Lecture Notes in Artificial Intelligence 9727

Subseries of Lecture Notes in Computer Science

LNAI Series Editors

Randy Goebel
  University of Alberta, Edmonton, Canada
Yuzuru Tanaka
  Hokkaido University, Sapporo, Japan
Wolfgang Wahlster
  DFKI and Saarland University, Saarbrücken, Germany

LNAI Founding Series Editor

Joerg Siekmann
  DFKI and Saarland University, Saarbrücken, Germany

More information about this series at http://www.springer.com/series/1244

João Silva · Ricardo Ribeiro
Paulo Quaresma · André Adami
António Branco (Eds.)

Computational Processing of the Portuguese Language

12th International Conference, PROPOR 2016
Tomar, Portugal, July 13–15, 2016
Proceedings

 Springer

Editors
João Silva
Universidade de Lisboa
Lisbon
Portugal

André Adami
Universidade de Caxias do Sul
Caxias do Sul
Brazil

Ricardo Ribeiro
ISCTE-IUL
Lisbon
Portugal

António Branco
Universidade de Lisboa
Lisbon
Portugal

Paulo Quaresma
Universidade de Évora
Évora
Portugal

ISSN 0302-9743   ISSN 1611-3349 (electronic)
Lecture Notes in Artificial Intelligence
ISBN 978-3-319-41551-2   ISBN 978-3-319-41552-9 (eBook)
DOI 10.1007/978-3-319-41552-9

Library of Congress Control Number: 2016942796

LNCS Sublibrary: SL7 – Artificial Intelligence

Printed on acid-free paper

This Springer imprint is published by Springer Nature
The registered company is Springer International Publishing AG Switzerland

Preface

The International Conference on the Computational Processing of Portuguese, PRO-POR, is the most important scientific event dedicated to the processing of spoken and written Portuguese, covering both basic and applied research. This event is hosted every other year, alternating between Brazil and Portugal. Previous events were held in Lisbon, Portugal (1993), Curitiba, Brazil (1996), Porto Alegre, Brazil (1998), Évora, Portugal (1999), Atibaia, Brazil (2000), Faro, Portugal (2003), Itatiaia, Brazil (2006), Aveiro, Portugal (2008), Porto Alegre, Brazil (2010), Coimbra, Portugal (2012), and São Carlos, Brazil (2014).

The community attending the conference is highly multidisciplinary and dynamic. The sharing of expertise, the communication of results, the promotion of methodologies and the exchanging of language resources and tools have been key contributions for the continuous scientific advance of the automated processing of the Portuguese language.

This 12th edition of PROPOR took place in Tomar, in the center of Portugal, and was hosted by the University of Lisbon and the NLX—Natural Language and Speech Group from its Department of Informatics at the Faculty of Sciences.

The event also featured the 4th edition of the MSc/MA and PhD Dissertation Contest, which selects the best new academic research work in the computational processing of Portuguese.

For the first time, PROPOR featured a Job-shop and Innovation Forum aiming at the creation of a fruitful environment to promote the exchange of results, talent, and partnerships between researchers in the field of language technologies and companies using and developing these technologies.

This year PROPOR had three workshops in addition to the main program, namely, (a) ASSIN, an evaluation forum for semantic similarity and textual entailment recognition; (b) LexSem+Logics, the Third Workshop on Logics and Ontologies and the First Workshop on Lexical Semantics for Lesser-Resourced Languages; and (c) CTPC, the Workshop on Corpora and Tools for Processing Corpora. It is noteworthy that the CTPC workshop was co-organized with the Portuguese Language Department of the Directorate-General for Translation of the European Commission.

Other additions to the main program were a Student Research Workshop and two tutorial sessions, one on "Gaussian Processes for Natural Language Processing" and another on "Translation Quality Estimation."

Two keynote speakers honored the event with their lectures: Dr. Hynek Hermansky (Johns Hopkins University, Baltimore, Maryland, USA and Brno University of Technology, Czech Republic) and Dr. Eneko Agirre (University of the Basque Country, San Sebastian, Spain).

In all, 52 submissions were received for the main event, totaling 134 authors from many countries worldwide, such as Belgium, Brazil, Denmark, Germany, Macao, Portugal, Spain, Switzerland, the UK, and the USA. This volume brings together a selection of the 39 best papers accepted at this meeting: 23 full papers and 14 short papers. The acceptance rate for full papers was 44 %. To these, the two papers corresponding to the winning submissions to the MSc/MA and PhD dissertation contest were added. In this volume, the papers are organized thematically and include the most recent developments in language applications, language processing, language resources, and speech processing.

We express our sincere thanks to every person and institution involved in the complex organization of this event, especially the members of the scientific committee of the main event, the dissertations contest and the associated workshops, the invited speakers, and the general organization staff.

We are also grateful to the agencies and organizations that supported and promoted the event, namely, the QTLeap Project, the Portuguese Language Department of the Directorate-General for Translation of the European Commission, the Hotel dos Templários, the Universidade de Lisboa and the Department of Informatics of its Faculty of Sciences, the Foundation of the Faculty of Sciences of the Universidade de Lisboa, and the Portuguese Foundation for Science and Technology.

May 2016

João Silva
Ricardo Ribeiro
André Adami
Paulo Quaresma
António Branco

Organization

General Chair

António Branco   Universidade de Lisboa, Portugal

Technical Program Chairs

André Adami   Universidade de Caxias do Sul, Brazil
Paulo Quaresma   Universidade de Évora, Portugal

Editorial Chairs

João Silva   Universidade de Lisboa, Portugal
Ricardo Ribeiro   ISCTE-IUL, Portugal

Demos Chair

Vládia Pinheiro   Universidade de Fortaleza, Brazil

Student Research Workshop Chairs

Pedro Balage   Universidade de São Paulo, São Carlos, Brazil
Fernando Batista   ISCTE-IUL, Portugal

Workshops Chair

Pablo Gamallo   Universidade de Santiago de Compostela, Spain

Tutorials Chairs

Fernando Perdigão   Universidade de Coimbra, Portugal
Margarita Correia   Universidade de Lisboa, Portugal
Lucia Specia   University of Sheffield, UK

Best MSc/MA and PhD Dissertation Contest Chair

Aline Villavicencio   UFRGS, Brazil

Jobshop and Innovation Forum Chair

Rosa Del Gaudio   Higher Functions, Portugal

Publicity and Sponsorship Chair

Daniela Braga   Defined crowd, USA
Ana Tavares   Universidade de Lisboa, Portugal

Organizing Committee

António Branco   Universidade de Lisboa, Portugal
Daniel Pereira   Universidade de Lisboa, Portugal
Ana Tavares   Universidade de Lisboa, Portugal

Steering Committee

Thiago Pardo   Universidade de São Paulo, Brazil
António Branco   Universidade de Lisboa, Portugal
Sara Candeias   Microsoft, Portugal
Nuno Mamede   Universidade de Lisboa, Portugal
Claudia Freitas   PUC-Rio de Janeiro, Brazil

Program Committee

Alberto Abad   Universidade de Lisboa, Portugal
Alberto Simões   Universidade do Minho, Portugal
Alexandre Rademaker   IBM, Brazil
Aline Villavicencio   UFRGS, Brazil
Amália Andrade   Universidade de Lisboa, Portugal
Amália Mendes   Universidade de Lisboa, Portugal
Ana Luís   Universidade de Coimbra, Portugal
Anabela Barreiro   INESC-ID, Portugal
André Adami   Universidade de Caxias do Sul, Brazil
Andreia Bonfante   UFMT, Brazil
Andreia Rauber   UCPEL, Brazil
Antonio Bonafonte   UPC, Spain
António Branco   Universidade de Lisboa, Portugal
António Serralheiro   Academia Militar, Portugal
António Teixeira   Universidade de Aveiro, Portugal
Ariani Di Felippo   UFSCAR, Brazil
Augusto Soares Silva   Universidade Católica, Portugal
Bento da Silva   UNESP, Brazil
Berthold Crysmann   CNRS, France
Brett Drury   USP-SC, Brazil
Carlos Prolo   UFRN, Brazil
Cicero Nogueira dos Santos IBM T.J. Watson Research Center, USA
Daniela Braga   defined crowd, USA
David Martins de Matos   Universidade de Lisboa, Portugal
Derek Wong   Universidade de Macau, China

Diamantino Freitas	Universidade do Porto, Portugal
Eraldo Rezende Fernandes	UFMS, Brazil
Eric Laporte	Université Paris-Est, France
Fabio Kepler	UNIPAMPA, Brazil
Fabio Violaro	UNICAMP, Brazil
Fernando Batista	ISCTE-IUL, Portugal
Fernando Perdigão	Universiade de Coimbra, Portugal
Fernando Resende Junior	UFRJ, Brazil
Gladis Almeida	UFSCAR, Brazil
Helena Caseli	UFSCAR, Brazil
Helena Moniz	INESC-ID, Portugal
Hugo Meinedo	Microsoft, Portugal
Hugo Oliveira	Universidade de Coimbra, Portugal
Irene Rodrigues	Universidade de Évora, Portugal
Isabel Falé	Universidade Aberta, Portugal
Isabel Trancoso	Universidade de Lisboa, Portugal
Ivandré Paraboni	USP, Brazil
João Balsa	Universidade de Lisboa, Portugal
João Luís Rosa	USP-SC, Brazil
João Silva	Universidade de Lisboa, Portugal
Joaquim Llisterri	Universitat Autònoma de Barcelona, Spain
Jorge Baptista	Universidade do Algarve, Portugal
José João Almeida	Universidade do Minho, Portugal
Laura Alonso Alemany	Universidad Nacional de Córdoba, Argentina
Leandro Oliveira	EMBRAPA, Brazil
Lucia Specia	University of Sheffield, UK
Magali Sanches Duran	USP-SC, Brazil
Margarita Correia	Universidade de Lisboa, Portugal
Maria das Graças Volpe Nunes	USP-SC, Brazil
Maria José Finatto	UFRGS, Brazil
Mário Silva	Universidade de Lisboa, Portugal
Nelson Neto	UFP, Brazil
Norton Roman	USP-EACH, Brazil
Nuno Cavalheiro Marques	Universidade Nova de Lisboa, Portugal
Nuno Mamede	Universidade de Lisboa, Portugal
Pablo Gamallo	Universidade de Santiago de Compostela, Spain
Palmira Marrafa	Universidade de Lisboa, Portugal
Paulo Gomes	Universidade de Coimbra, Portugal
Paulo Quaresma	Universidade de Évora, Portugal
Plínio Barbosa	UNICAMP, Brazil
Renata Vieira	PUC-RS, Brazil
Ricardo Ribeiro	ISCTE-IUL, Portugal
Rosa Del Gaudio	Higher Functions, Portugal
Ruy Luiz Milidiú	PUC-Rio, Brazil
Sandra Aluísio	USP-SC, Brazil

Contents

Language Processing – Full Papers

Language Processing – Short Papers

Speech Processing – Full Papers

Language Applications – Full Papers

Extending VITHEA in Order to Improve Children's Linguistic Skills

Vânia Mendonça[✉]

Instituto Superior Técnico, Porto Salvo, Portugal
vania.mendonca@tecnico.ulisboa.pt

Abstract. Autism Spectrum Disorder (ASD) often comprises difficulties in the acquisition of communication and language skills. Several researchers and companies have developed software to help individuals with ASD developing those skills; however, there is a lack of applications in Portuguese that are tailored to the individual needs of each child. In this context, we present VITHEA-Kids, a platform where caregivers can create exercises and customize the interaction between each child and the platform. We also developed a module for the automatic generation of multiple choice exercises, meant to be integrated in VITHEA-Kids. We evaluated this work with caregivers (which provided promising indicators), with a child (ongoing upon thesis delivery) and we also evaluated the generation of incorrect answers in multiple choice exercises (achieving acceptance rates between 61.11 % and 92.22 %).

Keywords: Autism Spectrum Disorder · Children learning · Language skills development · Automatic generation of content

1 Introduction

Autism Spectrum Disorder (ASD) is characterized by persistent deficits in social communication and interaction, as well as restricted, repetitive behaviors or interests since an early developmental period. Individuals with ASD often face difficulties in communication [2], which could be minimized by therapy. Therapeutic interventions might not be affordable, so the possibility of taking them to other settings or performing them in an affordable way could be helpful.

Given the interest that individuals with ASD display towards computers [6,15], some authors studied the use of software to teach academic skills to children with ASD [16]. More recently, both companies and researchers have developed educational mobile applications targeting children with impairments, thus easing the practice of verbal skills in diversified settings, with or without assistance. Despite the current large offer of such applications, there are issues left to be addressed: most applications are paid; there is a lack of applications available in Portuguese that take into account each user's characteristics and needs. Considering this situation, we developed a software platform where children with ASD can solve exercises to develop or improve linguistic skills regarding Portuguese, having in mind their individual characteristics – VITHEA-Kids. We also developed a module to generate multiple choice exercises.

© Springer International Publishing Switzerland 2016
J. Silva et al. (Eds.): PROPOR 2016, LNAI 9727, pp. 3–11, 2016.
DOI: 10.1007/978-3-319-41552-9_1

2   Background

According to the Diagnostic and Statistical Manual of Mental Disorders (DSM-V) [2], a diagnostic of ASD happens when "persistent deficits in social communication and social interaction" and "restricted, repetitive patterns of behaviour, interests" are present since an early developmental period, causing significant impairment. Often individuals diagnosed with ASD go through therapy in order to improve their communication skills. Most therapies are based in Applied Behaviour Analysis (ABA), a psychology approach where the focus is to modify a certain behavior considering the events that precede (antecedents) and follow (consequences) that behavior. An ABA based therapy comprises the following steps [13]: (1) Identification of the behaviors and the respective antecedents and consequences; (2) Selection of a target behavior as the focus of the treatment; (3) Measurement of the current level of the individual's target behavior (baseline); (4) Implementation of an intervention to improve the target behavior and measurement of this behavior during this process; (5) Assessment of whether the acquired skills were generalized across different settings, people and materials. The most common interventions are the reinforcement based ones, where the manipulated stimulus is something that pleases the individual, i.e., the individual's behavior is rewarded. These include several techniques, such as prompting, which consists in using an antecedent auxiliary stimulus that aims to elicit the desired response. An example of a prompt is to point at an object that the child was told to pick, in order to help the child performing such task [17].

3   Related Work

In this work, we reviewed works concerning the use of technology for children with Autism Spectrum Disorder (ASD) and the use of automatic generation of content in the context of education. Regarding technology for children with ASD, some authors performed surveys regarding aspects such as software features preferred by children with ASD and their caregivers, as well as feedback from users' relationship with technology, namely: previous experiences, main difficulties and reasons to abandon previously adopted technology [3,6,15]. The outcome of these surveys indicated that caregivers prioritize communication and social skills. Also, both children and caregivers display interest towards technology; however, caregivers have reported many issues, such as the lack of content customization based on the child's characteristics, among others.

Many researchers and companies have also developed software targeting children with ASD. Some researchers performed studies where they compared the outcome of using software *versus* non-computational solutions, regarding children's language development process, while others focused on evaluating the impact of a specific set of features. Studies reviewed include Heimann et al. [4], Moore and Calvert [12], Hetzroni and Tannous [5], and Massaro and Bosseler [8] and others (see Ramdoss et al. [16]). Most of these studies reported progress regarding the children's ability to learn new skills, but the applications utilized in these studies are either discontinued or paid and do not comprise customization options. We have also experimented several applications available in Google Play, and although they exist in a large amount, most of them are only available in English; additionally, many of them are paid (or include paid features) and, as for the free ones, some have several navigation bugs that negatively affect user experience; finally, the possibilities for creating content and the customization options were either limited or none at all. A more detailed analysis of the studies reviewed and applications experimented can be found in the dissertation document [10].

Regarding the use of automatic generation or processing features in the context of education, these can spare teachers and educators time while creating content, be adapted to each student's characteristics, etc. However, we only found one study where automatic generation of content was used in an application for children with ASD [7].

For the automatic generation of content, we analyzed resources and tools that we could use to extract words and images, as well as to assure the correctness of the resulting content. For the task of retrieving words and related images, the only resource that facilitates such task is ImageNet[1], which allows to browse through nested hierarchies of synsets (sets of synonyms) associated with a set of image URLs. As for the correction of the generated content, we experimented TreeTagger [18]: a tool for the Part of Speech (PoS) and lemma annotation of sequences of words that can be used for any language as long as a lexicon and a tagged training corpus are given, which is the case of Portuguese.

4 VITHEA-Kids: A Platform for Children with ASD and Their Caregivers

Recalling the goals defined in Sect. 1, we developed VITHEA-Kids: a platform where children can solve exercises created by their caregivers. This platform is based on the infrastructure of Virtual Therapist for Aphasia Treatment (VITHEA) [1], an awarded platform aiming to help patients with aphasia to recover their word naming skills by solving exercises in which they should orally reply to a question (e.g. to name the object in an image). VITHEA comprises two modules: the therapist's module, where therapists can create and manage the exercises for the patients to solve and the multimedia resources to be used in those exercises, as well as to manage each patient's information and statistics; the patient's module, where the patients can solve the kind of exercises described above. The exercise instruction and the feedback to the patient's answer are uttered by a talking animated character using a voice synthesizer [14]. The patient's answer is validated using automatic speech recognition [9] and it does not have to be an exact match of the correct answers provided by the therapist to be considered correct. When incorrect, the patient can try again for a user-defined number of attempts. However, the correct answer is never provided to the patient, and no helping cues are given during the exercises.

---

[1] http://image-net.org/.

Although VITHEA-Kids is based on the infrastructure of VITHEA, our target users are different and so is the purpose and functionality. One of the differences is that we use multiple choice exercises, since this kind of exercises has been used for children with ASD and might allow to work on skills such as vocabulary acquisition, word-picture association, and generalization. These exercises are composed by a question (e.g., "What is the name of this object?"), an optional image or textual stimulus (e.g., the picture of a fork), and a set of textual or image possible answers, respectively ("Fork", "Spoon", "Cup", "Bowl"), in which only one of the answers is correct (in this case, "Fork"). Each exercise can contain from zero to three distractors, easing the task of creating several exercises with small variations in content and difficulty.

VITHEA-Kids is composed of two modules: the caregiver's module and the child's module, based on VITHEA's therapist's and patient's modules respectively. The caregiver's module allows to create and manage the exercises described above (see Fig. 1a), as well as to upload and manage image files to be featured in the exercises, and create and manage child users (whose info differ from VITHEA's patients'). Unlike VITHEA, it also allows for the caregiver to customize the child's module, namely the utterances for the animated character and the reinforcement images to display when the child correctly solves an exercise.

As for the child's module, in each exercise the animated character utters the question, and the exercise area is filled with the stimulus and the possible answers in a random order (see Fig. 1b). The child should then tap over the answer they think to be correct. Selecting the correct answer on the first attempt will lead to a reinforcement image. Selecting any other answer will prompt the child to pick the correct answer: the selected distractor disappears, the correct answer is highlighted and the remaining answers are uttered by the animated character. If, after that, the child selects the correct answer, a weaker reinforcement screen is shown. The current exercise can be skipped at any time. When the exercise session ends, information about child's performance is shown. There is also a repeat button to have the animated character repeating her last utterance.

5   Automatic Generation of Exercises

Besides VITHEA-Kids, we also developed a module for automatic generation of exercises, aiming to ease the task of creating exercises, either in VITHEA-Kids or in other contexts. This module generates content for multiple choice exercises of the two variants presented in Sect. 4, taking as input the exercise topic (e.g., "Emoções" – "Emotions"), a user-provided template for the question generation, composed by an immutable part and a variable whose value depends on the exercise's topic ("Que <%variavel> é este?" – "Which <%variable> is this?"), information regarding the number of distractors (e.g., "3") and whether the distractors should be related to the given topic (e.g., "true"). The output is an exercise composed by a question/instruction (e.g., "Que emoçã é esta?" – "Which emotion is this?"), the correct answer to the question (e.g., "alegria" – "happiness", along

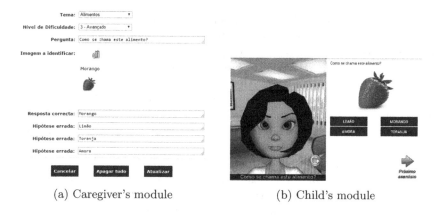

(a) Caregiver's module   (b) Child's module

Fig. 1. VITHEA-Kids's modules

with an image of a happy person) and a set of distractors ("tristeza" – "sadness", "raiva" – "anger", and "medo" – "fear"). The words and images are obtained from a hierarchy of synsets similar to ImageNet (recall Sect. 3).

The exercise generation process comprises three main tasks: (1) Generating the correct answer and the corresponding stimulus, given a certain topic; (2) Generating a question given the exercise's topic, the correct answer generated in 1 and the question template; (3) Generating a set of distractors given the exercise's topic (if relevant), the correct answer and the number of distractors. Generating a correct answer consists in selecting an hyponym of the topic provided (upon being converted to singular). As for the question generation, it involves the following steps (unless the template provided does not include a variable, in which case the template remains unaltered):

1. The template is searched for the variable (a word surrounded by <% and >);
2. If the variable was found, it is replaced by the word selected within the next steps. Otherwise, the process ends here and the template is returned as the final question;
3. The first hyperonym of the correct answer that contains the topic is selected (if none contains the topic, a random one is selected);
4. The variable is replaced with a word randomly selected from the hyperonyms list obtained in the previous step;
5. The resulting sentence is PoS tagged using TreeTagger;
6. The resulting tagged sentence is checked for inconsistencies regarding number and gender agreement between the variable's replacement and the rest of the sentence (number and gender information for each word was provided by TreeTagger in the previous step).
7. If an inconsistency is detected, all the words that should agree in gender and/or number with the variable but fail this criterion are corrected using a list of substitutions. Each substitution has the format `regex before after` – if the word to correct matches the regular expression (`regex`), its suffix (`before`) is replaced by the suffix `after`.

In Step 6, we check gender and number agreement for the following patterns: the variable is a noun preceded by a demonstrative determinant; the variable is a noun preceded by an article; the variable is a noun preceded by a demonstrative determinant and a verb; the variable is a noun preceded by an article and a verb; the variable is a noun preceded by a demonstrative determinant, a verb and an interrogative pronoun; the variable is a noun preceded by an article, a verb and an interrogative pronoun; the variable is a noun followed by a demonstrative determinant; the variable is a noun followed by a verb demonstrative determinant.

Regarding the generation of textual distractors, if the distractor should belong to the given topic, it is selected from the leaf hyponyms of the topic. Otherwise, it is selected from all the leaf synsets. The selected distractor cannot be a synonym of the correct answer (i.e., it cannot belong to the same synset) nor match any of the remaining distractors for the current exercise. This process repeats itself for the number of distractors specified. For image distractors, the selected synsets must also include images.

6   Evaluation

We individually evaluated each of the three modules delivered. Regarding VITHEA-Kids, the caregiver's module was evaluated with seven caregivers, including parents and therapists of children with special needs. Participants had to perform 5 tasks and rate them using a scale of 1 to 5 in a questionnaire which also included a set of statements about the overall experience (to be rated using the same scale), and open questions for further observations and suggestions, as well as a question to assess the interest in automatically generated exercises. Most of the tasks were rated with 4 or 5 in terms of how easily/fast they could be performed. All the participants strongly agreed that the platform was easy to browse and makes use of a discourse easy to understand. However, some participants disagreed concerning the clarity of the feedback given in error or confirmation messages. The majority of the participants (6 in 7) found the exercises in VITHEA-Kids useful for their children and would like to use the application again. Furthermore, all the participants were interested in the automatic generation of exercises.

The child's module was evaluated using a Single Subject Design, since each individual with ASD has a unique set of symptoms, characteristics and needs, making it difficult to generalize results across participants. The variant chosen comprises baseline and intervention phases (recall Sect. 2), and also a final follow-up phase. The application was tested with a seven year old male child, diagnosed with ASD, who is currently learning how to read, in his classroom, with the help of a therapist. Our goal was to assess whether the child could learn a new word taking advantage of the application's prompting and customized reinforcement. In the baseline phase (using zero distractors), the child never selected the correct answer, so this phase ended after four sessions (the minimum number of sessions required to consider that a stable pattern was observed). In the intervention phase, which overlapped the delivery of this thesis, the child had to solve exercises in which the number of distractors was initially zero and progressively incremented (up to three). This phase ended when the child selected the correct answer during two consecutive sessions, without prompting. In the follow-up phase, prompting and reinforcement were deactivated in order to assess whether the child had learned the target word. Once again, the child selected the correct answer during two consecutive sessions (always using three distractors).

The evaluation of the exercise generation module was focused on the generation of distractors. Thirty participants replied to a questionnaire to generate and rate a set of distractors for each pair of question and correct answer to that question. For each distractor, the participants should indicate whether they think the distractor makes sense in the context of the given input or not (i.e., whether it would be possible to solve an exercise if it was composed of the given question and a set of answers containing the given correct answer and the distractor being evaluated). A total of 180 word distractors and 180 image distractors were generated. Out of the word distractors, 79.44 % were marked as making sense, and the average quality rating was 3.75 out of 5 (std. dev. = 1.41). As for the image distractors, 76.67 % were marked as making sense and the average quality rating was 3.82 out of 5 (std. dev. = 1.44).

7   Conclusions

In this work, we are taking the first steps into addressing the issues presented by the currently available software for individuals with ASD, by presenting a customizable, free of charges platform for the development of linguistic and generalization skills regarding Portuguese, where caregivers can create content having in mind each child of whom they take care. We also present a module that aims to save time to the caregivers by allowing to automatically generate multiple choice exercises. The work described in this thesis originated a publication in 17th International ACM SIGACCESS Conference on Computers & Accessibility – ASSETS 2015[2], under the title: *VITHEA-Kids: a platform for improving language skills of children with Autism Spectrum Disorder* [11].

As for future work, one of our goals is the integration of the exercise generation module in VITHEA-Kids. We also intend to add some features that were suggested by caregivers during evaluation (specially regarding customization) and expand the variety of exercises available. Regarding the automatic generation of exercises, it would be interesting to allow the generation of paraphrases given a question, since it could contribute to improve the child's ability to generalize different formulations of the same question.

Acknowledgments. This work was supported by national funds through Fundação para a Ciência e a Tecnologia (FCT) with reference UID/CEC/50021/2013, and under project CMUP-ERI/HCI/0051/2013. The author is deeply thankful to her parents and supervisors, as well as to her friends, colleagues and contacted caregivers for having

---

[2] http://www.sigaccess.org/assets.

tested the applications developed, proof-read the dissertation documents, provided help in debugging tasks, and/or for emotional support.

References

1. Abad, A., Pompili, A., Costa, A., Trancoso, I., Fonseca, J., Leal, G., Farrajota, L., Martins, I.P.: Automatic word naming recognition for an on-line aphasia treatment system. Comput. Speech Lang. **27**(6), 1235–1248 (2013)
2. American Psychiatric Association. Diagnostic and Statistical Manual of Mental Disorders: DSM-5, 5th edn. American Psychiatric Association Arlington, VA (2013)
3. Dawe, M.: Desperately seeking simplicity: how young adults with cognitive disabilities and their families adopt assistive technologies. In: Proceedings of the SIGCHI Conference on Human Factors in Computing Systems, CHI 2006, pp. 1143–1152. ACM, New York (2006)
4. Heimann, M., Nelson, K.E., Tjus, T., Gillberg, C.: Increasing reading and communication skills in children with autism through an interactive multimedia computer program. J. Autism Dev. Disord. **25**(5), 459–480 (1995)
5. Hetzroni, O.E., Tannous, J.: Effects of a computer-based intervention program on the communicative functions of children with autism. J. Autism Dev. Disord. **34**(2), 95–113 (2004)
6. Lehman, J.F.: A Feature-based Comparison of Software Preferences in Typically-Developing Children versus Children with Autism Spectrum Disorders (1998)
7. Lehman, J.F.: Toward the use of speech and natural language technology in intervention for a language disordered population. Behav. Inf. Technol. **18**(1), 45–55 (1999)
8. Massaro, D.W., Bosseler, A.: Read my lips: the importance of the face in a computer-animated tutor for vocabulary learning by children with autism. Autism Int. J. Res. Pract. **10**(5), 495–510 (2006)
9. Meinedo, H., Caseiro, D.A., Neto, J.P., Trancoso, I.: AUDIMUS.MEDIA: a broadcast news speech recognition system for the european portuguese language. In: Mamede, N.J., Baptista, J., Trancoso, I., Nunes, M.G.V. (eds.) PROPOR 2003. LNCS, vol. 2721, pp. 9–17. Springer, Heidelberg (2003)
10. Mendonça, V.: Extending VITHEA in order to improve children's linguistic skills. Master's thesis, Instituto Superior Técnico (2015)
11. Mendonça, V., Coheur, L., Sardinha, A.: Vithea-kids: a platform for improving language skills of children with autism spectrum disorder. In: Proceedings of the 17th International ACM SIGACCESS Conference on Computers and Accessibility, ASSETS 2015, pp. 345–346. ACM, New York (2015)
12. Moore, M., Calvert, S.: Brief report: vocabulary acquisition for children with autism: teacher or computer instruction. J. Autism Dev. Disord. **30**(4), 359–362 (2000)
13. Naoi, N.: Intervention and treatment methods for children with autism spectrum disorders. In: Matson, J.L. (ed.) Applied Behavior Analysis for Children with Autism Spectrum Disorders, Chap. 4, pp. 67–81. Springer, New York (2009)
14. Paulo, S., Oliveira, L.C., Mendes, C., Figueira, L., Cassaca, R., Viana, C., Moniz, H.: DIXI – a generic text-to-speech system for european portuguese. In: Teixeira, A., de Lima, V.L.S., de Oliveira, L.C., Quaresma, P. (eds.) PROPOR 2008. LNCS, vol. 5190, pp. 91–100. Springer, Heidelberg (2008)

15. Putnam, C., Chong, L.: Software and technologies designed for people with autism: what do users want? In: Proceedings of the 10th International ACM SIGACCESS Conference on Computers and Accessibility, ASSETS 2008, pp. 3–10. ACM, New York (2008)
16. Ramdoss, S., Mulloy, A., Lang, R., O'Reilly, M., Sigafoos, J., Lancioni, G., Didden, R., El Zein, F.: Use of computer-based interventions to improve literacy skills in students with autism spectrum disorders: A systematic review. Res. Autism Spectr. Disord. **5**(4), 1306–1318 (2011)
17. Ringdahl, J.E., Kopelman, T., Falcomata, T.S.: Applied behavior analysis and its application to autism and autism related disorders. In: Matson, J.L. (ed.) Applied Behavior Analysis for Children with Autism Spectrum Disorders, Chap. 2, pp. 15–32. Springer, New York (2009)
18. Schmid, H.: Probabilistic part-of-speech tagging using decision trees. In: Proceedings of the International Conference on New Methods in Language Processing, pp. 44–49 (1994)

Automatic Classification of the Complexity of Nonfiction Texts in Portuguese for Early School Years

Nathan Hartmann[1]($\boxtimes$), Livia Cucatto[2], Danielle Brants[2], and Sandra Aluísio[1]

[1] Interinstitutional Center for Computational Linguistics (NILC),
Institute of Mathematical and Computer Sciences,
University of São Paulo, São Carlos, Brazil
{nathansh,sandra}@icmc.usp.br
[2] Guten Educação e Tecnologia Ltda., São Paulo, Brazil
liviacucatto@gmail.com, dbrants@gutennews.com.br

Abstract. Recent research shows that most Brazilian students have serious problems regarding their reading skills. The full development of this skill is key for the academic and professional future of every citizen. Tools for classifying the complexity of reading materials for children aim to improve the quality of the model of teaching reading and text comprehension. For English, Feng's work [11] is considered the state-of-art in grade level prediction and achieved 74 % of accuracy in automatically classifying 4 levels of textual complexity for close school grades. There are no classifiers for nonfiction texts for close grades in Portuguese. In this article, we propose a scheme for manual annotation of texts in 5 grade levels, which will be used for customized reading to avoid the lack of interest by students who are more advanced in reading and the blocking of those that still need to make further progress. We obtained 52 % of accuracy in classifying texts into 5 levels and 74 % in 3 levels. The results prove to be promising when compared to the state-of-art work.

Keywords: Automatic readability assessment · Early grade reading · Methods for selecting reading material

1 Introduction

According to data collected by the Organisation for Cooperation and Economic Development (OECD) in the Programme for International Student Assessment (PISA)[1], Brazilian students have serious problems regarding their reading skills. The most recent survey, carried out in 2012, showed results for Brazil below the average of the countries surveyed. 49.5 % of Brazilian students did not reach the levels considered minimum in reading, which means that, at best, they can only recognize themes of simple and familiar texts. Furthermore, only 0.5 % of Brazilian students reached maximum reading levels, which means that only one

---

[1] Available at oecd.org/education/PISA-2012-results-brazil.pdf.

© Springer International Publishing Switzerland 2016
J. Silva et al. (Eds.): PROPOR 2016, LNAI 9727, pp. 12–24, 2016.
DOI: 10.1007/978-3-319-41552-9_2

in every 200 young people in Brazil is able to deal with complex texts and perform in-depth analysis on such texts. More negative numbers were seen in the Brazilian National High School Exam (ENEM – *Exame Nacional do Ensino Médio*) in 2014: from the 6.1 million students who did the exam, 529 flunked the composition. Experts stated that most students do not even understand the wording of the question. Only 250 students, equivalent to 0.004 %, aced the composition.

The development of reading skills has long been related to success in future academic and professional activities. Aimed at raising the quality of the teaching model for reading and text comprehension in this country and trying to close some gaps in Brazilian public policies for education, many features and computer systems for the Brazilian Portuguese have been launched recently. An example is the First Book Project (*Projeto Primeiro Livro*)[2], which helps children and young people from public schools to learn grammar, spelling and develop narratives. Another example is the Victor Civita Foundation, sponsored by the publishing house Abril, which supports teachers, school managers and public policy makers of Elementary Education with lesson plan search engines, social network for educators to exchange experience and share knowledge, and a resource bank for classes[3].

Currently, in Brazil, the elementary school is divided into two stages - 1st to 5th year, and 6th to 9th year. The National Curriculum Parameters (1998), however, divide these two stages into four cycles. In this article, we focus on the end of the first cycle - 3rd year -, and the second and third cycles - 4th/5th and 6th/7th years because they are fundamental for students to achieve adult reading comprehension.

There are some tools for Brazilian Portuguese such as the Flesch Index [30], which is adapted for Portuguese and used in the Microsoft Word, and mainly the Coh-Metrix-Port and AIC, developed in the PorSimples project [3], whose goal is to simplify Web texts for people with poor literacy levels. These tools, however, do not meet the needs of educators in the classroom: there are no classifiers able to discriminate the level of complexity of each year focus of this study – 3rd to 7th years, using metrics of the many language levels.

For the English language, there are tools for classifying reading materials for children used in US schools, based on both quantitative data such as Lexile[4] [25,39] and better informed such as Text Easability Assessor (TEA)[5] that uses Coh-Metrix [17,18] metrics.

In this article, we present the process of features development and training of a classifier based on machine learning to automatically distinguish five levels of textual complexity to support the selection of texts for students of a given class. Here, we use grade levels, which indicate the number of years of education required to completely understand a text, as a proxy for reading difficulty, the

---

[2] Available at primeiro-livro.com.
[3] Available at rede.novaescolaclube.org.br.
[4] Available at lexile.com.
[5] Available at tea.cohmetrix.com.

same way as [11]. However, we understand that there can be a great diversity of competences, abilities and background knowledge regarding reading in a same classroom.

In Sect. 2 we present some recent work on automatic readability assessment of grade levels. In Sect. 3 we present the manual annotation criteria and the process of manual annotation of our corpus. In Sect. 4 we present the experiments carried out and the results obtained on 5 grade levels and on combining adjacent levels, achieving best results on 3 classes. Finally, in Sect. 5 we present our final remarks and future work.

2   Related Work

In recent years, the interest in building automatic classifiers of text complexity has increased. Although the English language is a highlight in this topic [8,17,26,38], it has served as base for other languages to develop their own classifiers, such the French [14], Italian [10], Spanish [36], German [19,41], Arabic [13] and Portuguese [1,9]. Automatic classifiers of text complexity have various applications, as follows: teaching a second language [9], reading and comprehension for poor literacy readers [3], legal and scientific texts and as a first step in building Text Simplification Systems [1].

Readability studies are an area of great interest for language teaching, particularly in building materials for reading and learning vocabulary. The studies in this area allow to establish a scale of difficulty levels of texts used to assess students. Generally, in elementary levels of education, teachers acknowledge that giving reading materials not suitable for the students' level impairs their learning, discouraging them [15].

Curto [9] developed a system to extract linguistic features and a text classifier to teach Portuguese as a second language. The motivation presented by the author is the need of selecting texts for language teaching, which is done manually.

The Coh-Metrix-Port 2.0[6], an adaptation of the Coh-Metrix developed in the PorSimples project [1], currently provides 48 metrics that enable the analysis of lexical, morphosyntactic, syntactic (chunking), semantic and discursive features [37]. The AIC tool, with 39 metrics [31], covers the lack of syntactic analysis (full parsing) in the Coh-Metrix-Port. Scarton and Aluísio [37] evaluated the first version of the Coh-Metrix-Port tool (with 38 metrics) comparing written texts for adults with written texts for children, considering only two levels: simple texts and complex ones related to the journalistic and scientific dissemination genre. It is worth noting that a simple measure such as the Flesch Index and its components results in a SVM classifier with polynomial kernel with 82.5 % accuracy, while the Coh-Metrix-Port increased accuracy to 92 % and the measures altogether resulted in 93 % of accuracy.

The work most related to ours is for the English language [11] and classifies textual complexity using a corpus of magazines for elementary and high

---

[6] Available at nilc.icmc.usp.br/coh-metrix-port.

school students (Weekly Reader Corpus[7] that has texts for elementary school students labeled with grade levels, which range from 2 to 5). Their best results were obtained by group-wise add-one-best feature selection, resulting in 74 % classification accuracy, with 273 features selected, including language modeling features, syntactic features, PoS features, traditional readability metrics, and out-of-vocabulary features.

3   Corpus and Manual Annotation on Grade Levels

3.1   Description of Grade Levels and the Problem

In recent years, the Brazilian government has been working on a systematization of the education policy in an attempt to unify the curricula methods and content for schools and teachers all over Brazil to speak the same language. The *Provinha Brasil*[8], the state assessment tests (e.g., SARESP[9] in the state of São Paulo) and even the ENEM (National High School Exam) are attempts to direct education professionals to the same educational setting. However, it is still not clear for teachers, especially for elementary school ones, how to distribute such content by school year, especially when it comes to reading. In addition, in Brazil, there is an extremely diverse learning scenario in the same grade. The insertion of dictionaries in grade levels by the National Textbook Program (PNLD) [23] since 2006 shows a change, albeit slow, in the Brazilian educational system.

Building a five-level classifier is in line with this emerging educational scenario. For the 3rd, 4th and 5th years (*Ensino Fundamental I*) and the 6th and 7th years of the elementary school (*Ensino Fundamental II*), we can measure the complexity of texts and, thus, meet the diversity in reading comprehension.

The creation basis was: the National Curriculum Parameters (PCNs) (1998), the descriptors of Prova Brasil[10], analysis of textbooks, articles in the psycholinguistics area [7,12,16,27–29,32,33,35] and language acquisition [21,22], and the knowledge of linguists with experience in Education and the Portuguese language (phonology, morphology, syntax, semantics and discourse).

With respect to PCNs, one way to measure these skills was to create descriptors that synthesized the competencies and skills. Such descriptors are used as reference matrix for Prova Brasil. The Portuguese language test assesses only reading skills, represented by 21 descriptors for the 9th year and by 15 descriptors for the 5th year, divided into six groups: (1) Reading procedures; (2) implications of support, gender and/or enunciator in the text comprehension; (3) Relationship between texts; (4) Coherence and cohesion in text processing; (5) Relations between expressive features and effects of meaning; and (6) Linguistic variation.

---

[7] Available at www.weeklyreader.com.

[8] *Provinha Brasil* is a test to evaluate how much children have learned about Portuguese and Mathematics subjects. Available at provinhabrasil.inep.gov.br.

[9] Available at http://www.educacao.sp.gov.br/saresp.

[10] *Prova Brasil* is a test to evaluate the quality of the educational brazilian system. Available at http://portal.mec.gov.br/prova-brasil.

However, neither the PCNs nor the descriptors distinguish five levels. On the other hand, it is known that each grade level has a specific curriculum and, therefore, its difficulties and expected progress. One way to obtain a more objective division by grade levels was to resort to textbooks. All of them indicate the content to be taught and bring nonfiction texts.

3.2   Corpus and Selection of Texts for Annotation

In order to build the corpus, we search for pre-selected texts in terms of complexity levels, using the following sources: SARESP and textbooks. We obtained only 72 texts, distributed in five levels, from SARESP tests, given limitations such as they do not cover all school years; they are generally applied once a year; the test contains several textual genres – that is, there are few informative texts; and, above all, not all texts are available online. Considering the difficulties above and knowing the importance of a large amount of data to machine learning techniques, we turned to textbooks as our main source of texts. Experts selected 178 informative texts from Portuguese language textbooks. Therefore, we equally distributed 50 texts in each level, totaling 250.

Because of the small amount of texts which had some level information, new sources, not previously classified, were included in the corpus: NILC corpus[11], *Ciência Hoje das Crianças* (CHC)[12], *Folhinha*[13], *Para Seu Filho Ler*[14] and *Mundo Estranho*[15], which currently contains 7,645 texts compiled, whose sources distribution is shown in Table 1. Among the seven sources, the one that presents great diversity of textual type and gender is textbooks, since the purpose of this type of source is to present the student with all existing genres and types – we found from simple expository texts to more complex structures such as argumentative texts very common in the editorial genre; the same textual amplitude is seen in SARESP tests[16]. Although the NILC corpus is also composed of textbooks, its texts generally have three text types: descriptive, narrative and expository. However, CHC, *Folhinha* and *Mundo Estranho* are similar: they present, in most cases, dialogues; varied text types in the same text; and the predominance of a particular type. These different possibilities of textual occurrence increase the challenge of building the curricula (see Sect. 3.3) and, therefore, the classification system. So far, 1,456 texts have been annotated by a sole linguist.

3.3   Annotation Criteria

The first annotation grid built relied on textbook curricula, which has linguistic phenomena organized by grade levels. From this basis, the contact with texts

---

[11] Available at nilc.icmc.usp.br/nilc/images/download/corpusNilc.zip.

[12] Available at chc.cienciahoje.uol.com.br.

[13] Available at www.folha.uol.com.br/folhinha.

[14] Available at zh.clicrbs.com.br/rs.

[15] Available at mundoestranho.abril.com.br.

[16] Available at sites.google.com/site/provassaresp.

Table 1. Distribution of texts by source.

Textbooks	NILC corpus	SARESP tests	Ciência Hoje das Crianças	Folhinha issue of Folha de São Paulo	Para Seu Filho Ler issue of Zero Hora	Mundo Estranho
492	262	72	2.589	308	166	3.756

targeted to school years and the knowledge of linguists, we kept on improving the grid. We should emphasize that although the school introduces linguistic elements in certain years, children can already understand and produce them long before being exposed to them in the educational system. Hence, the need to link different sources of knowledge.

Another challenge lies in the text type diversity found in informative texts, namely: narrative, descriptive, injunctive, expository and argumentative [4]. Such text types have different structures, but they may still be in the same reading comprehension level. Thus, for example, a mostly injunctive text may have the same level of complexity as a text that is mostly descriptive. Structural possibilities were and are still considered in the grid detailing.

Linguistic and non-linguistic elements are divided into six groups: morphological, lexical, syntactic, textual, punctuation and semantic and reader's commonsense knowledge. The first one corresponds to linguistic elements in the morphological level such as verb endings, affixes and grammatical categories; the second brings together linguistic phenomena connected to vocabulary and semantic relationships such as synonymy, antonymy, polysemy, among others; the syntactic group highlights the types of clauses present in the texts, how they are organized within the sentence, the paragraph, the order and size of constituents; with regard to text metrics, the main focus is cohesion: the type of cohesion used and the elements used for this end. The Punctuation and Semantic and reader's commonsense knowledge complement the previous ones: this maps the punctuation richness and the other is an attempt to capture the semantic and world knowledge of the reader, so far, by means of named entities.

4 Experiments

4.1 Preliminary Experiments: Using Language Independent Features

The manual annotation process started focusing on a balanced sample of 971 texts in 5 levels of textual complexity, from the 3rd to 7th grade levels, mapped here from level 1 to 5. The distribution of our initial data set is as follows: 208 texts of level 1, 185 texts of level 2, 196 texts of level 3, 191 texts of level 4 and 191 texts of level 5. For this set of texts, we extracted the following 10 features list we call "simple statistics feature": Flesch-Kincaid Grade Level index, the average sentences per paragraph, average words per sentence, number of paragraphs, number of sentences, number of words in the text, type-token ratio, number of simple words matching the dictionary of simple words to youngsters [6], incidence of punctuation and diversity of punctuation. All of these features are independent of language, except for the dictionary of simple words, but it is easy to find it for many languages. When performing a 10-fold cross-validation experiment on the initial data set, with an SVM classifier[17] with linear kernel and $C = 1$, we obtained 52 % of accuracy $(+/- 14)$. It is worth noting that the 3 features best classified by the recursive feature elimination (RFE) process for selecting features were the Flesch-Kincaid, the number of paragraphs in the text and the diversity of punctuation.

4.2   Increasing the Number of Features and Data

Keeping the size of the initial corpus, we decided to increase our features set to better represent differences among the textual levels. Table 2 maps the features implemented in 6 linguistic categories used for corpus annotation, described in Sect. 3.3. Table 2 shows a total of 108 features: (i) 52 Coh-Metrix-Port features 2.0[18], (ii) 32 AIC Features, (iii) two features based on the lists of positive and negative words of the LIWC - Dictionary for Sentiment Analysis[19], 14 features about Named Entities, calculated on the flat output of the PALAVRAS parser [5], and (v) 8 new features on Verbs Incidence implemented especially for this work comprising Portuguese verb tenses and moods. Some features were duplicated on Table 2 because they use information from many linguistic categories.

By repeating the experiment with the same fold and SVM settings for the new set of 108 features, we obtained 56 % of accuracy $(+/-13)$. We know it is difficult to have statistical learning in a small dataset such as the initial dataset. Therefore, we use the Active Learning Approach [40] for selecting new instances for annotation, so that the new instances are those that are most difficult for our classifier to label. Thus, we use the distance of texts from SVM separating hyperplanes as criteria for selecting instances for annotation. The closer an instance is from the separating hyperplanes, there is greater indecision in classifying that instance. Therefore, when we label this text manually, we believe we are helping the classifier to better define the existing limits between classes.

We performed four steps to select texts for annotation, where each step selected the 100 most complex texts for SVM. The texts that could not be processed due to parsing problems were removed. The results are shown in Table 3. They show that even when we select the texts in which the classifier has greater indecision in classifying, the SVM has not yet been able to define a boundary between the classes, which led to lower accuracy in classifying data. This shows that there is a mix between classes so that the 108 current features are not able to correctly distinguish the five levels manually annotated. Finally, we conducted a stage of selecting the 100 most easily annotated texts (those with

---

[17] It was used a libsvm implementation of SVM classifier.

[18] Available at http://143.107.183.175:22680.

[19] Available at http://143.107.183.175:21380/portlex/index.php/en/liwc.

Table 2. Full set of 108 features currently been used.

Morphological Features

Inc. of Indicative mood (preterite perfect tense)	Mean syllables per content word	Inc. of Imperative mood
Inc. of Indicative mood (imperfect tense)	Inc. of Indicative mood (future tense)	Inc. of Subjunctive mood
Inc. of Indicative mood (pluperfect tense)	Inc. of Indicative mood (present tense)	Flesch index
Inc. of Indicative mood (future of the past tense)		

Lexical Features

Adjective incidence	Adverb incidence	Content word incidence
Flesch index	Function word incidence	Mean words per sentence
Noun incidence	Number of Words	Verb incidence
Content words frequency (BP)	Min among content words freq	Mean hypernyms per verb
Brunet Index	Honore Statistic	Mean pronouns per noun phrase
Type to token ratio	Ambiguity of adjectives	Ambiguity of adverbs
Ambiguity of nouns	Ambiguity of verbs	Words before Main Verb
Inc. of Prepositions Per Clauses	Inc. of Prepositions Per Sentence	

Syntactic Features

Mean Clauses per Sentence	Mean pronouns per noun phrase	Modifiers per Noun Phrase
Noun Phrase Inc.	Mean Adverbial Adjunct Per Phrase	Inc. of Coordinate Clauses
Mean Apposition Per Clause	Inc. of Gerund Verbs	Inc. of Infinitive Verbs
Inc. of Verbals	Inc. of Coordinate Clauses	Mean of Clauses Per Sentence
Inc. of Initiating Subordinate Clauses	Inc. of Participle Verbs	Inc. of Passive Sentences
Inc. of Prepositions Per Clauses	Inc. of Prepositions Per Sentence	Inc. of Relative Clauses
Inc. of Sentences With 5 Clauses	Inc. of Sentences With Four Clauses	Inc. of Sentences With 1 Clause
Inc. of Sentences With 7 or More Clauses	Inc. of Sentences With 6 Clauses	Inc. of Sentences With 3 Clauses
Inc. of Sentences With 2 Clauses	Inc. of Sentences With Zero Clauses	Inc. of Subordinate Clauses
Inc. of Imperative mood	Inc. of Subjunctive mood	Inc. of Indicative mood (future tense)
Inc. of Indicative mood (preterite tense)	Inc. of Indicative mood (pluperfect tense)	Inc. of Indicative mood (present tense)
Inc. of Indicative mood (preterite perfect tense)	Inc. of Indicative mood (future of the past tense)	

Textual Features

Inc. of ANDs	Inc. of IFs	Inc. of ORs
Inc. of negations	Logic operators Inc.	Inc. of connectives
Inc. of additive negative connec.	Inc. of additive positive connec.	Inc. of causal negative connec.
Inc. of causal positive connec.	Inc. of logical negative connec.	Inc. of logical positive connec.
Inc. of temporal negative connec.	Inc. of temporal positive connec.	Adjacent anaphoric references
Anaphoric references	Adjacent argument overlap	Argument overlap
Adjacent stem overlap	Stem overlap	Adjacent content word overlap
Inc. of Ambiguous Discourse Markers	Inc. of Discourse Markers	Incidence of Pronouns
Inc. of 1st Person Poss. Pronouns	Inc. of 1st Person Pronouns	Inc. of 2nd Person Poss. Pronouns
Inc. of 2nd Person Pronouns	Inc. of 3th Person Poss. Pronouns	Inc. of 3th Person Pronouns

---

Punctuation Features

Punctuation diversity in a text Number of Paragraphs in a text Punctuation incidence in a text
Number of sentences in a text   Flesch index

Semantic and reader's commonsense knowledge

Inc. of LIWC Negative Words	Inc. of LIWC Positive Words
Inc. of Concrete Moving Entities in Sentences	Inc. of Concrete Moving Entities in Text
Inc. of Concrete Non-Moving Entities in Sentences	Inc. of Concrete Non-Moving Entities in Text
Inc. of Human Named Entities in Sentences	Inc. of Human Named Entity Sentence
Inc. of Named Entities in Sentences	Inc. of Named Entities in Text
Inc. of Non-Human Anim. Moving Entities in Sentences	Inc. of Non-Human Anim. Moving Entities in Text
Inc. of Non-Human Anim. Non-Moving Entities in Sentences	Inc. of Non-Human Anim. Non-Moving Entities in Text
Inc. of Topological Entities in Sentences	Inc. of Topological Entities in Text

---

greater distance from SVM separating hyperplanes) in order to contrast with the current distribution of data and the accuracy obtained. We obtained a set of 1,456 texts with the following distribution: 242 texts of level 1, 313 texts of level 2, 338 texts of level 3,287 texts of level 4 and 276 texts of level 5. The accuracy obtained when performing a 10-fold cross-validation experiment with linear kernel SVM and $C = 1$ was 52 % (+/−15).

This slight improvement in performance shows us that, in fact, there is a set of complex texts that the classifier cannot handle: due to either lack of discriminative features or lack of data for training (see confusion matrix on Table 4).

Table 3. Selection of texts via Active Learning and accuracy obtained from SVM

Step	Texts	Accuracy
First	1,070	53 (+11)
Second	1,169	50 (+14)
Third	1,268	51 (+13)
Forth	1,364	50 (+15)

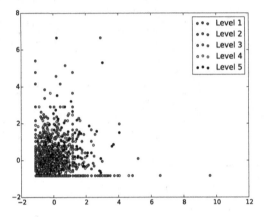

Fig. 1. $\mathbb{R}^2$ distribution of our 1,456 texts with the 2 most significant features. X-axis represents Incidence of Indicative mood (Preterit perfect tense) and Y-axis Incidence of additive negative connectives. Data scaling with mean 0 and standard deviation 1. (Color figure online)

The problem can also consist in human annotation errors. To evaluate that we performed a double-blind annotation of a random sampling of 100 texts. We obtained a Kappa score of 0.528 that represents a moderate agreement on Landis and Koch scale [24]. This agreement suggests that the manual annotation process and the labeled data should be reviewed because, as Hovy and Lavid says, "if humans can agree on something at N%, systems will achieve (N−10)%" [20]. In addition to the confusion matrix, we can see in Fig. 1 the axes that represent the two most discriminative features of the 44 selected by the RFE method of feature selection, and that there is, in fact, a mixture in the features space, particularly between the 2–3, 3–4–5, and 4–5 levels. This scenario will be hardly separated by SVM.

Feng's work [11] addresses 4 levels of difficulty, reaching the state-of-art 74 % of accuracy in English. Our experiments with fewer classes showed that, when joining classes 2 and 3, we achieved 65 % (+/−15) of accuracy, and by joining classes 4 and 5, we achieved 63 % (+/−11) of accuracy. By simultaneously joining class 2 with class 3 and 4 with 5, we reached the 74 % of accuracy achieved by the state of art. This division of grade levels better reflects the division into cycles indicated by the PCNs (1998).

Table 4. Confusion matrix of a 10-fold cross-validation experiment on our dataset.

	Level 1	Level 2	Level 3	Level 4	Level 5
Level 1	182	45	9	4	2
Level 2	36	160	102	14	1
Level 3	11	99	170	39	19
Level 4	6	13	79	118	71
Level 5	3	5	28	60	180

5   Discussion and Future Work

Our work presents the first efforts to automatically classify Portuguese texts into 5 close grade levels. The literature shows that this task is complex and, in this sense, our results are promising. We also understand that, despite the number of features used is 40 % of the 273 features used in the state-of-art work for the English language [11], there is a high rate of mixed data, especially in the central levels 4–6. Our selection of features brought 44 of the 108 features used in this work, obtaining 52 % (+/−15) of accuracy. This selection brings features to meet 5 out of 6 linguistic groups that model the manual annotation, for example: Flesch Index for the Morphological category; Ambiguity of adjectives and Incidence of Adverbs for the Lexical category; Mean Apposition Per Clause for the Syntactic category; Adjacent content word overlap and Incidence of Negative Additive Connective for the Textual category; Incidence of Human Named Entity in Text for the Semantic and reader's commonsense knowledge. By reducing the classification to 3 levels of textual complexity, we achieved 74 % of accuracy - as obtained by the state-of-art work for the English language that focuses on 4 levels.

As future work, we indicate two fronts of efforts: (i) the re-annotation of the corpus by a second annotator, using the manual annotation developed to check discrepancies; (ii) the addition of features in the six categories of linguistic elements that were used for manual classification of texts. We will replicate 6 out-of-vocabulary features described in [11]. For each text in our final corpus, these 6 features are computed using the most common 100, 200 and 500 word tokens and types based on texts from 3th grade. Also, we will implement successful features for the English language, cited by [34], such as average sentence length and features from the language model of our corpus. Moreover, and more importantly, we will implement a text type classifier to distinguish the text types occurring in our corpus. As the features of each text in our corpus are being annotated and there is a corpus annotated with text types in the Láicio-Web project [2] we will be able to better understand the correlations between text types and the others features for readability assessment in our project.

References

1. Aluisio, S., Specia, L., Gasperin, C., Scarton, C.: Readability assessment for text simplification. In: Proceedings of the NAACL HLT 2010 Fifth Workshop on Innovative Use of NLP for Building Educational Applications, pp. 1–9. Association for Computational Linguistics (2010)
2. Aluísio, S.M., Pinheiro, G.M., Manfrin, A.M., de Oliveira, L.H., Genoves Jr., L.C., Tagnin, S.E.: The lácio-web: corpora and tools to advance brazilian portuguese language investigations and computational linguistic tools. In: Proceedings of LREC, pp. 1779–1782 (2004)
3. Aluísio, S.M., Gasperin, C.: Fostering digital inclusion and accessibility: the porsimples project for simplification of portuguese texts. In: Proceedings of the NAACL HLT 2010 Young Investigators Workshop on Computational Approaches to Languages of the Americas, pp. 46–53. Association for Computational Linguistics (2010)
4. Bakhtin, M.: Estética da criação verbal. Livraria Martins Fontes, São Paulo (2003)
5. Bick, E.: The Parsing System "Palavras": Automatic Grammatical Analysis of Portuguese in a Constraint Grammar Framework. Aarhus University Press, Aarhus (2000)
6. Biderman, M.T.C.: Dicionários do português: da tradição à contemporaneidade. ALFA: Revista de Linguística **47**(1) (2003)
7. Cimadon, É.: Funções executivas em crianças com dificuldade de leitura (2012)
8. Collins-Thompson, K., Bennett, P.N., White, R.W., de la Chica, S., Sontag, D.: Personalizing web search results by reading level. In: Proceedings of the 20th ACM International Conference on Information and Knowledge Management, pp. 403–412. ACM (2011)
9. Curto, P.: Classificador de textos para o ensino de português como segunda língua. Master's thesis, Universidade Técnico Lisboa, Portugal (2014)
10. Dell'Orletta, F., Venturi, G., Cimino, A., Montemagni, S.: T2k2: system for automatically extracting and organizing knowledge from texts. In: Proceedings of the 9th International Conference on Language Resources and Evaluation (LREC 2014) (2014)
11. Feng, L., Jansche, M., Huenerfauth, M., Elhadad, N.: A comparison of features for automatic readability assessment. In: Proceedings of the 23rd International Conference on Computational Linguistics: Posters, COLING 2010, pp. 276–284. Association for Computational Linguistics (2010)
12. Flor, M., Klebanov, B.B.: Associative lexical cohesion as a factor in text complexity. Int. J. Appl. Linguist. **165**(2), 223–258 (2014)
13. Forsyth, J.N.: Automatic Readability Detection for Modern Standard Arabic. Master's thesis, Brigham Young University, United States
14. François, T.: An analysis of a french as a foreign language corpus for readability assessment. NEALT Proc. Ser. **22**, 13–32 (2014)
15. Fulcher, K.Y., White, P.D.: Randomised controlled trial of graded exercise in patients with the chronic fatigue syndrome. BMJ **314**(7095), 1647–1652 (1997)
16. Giangiacomo, M.C.P.B., Navas, A.L.G.P.: A influência da memória operacional nas habilidades de compreensão de leitura em escolares de 4ª série influence of working memory in reading comprehension in 4th grade students. Sociedade Brasileira de Fonoaudiologia **13**(1), 69–74 (2008)

17. Graesser, A.C., McNamara, D.S., Kulikowich, J.M.: Coh-metrix providing multi-level analyses of text characteristics. Edu. Res. **40**(5), 223–234 (2011)
18. Graesser, A.C., McNamara, D.S., Louwerse, M.M., Cai, Z.: Coh-metrix: analysis of text on cohesion and language. Behav. Res. Methods, Instrum. Comput. **36**(2), 193–202 (2004)
19. Hancke, J., Vajjala, S., Meurers, D.: Readability classification for german using lexical, syntactic, and morphological features. In: Proceedings of COLING, pp. 1063–1080 (2012)
20. Hovy, E., Lavid, J.: Towards a 'science'of corpus annotation: a new methodological challenge for corpus linguistics. Int. J. Transl. **22**(1), 13–36 (2010)
21. Kato, M.: O aprendizado da leitura. Martins Fontes, São Paulo (1985)
22. Kato, M.A.: No mundo da escrita: uma perspectiva psicolingüística, vol. 9. Editora Ática (1986)
23. da Graça Krieger, M.: Dicionários para o ensino de língua materna: princípios e critérios de escolha. Revista Língua & Literatura **7**(10-11), 101–112 (2012)
24. Landis, J.R., Koch, G.G.: The measurement of observer agreement for categorical data. Biometrics **33**(1), 159–174 (1977)
25. Lennon, C., Burdick, H.: The lexile framework as an approach for reading measurement and success. Electronic publication on www.lexile.com (2004)
26. LoPucki, L.M.: System and method for enhancing comprehension and readability of legal text (2014). US Patent 8, 794–972
27. Maia, M.: Gramática e parser. Boletim da ABRALIN **1**(26), 288–291 (2001)
28. Maia, M.: Efeitos do status argumental e de segmentação no processamento de sintagmas preposicionais em português brasileiro. Cadernos de Estudos Lingüísticos **50**(1) (2011)
29. Maia, M., Finger, I.: Processamento da linguagem. Educat, Pelotas (2005)
30. Martins, T.B., Ghiraldelo, C.M., Nunes, M.d.G.V., de Oliveira Jr., O.N.: Readability formulas applied to textbooks in brazilian portuguese. Icmsc-Usp (1996)
31. Maziero, E.G., Pardo, T.A.S., Aluísio, S.M.: Ferramenta de análise automática de inteligibilidade de córpus (aic). Technical report (2008)
32. Navas, A.L.G.P., Pinto, J.C.B.R., Dellisa, P.R.R.: Avanços no conhecimento do processamento da fluência em leitura: da palavra ao texto improvements in the knowledge of the reading fluency processing: from word to text. Sociedade Brasileira de Fonoaudiologia **14**(3), 553–9 (2009)
33. O'Reilly, T., Sinclair, G., McNamara, D.S.: istart: a web-based reading strategy intervention that improves students's science comprehension. In: CELDA, pp. 173–180 (2004)
34. Petersen, S.E., Ostendorf, M.: A machine learning approach to reading level assessment. Comput. Speech Lang. **23**(1), 89–106 (2009)
35. de Salles, J.S.F., Parente, M.A.d.M.P.: Heterogeneidade nas estratégias de leitura/escrita em crianças com dificuldades de leitura e escrita. Psico **37**(1), 83–90
36. San Norberto, E.M., Gómez-Alonso, D., Trigueros, J.M., Quiroga, J., Gualis, J., Vaquero, C.: Readability of surgical informed consent in spain. Cirugía Española **92**(3), 201–207 (2014)
37. Scarton, C., Aluísio, S.: Análise da Inteligibilidade de textos via ferramentas de Processamento de Língua Natural: adaptando as métricas do Coh-Metrix para o Português. Linguamática **2**(1), 45–62 (2010)
38. Sheehan, K.M., Flor, M., Napolitano, D.: A two-stage approach for generating unbiased estimates of text complexity. In: Proceedings of the Workshop on Natural Language Processing for Improving Textual Accessibility, pp. 49–58 (2013)

39. Stenner, A.J.: Measuring reading comprehension with the lexile framework (1996)
40. Tong, S., Koller, D.: Support vector machine active learning with applications to text classification. J. Mach. Learn. Res. **2**, 45–66 (2002)
41. Vajjala, S., Meurers, D.: Readability assessment for text simplification: from analysing documents to identifying sentential simplifications. Int. J. Appl. Linguist. **165**(2), 194–222 (2014)

Improving Question-Answering for Portuguese Using Triples Extracted from Corpora

Ricardo Rodrigues[1,2(✉)] and Paulo Gomes[1]

[1] Centre for Informatics and Systems of the University of Coimbra,
Coimbra, Portugal
{rmanuel,pgomes}@dei.uc.pt
[2] College of Education of the Polytechnic Institute of Coimbra, Coimbra, Portugal

Abstract. We present here an evolution of a QA system for Portuguese that uses *subject-predicate-object* triples extracted from sentences in a corpus. The system is supported by indices that store those triples, related sentences and documents. It processes the questions and retrieves answers based on the triples.

For purposes of testing and evaluation, we have used the CHAVE corpus, used in multiple editions of the CLEF multilingual QA tracks. The questions from those editions were used to query and benchmark our system. Currently, the system manages to answer up to 42 % of those questions. This document describes the modules that compose the system and how they are combined, providing a brief analysis on them, and also current results, as well as some expectations regarding future work.

Keywords: Question Answering · Open information extraction · Triple extraction · Portuguese

1 Introduction

The quest for information is a quintessential human endeavour. And as soon as computers came into play, they immediately started to be used for storing and retrieving information, most of which in the form of natural language. Not long after, there were attempts to use computers in tasks related to natural language processing (NLP), trying to make sense of all the data described using natural language, which keeps increasing by the day. However, it is not enough to store and retrieve documents, being needed tools that can process them in order to retrieve just what the user wants or needs, instead of just a list of documents.

This issue is addressed by question answering (QA) systems [25], which allow the user to interact with those systems by means of natural language, and process documents whose contents are specified also using natural language.

In this context, we present RAPPORT, a system that addresses QA for Portuguese that uses triples extracted from sentences in a corpus, much like open information extraction performs, that are then used to present "short answers" (passages), alongside the sentences and documents they belong to.

© Springer International Publishing Switzerland 2016
J. Silva et al. (Eds.): PROPOR 2016, LNAI 9727, pp. 25–37, 2016.
DOI: 10.1007/978-3-319-41552-9_3

In the remaining document, we present a brief contextualization on QA, address related work, describe the overall used approach and each of its modules, and draw some conclusions and reflections about future work.

2   Question Answering

QA, much like other subfields of information retrieval (IR), may include techniques such as: named entity recognition (NER) or semantic classification of entities, relation extraction between entities, and selection of semantically relevant sentences or chunks [16], beyond customary sentence splitting, tokenization, lemmatization, and part-of-speech (POS) tagging. QA can also address a restricted set of topics, in a closed domain, or forgo that restriction, operating in an open domain. Focusing specifically on open domain QA, it can consist of fundamentally two distinct approaches: IR-based QA or knowledge-based QA [11].

QA systems based on IR typically follow the framework depicted in Fig. 1, where the processing stages are made at run-time, except for document indexing. Knowledge-based QA systems, although sharing some similarities, tend to adopt logical representations of facts, for instance, through the use triples (*subject*, *predicate* and *object*) backed up by ontologies, often implemented by means of RDF triple stores, using SPARQL to query them [26], or similar data repositories.

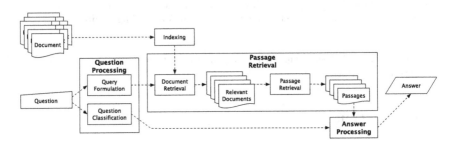

Fig. 1. A typical framework for a IR-based QA system (reproduced [11])

Regarding specific approaches to Portuguese, we present next the most relevant works whose results are compared against our work later on this document.

2.1   Senso

The Senso Question Answering System [22] (alias PTUE [20]) uses a local knowledge base, providing semantic information for text search terms expansion. It is composed of five major modules: *query* (for question analysis), *libs* (for corpora management), *ontology* (for knowledge representation), *solver* (for answer searching), and *web interface*. After all modules are used, the results are merged for answer list validation, to filter and adjust answers weight, ranking them.

2.2   Esfinge

Esfinge [5] is a general domain QA system that tries to take advantage of the great amount of information existing in the Web. Esfinge relies on pattern identification and matching. For each question, a tentative answer beginning is created. Then the probable answer beginning is used to search the corpus, through a search engine, in order to find possible answers that match the same pattern. In the remaining stages of the process, *n-grams* are scored and NER is performed in order to improve the performance of the system.

2.3   RAPOSA

The RAPOSA Question Answering System [24] tries to provide a continuous online processing chain from question to answer, combining stages from information extraction and retrieval. The system involves expanding queries for event-related or action-related factoid questions, using a verb thesaurus automatically generated using information extracted from large corpora. RAPOSA consists of six modules more or less typical on QA systems: a *question parser*, a *query generator*, a *snippet searcher*, an *answer extractor*, *answer fusion*, and an *answer selector*. It deals with two categories of questions: definitions and factoids.

2.4   IdSay

IdSay: Question Answering for Portuguese [3, 4] uses mainly techniques from the area of IR, where the only external information that it uses, besides the text collections, is lexical information for the Portuguese language. IdSay uses a conservative approach to QA, being its main stages: *question analysis, set Wikipedia answer (SWAN), document retrieval, passage retrieval, answer extraction* and *answer validation*. IdSay starts by performing document analysis and then proceeding to entity recognition. After that, the system makes use of patterns to define the type of the questions and expected answers. However, contrary to most QA systems, it does not store passages in the IR module, but documents, with the passages being extracted in real time, allowing for more flexibility.

2.5   QA@L²F

QA@L²F [15], the QA system from L²F, INESC-ID, is a system that relies on three main tasks: *information extraction, question interpretation* and *answer finding*. The system starts by processing and analyzing the text sources in order to extract potentially relevant information (such as named entities or relations between concepts), which is stored into a knowledge base. Then, the questions are also processed and analyzed, selecting which terms should be used to build a query to search the database. Finally, the retrieved records are then processed, selecting the answer according to the question type and other strategies.

2.6 Priberam

Priberam's Question Answering System for Portuguese [2] is divided in five major modules: *indexation, question analysis, document retrieval, sentence retrieval,* and *answer extraction*. It starts by processing the documents, mainly at sentence level, and storing related data (lemmas, heads of derivation, named entities and fixed expressions, question categories and ontology domains) in different indices. Then each question is processed, extracting and expanding pivots for querying the indices. The resulting queries are used first for retrieving documents based on their scores (using lexical frequency, document frequency, and weighted POS tags) and then for selecting the sentences and extracting the answers, according to matches against the pivotal words in the questions.

2.7 GistSumm

Brazil's Núcleo Interinstitucional de Lingüística Computacional (NILC) had built previously a summarization system, dubbed GistSum [19], that has been adapted for use in the task of monolingual QA for Portuguese texts. NILC's system comprises three main processes: *text segmentation, sentence ranking,* and *extract production* [6], associating sentences to a topic. The questions are then matched against the sentences and associated summaries, with the highest scored sentences being used to produce an answer.

3 RAPPort

Our system adheres to most of the typical framework for a QA system, combining aspects from both IR-based QA and knowledge-based QA. It does also improve on some techniques that differ from other approaches to Portuguese.

One of the most identifying elements of RAPPort is the use of triples as the basic unit of information regarding any topic, represented by a *subject*, a *predicate* and an *object*, and then using those triples as a basis for answering questions. This approach also possesses somes characteristics from open information extraction, regarding the extraction and storage of information in triples [8].

The system depends on a combination of four major modules for addressing information extraction, storage, querying and retrieving, namely:

- triple extraction (performed offline);
- triple storage (performed offline);
- data querying (performed online);
- and answer retrieving (performed online).

Each of these modules is described next, specifying the main tasks that compose them. An overview of the modules can also be seen in Fig. 2.

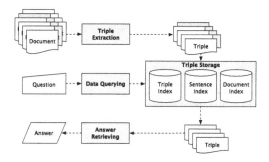

Fig. 2. Our system's general approach

3.1 Triple Extraction

This module processes the contents of the corpus, picking each of the documents, selecting sentences and extracting triples. It includes multiple tasks, namely sentence splitting, phrase chunking, tokenization, POS tagging, lemmatization, dependency parsing, and NER. Except for lemmatization and dependency parsing, these tasks are done using the *Apache OpenNLP toolkit*[1], with some minor tweaks for better addressing Portuguese, and with the models used for chunking and in being specifically created, as there was no available pre-built models.

For the lemmatization process, *LemPORT* [21], a Portuguese specific lemmatizer was used. For dependency parsing, it was used *MaltParser* [18], with a model trained on Bosque 8.0[2] [1]. The output of MaltParser is also further processed in order to group the tokens around the *main* dependencies, such as: subject, root (verb), and objects, among others.

Triple extraction is performed using two complementary approaches, both depending on named entities for determining which triples are of use. The triples are defined by *subject*, *predicate*, and *object*, that are obtained either through the proximity relations between phrase chunks, or through the analysis of the dependencies in sentences. Only the triples with entities in the *subject* or in the *object* are stored for future querying. Also, the predicate has the verb stored in its lemmatized form in order to facilitate later matches.

In the triples that are based on the proximity between chunks, most of the predicates comprehend, but are not necessarily limited to, the verbs *ser* (to be), *pertencer* (to belong), *haver* (to have), and *ficar* (to be located). For instance, if two noun phrase (NP) chunks are found sequentially, and the first chunk contains a named entity, it is highly probable that it is further characterized by the second chunk. If the second chunk starts with a determinant or a noun, the predicate of the future triple is set to *ser*; if it starts with the preposition *em* (in), it is used the verb *ficar*; if it starts with the preposition *de* (of), it is used the verb *pertencer*; and so on. An algorithm describing the process is found in Algorithm 1.

Data: Corpus documents
Result: Triple list
Read documents;
foreach *document* **do**
    Split sentences;
    foreach *sentence* **do**
        Tokenize, POS tag, lemmatize;
        Extracts phrase chunks and dependency chunks;
        Extract named entities;
        foreach *phrase chunk* **do**
            if *chunk contains any entity* **then**
                if *neighbouring chunk has a specific type* **then**
                    Create *triple* relating both chunks, depending on the
                    neighbouring chunk type and contents;
                    Add it to the triple list;
                end
            end
        end
        foreach *dependency chunk* **do**
            if *chunk contains any entity and is a subject or an object* **then**
                Create *triple* using the subject or object, the root, and
                corresponding object or subject, respectively;
                Add it to the triple list;
            end
        end
    end
end

Algorithm 1. Triple Extraction Algorithm

As an example, the sentence "Mel Blanc, o homem que deu a sua voz a o coelho mais famoso de o mundo, Bugs Bunny, era alérgico a cenouras."[3] yields distinct triples, such as: "{*Bugs Bunny*} {*ser*} {*o coelho mais famoso do mundo*}" and "{*Mel Blanc*} {*ser*} {*o homem que deu a sua voz ao coelho mais famoso do mundo*}", both using the proximity approach, and "{*Mel Blanc*} {*ser*} {*alérgico a cenouras*}", using the dependency approach.

3.2 Triple Storage

After triple extraction is performed, *Lucene* [14] is used for storing the triples, the sentences where the triples are found, and the documents that, by their turn, contain those sentences. For that purpose, three indices were created:

– the triple index stores the triples (subject, predicate and object), their *ids*, and the *ids* of the sentences and documents that contain them;

---

[3] Loosely translated as: "Mel Blanc, the man who lent his voice to the world's most famous rabbit, Bugs Bunny, was allergic to carrots.".

- the sentence index stores the sentences *ids* (a sequential number representing their order within the document), the tokenized text, the lemmatized text and the documents *ids* they belong to;
- the document index stores the data describing the document, as found in CHAVE (number, *id*, date, category, author, and original text);

Although each index is virtually independent from the others, they can refer one another by using the *ids* of the sentences and of the documents. That way, it is easy to determine the relations between documents, sentences, and triples. These indices (mainly the sentence and the triple indices) are then used in the next steps of the presented approach.

3.3 Data Querying

In a similar way to the sentences in the corpus, the questions are processed in order to extract tokens, lemmas and named entities, and identify their types, categories and targets (although the last three tasks are not currently performed).

For building the queries, the system starts by performing NER and lemmatizing the questions. The lemmas are useful for broadening the matches and results that could be found only by using the tokens. The queries are essentially built on the lemmas found in the questions. All the query elements are, by default, optional, except for named entities. If no entities are present in the questions, proper nouns are made mandatory; by its turn, if there are also no proper nouns, (common) nouns replace them as mandatory keywords in the queries.

For instance, in order to retrieve the answer to the question *"A que era alérgico Mel Blanc?"*[4], the Lucene query will end up being defined by five terms: *"+Mel_Blanc a que ser alérgico"*. We have chosen to keep all the lemmas because Lucene scores higher the hits with the optional lemmas, and virtually ignores them if they are not present. The query is then applied to the sentence index. When a match occurs, the associated triples are retrieved, along with the document data. In the same step, when applicable, and for the moment, just synonyms for the verb are added to query as optional items, using the synonymy relations defined in PAPEL [10].

The triples that are related to the sentence are then processed, checking for the presence of the question entities in either the *subject* or the *object* of the triples, for selecting which triples are of interest.

3.4 Answer Retrieving

After a sentence matches a query, as stated before, the associated triples and document data are retrieved — and this goes for all the sentences matching that query. As the document data is only used for better characterizing the answers, let us focus on the triples.

For each triple, it is retrieved each of its components: if the best match against the query is found in the *subject*, the *object* is returned as being the

---

[4] Loosely translated as: "What was Mel Blanc allergic to?".

answer; if, on the other hand, the best match is found against the *object*, it is the *subject* that is returned. This candidate answer, before being presented to the user, is ordered against other candidate answers. For that, the triples are used once again, as the candidate answers are ordered against the number of triples they are found in. An algorithm describing both data querying and this process is found in Algorithm 2.

Data: Question &Indices
Result: Answers
Create query using *named entities* (or, if inexistent, *proper nouns*, or *nouns*) as mandatory, and the remaining lemmas from the *question* as optional;
Run *query* against *sentence index*;
foreach *sentence hit* **do**
    Retrieve triples related to the sentence hit;
    foreach *triple* **do**
        if *subject contains named entities from question* **then**
            Add *object* to *answers* and retrieve sentence and document associated with the triple;
        end
        else if *object contains named entities from question* **then**
            Add *subject* to *answers* and retrieve sentence and document associated with the triple;
        end
    end
end
Order *answers* based in the number of triples they belong to;

Algorithm 2. Answer Retrieval Algorithm

Continuing with the example provided earlier, after the correct sentence is retrieved, of the three corresponding triples, the one that best matches the question is "{*Mel Blanc*} {*ser*} {*alérgico a cenouras*}" — there is a match on the *predicate* and the named entity is found in the *subject*. Removing from the triple the terms found in the question, what remains must yield the answer: "[a] cenouras". Besides that, as the named entity, Mel Blanc, is found in the *subject* of the triple, the answer is most likely to be found in the *object*, and so retrieved.

4   Experimentation Results

For the experimental work, we have used the CHAVE corpus [23], a collection of 1456 editions of newspapers "Público" and "Folha de São Paulo", from 1994 and 1995, with each of the editions comprehending about one hundred articles, identified by *id*, number, date, category, author, and the text of article itself.

CHAVE was used in the Cross Language Evaluation Forum (CLEF) multilingual QA tracks for Portuguese [7,9,12,13,27], although in the editions of 2007 and 2008 a dump of the Portuguese Wikipedia was also used in addition — that is the reason, in the present paper, for just being addressing the 2004, 2005 and 2006 campaigns for evaluation purposes.

Nearly all of the questions used in each of the CLEF editions (200 for each language), and respective answers, are known. It is also known the results of each of the contestant systems. The questions used in CLEF adhere to the following criteria [13]: they can be *list* questions, *embedded* questions, *yes/no* questions (although none was found in the questions used for Portuguese), *who, what, where, when, why,* and *how* questions, and definitions.

For reference, in Table 1 there is a summary of the best results for the Portuguese QA tasks on CLEF from 2004 to 2008 (abridged [7,9,12,13,27]), alongside with the arithmetic mean for each system comprehending the editions where they were contenders. At the end of the table, it is also shown the current results of our system, for a maximum of ten answers per question.

Table 1. Comparison of the Results at CLEF 2004 to 2008

Approach	Overall Accuracy (%)						
	2004	2005	2006	2007	2008	*(2004–06 Avg)*	*(2004–08 Avg)*
Esfinge	15.08	23.00	24.5	8.0	23.5	*(20.86)*	*(18.82)*
Senso	28.54	25.00	—	42.0	46.5	*(26.77)*	*(35.51)*
Priberam	—	64.50	67.0	50.5	63.5	*(65.75)*	*(61.34)*
NILC	—	—	1.5	—	—	*(1.5)*	*(1.5)*
RAPOSA	—	—	13.0	20.0	14.5	*(13.0)*	*(18.83)*
QA@L^{2}F	—	—	—	13.0	20.0	—	*(16.5)*
IdSay	—	—	—	—	32.5	—	*(32.5)*
RAPPORT	41.21	45.00	38.50	—	—	*(41.57)*	—

As already mentioned, we are only addressing the questions for Portuguese used in CLEF in 2004, 2005 and 2006. As such, a grand total of 599 questions[5] were used for testing our system, of which 10 % don't have an answer in the corpus — being 'NIL' the expected answer in that case. That is the reason for considering the average result of our system in Table 1 just for the years 2004 to 2006, and omitting the results for the years 2007 and 2008.

For verifying if the retrieved answers match the expected answers, the answers must contain the already known answers, and the corresponding document *ids* must also match those of the known answers.

Using the set of questions from 2004 to 2006, which were known to have their answers found on CHAVE, we were able to find the answers to 41.57 % of the questions (249 in 599), grouping all the question from the already identified editions of CLEF, with a limit of ten answers for each question. (If that limit is

---

[5] In 2004, one of the questions was unintentionally duplicated, hence 599 and not 600.

relinquished, the number of answered questions rises to 67.61 %, which may lead to the conclusion that one of the big issues to be further addressed is to improve the ordering and selection of the answers.)

For comparison purposes, a previous version of RAPPORT (whose main differences to the current version was not using verb synonyms, and mainly the ranking of the answers, which was then directly related to the score of each Lucene match of the sentences housing the triples against the query generated from the question), for a limit of ten answers per question, achieved 20.75 % of right answers, and without a limit, 43.33 % of right answers.

On the answers that have not been found, we have determined that in a few cases the fault is due to questions depending on information contained in other questions or their answers. There are certainly also many shortcomings in the creation of the triples, mainly on the phrase chunks that are close together, as opposed to the dependency chunks, that should and must be addressed, in order to improve and create more triples. Furthermore, there are questions that refer to entities that fail to be identified as such by our system, an so no triples were created for them when processing the sentences.

5   Conclusions and Future Work

We have come to the conclusion that using triples as a means of representing and storing information found in corpora has strong advantages, besides allowing the exploration of a different way of supporting QA systems for Portuguese.

Firstly, the use of triples, restricting them to those containing named entities, provides a way of selecting which information should really be stored (just the triples and associated sentences), instead of, for instance, storing and indexing all text as a source for providing answers, having to processed the text later. Secondly, triples, being composed mostly of small chunks, already contain in themselves (in the *subject*, *predicate* or *object*) the passage that will be used as the short answer to a question.

Earlier experiments have shown us that trying to store every single bit of information regarding texts in corpora — such as using ontologies for storing syntactic and semantic data, or indexing and storing all and whole sentences — creates considerable overhead and noise, besides having its toll on performance. Using triples in the way described here helps to mitigate these problems.

Although the proposed system scores a strong second place for the three years considered (using solely CHAVE), the use of triples keeps proving to be a promising way of selecting the right and shorter answers to most of the questions addressed. However, there is still a lot that can be improved.

Triples could be improved, namely those that are built from the relations of proximity between chunks, so the system is able to have a number of retrieved triples on par with the sentences that contain the answers (and the triples). Another boost to the approach would be to properly differentiate the queries accordingly to the types of the named entities found in the questions, and improve NER, both on questions an on corpus sentences.

Another aspect that should be considered is the use of coreference resolution in order to increase the number of extracted triples by means of replacing, for instance, pronouns with the corresponding, if any, named entities.

And the system has yet to properly address NIL answers, as it currently provides almost always an answer, even if the match when querying the indices has an extremely low score.

We believe that expanding the queries using the above techniques, together with the creation of better models to extract triples and coreference resolution, will achieve better results in a short time span.

Finally, the next major goal is to use the Portuguese Wikipedia as a repository of information, either alongside CHAVE, to address the latter editions of CLEF, or by itself, as it has happened in Págico [17].

References

1. Afonso, S., Bick, E., Haber, R., Santos, D.: Floresta sintá(c)tica: a treebank for portuguese. In: Rodríguez, M.G., Araujo, C.P.S. (eds.) Proceedings of LREC 2002, The Third International Conference on Language Resources and Evaluation, pp. 1698–1703. ELRA, Paris (2002)
2. Amaral, C., Figueira, H., Martins, A., Mendes, A., Mendes, P., Pinto, C.: Priberam's question answering system for portuguese. In: Peters, C., Gey, F.C., Gonzalo, J., Müller, H., Jones, G.J.F., Kluck, M., Magnini, B., de Rijke, M., Giampiccolo, D. (eds.) CLEF 2005. LNCS, vol. 4022, pp. 410–419. Springer, Heidelberg (2006)
3. Carvalho, G., de Matos, D.M., Rocio, V.: IdSay: question answering for portuguese. In: Peters, C., Deselaers, T., Ferro, N., Gonzalo, J., Jones, G.J.F., Kurimo, M., Mandl, T., Peñas, A., Petras, V. (eds.) CLEF 2008. LNCS, vol. 5706, pp. 345–352. Springer, Heidelberg (2009)
4. Carvalho, G., Matos, D.M., Rocio, V.: Robust Question Answering. In: PhD and MSc/MA Dissertation Contest of the of the 10th International Conference on Computational Processing of the Portuguese Language (PROPOR 2012), Coimbra, Portugal, April 2012
5. Costa, L.F.: Esfinge – a question answering system in the web using the web. In: Proceedings of the Demonstration Session of the 11th Conference of the European Chapter of the Association for Computational Linguistics, pp. 410–419. Association for Computational Linguistics, Trento, Italy, April 2006
6. Filho, P.P.B., de Uzêda, V.R., Pardo, T.A.S., das Graças Volpe Nunes, M.: Using a Text Summarization System for Monolingual Question Answering. In: CLEF 2006 Working Notes (2006)
7. Forner, P., Peñas, A., Agirre, E., Alegria, I., Forăscu, C., Moreau, N., Osenova, P., Prokopidis, P., Rocha, P., Sacaleanu, B., Sutcliffe, R., Tjong Kim Sang, E.: Overview of the CLEF 2008 multilingual question answering track. In: Peters, C., Deselaers, T., Ferro, N., Gonzalo, J., Jones, G.J.F., Kurimo, M., Mandl, T., Peñas, A., Petras, V. (eds.) CLEF 2008. LNCS, vol. 5706, pp. 262–295. Springer, Heidelberg (2009)
8. Gamallo, P.: An overview of open information extraction. In: Pereira, M.J.V., Leal, J.P., Simões, A. (eds.) Proceedings of the 3rd Symposium on Languages, Applications and Technologies (SLATE 2014), pp. 13–16. Schloss Dagstuhl – Leibniz-Zentrum für Informatik Dagstuhl Publishing, Germany (2014)

9. Giampiccolo, D., Forner, P., Herrera, J., Peñas, A., Ayache, C., Forascu, C., Jijkoun, V., Osenova, P., Rocha, P., Sacaleanu, B., Sutcliffe, R.F.E.: Overview of the CLEF 2007 multilingual question answering track. In: Peters, C., Jijkoun, V., Mandl, T., Müller, H., Oard, D.W., Peñas, A., Petras, V., Santos, D. (eds.) CLEF 2007. LNCS, vol. 5152, pp. 200–236. Springer, Heidelberg (2008)

10. Oliveira, H.G., Santos, D., Gomes, P., Seco, N.: PAPEL: a dictionary-based lexical ontology for portuguese. In: Teixeira, A., de Lima, V.L.S., de Oliveira, L.C., Quaresma, P. (eds.) PROPOR 2008. LNCS (LNAI), vol. 5190, pp. 31–40. Springer, Heidelberg (2008)

11. Jurafsky, D., Martin, J.H.: Speech and Language Processing, 2nd edn. Pearson Education International Inc., Upper Saddle River (2008)

12. Magnini, B., Giampiccolo, D., Forner, P., Ayache, C., Jijkoun, V., Osenova, P., Peñas, A., Rocha, P., Sacaleanu, B., Sutcliffe, R.F.E.: Overview of the CLEF 2006 multilingual question answering track. In: Peters, C., Clough, P., Gey, F.C., Karlgren, J., Magnini, B., Oard, D.W., de Rijke, M., Stempfhuber, M. (eds.) CLEF 2006. LNCS, vol. 4730, pp. 223–256. Springer, Heidelberg (2007)

13. Magnini, B., Vallin, A., Ayache, C., Erbach, G., Peñas, A., de Rijke, M., Rocha, P., Simov, K.I., Sutcliffe, R.F.E.: Overview of the CLEF 2004 multilingual question answering track. In: Peters, C., Clough, P., Gonzalo, J., Jones, G.J.F., Kluck, M., Magnini, B. (eds.) CLEF 2004. LNCS, vol. 3491, pp. 371–391. Springer, Heidelberg (2005)

14. McCandless, M., Hatcher, E., Gospodnetić, O.: Lucene in Action. Manning Publications Co., Greenwich (2010)

15. Mendes, A., Coheur, L., Mamede, N.J., Ribeiro, R., Batista, F., de Matos, D.M.: QA@L^{2}F, first steps at QA@CLEF. In: Peters, C., Jijkoun, V., Mandl, T., Müller, H., Oard, D.W., Peñas, A., Petras, V., Santos, D. (eds.) CLEF 2007. LNCS, vol. 5152, pp. 356–363. Springer, Heidelberg (2008)

16. Moens, M.F.: Information Extraction: Algorithms and Prospects in a Retrieval Context. Springer, Heidelberg (2006)

17. Mota, C.: Resultados Págicos: Participação, Resultados e Recursos. Linguamática 4(1), April 2012

18. Nivre, J., Hall, J., Nilsson, J., Chanev, A., Eryiğit, G., Kübler, S., Marinov, S., Marsi, E.: MaltParser: a language-independent system for data-driven dependency parsing. Nat. Lang. Eng. **13**(2), 95–135 (2007)

19. Pardo, T.A.S., Rino, L.H.M., Nunes, M.G.V.: GistSumm: a summarization tool based on a new extractive method. In: Mamede, N.J., Baptista, J., Trancoso, I., Nunes, M.G.V. (eds.) PROPOR 2003. LNCS, vol. 2721, pp. 210–218. Springer, Heidelberg (2003)

20. Quaresma, P., Quintano, L., Rodrigues, I., Saias, J., Salgueiro, P.: The University of Évora approach to QA@CLEF-2004. In: CLEF 2004 Working Notes (2004)

21. Rodrigues, R., Gonçalo-Oliveira, H., Gomes, P.: LemPORT: a high-accuracy cross-platform lemmatizer for portuguese. In: Pereira, M.J.V., Leal, J.P., Simões, A. (eds.) Proceedings of the 3rd Symposium on Languages, Applications and Technologies (SLATE 2014). pp. 267–274. Germany (2014)

22. Saias, J., Quaresma, P.: The senso question answering approach to portuguese QA@CLEF-2007. In: Nardi, A., Peters, C. (eds.) Working Notes for the CLEF 2007 Workshop, Budapest, Hungary, September 2007

23. Santos, D., Rocha, P.: The key to the first CLEF with portuguese: topics, questions and answers in CHAVE. In: Peters, C., Clough, P., Gonzalo, J., Jones, G.J.F., Kluck, M., Magnini, B. (eds.) Multilingual Information Access for Text, Speech and Images. LNCS, vol. 3491, pp. 821–832. Springer, Heidelberg (2005)

24. Sarmento, L., Oliveira, E.: Making RAPOSA (FOX) smarter. In: Nardi, A., Peters, C. (eds.) Working Notes for the CLEF 2007 Workshop, Budapest, Hungary, September 2007
25. Strzalkowski, T., Harabagiu, S. (eds.): Advances in Open Domain Question Answering, Text, Speech and Language Technology, vol. 32. Springer, Heidelberg (2006)
26. Unger, C., Bühmann, L., Lehmann, J., Ngomo, A.C.N., Gerber, D., Cimiano, P.: Template-based question answering over RDF data. In: Proceedings of the 21st International Conference on World Wide Web (WWW 2012), pp. 639–648. ACM Press, Lyon, France, April 2012
27. Vallin, A., Magnini, B., Giampiccolo, D., Aunimo, L., Ayache, C., Osenova, P., Peñas, A., de Rijke, M., Sacaleanu, B., Santos, D., Sutcliffe, R.F.E.: Overview of the CLEF 2005 multilingual question answering track. In: Peters, C., Gey, F.C., Gonzalo, J., Müller, H., Jones, G.J.F., Kluck, M., Magnini, B., de Rijke, M., Giampiccolo, D. (eds.) CLEF 2005. LNCS, vol. 4022, pp. 307–331. Springer, Heidelberg (2006)

Applying Lexical-Conceptual Knowledge for Multilingual Multi-document Summarization

Ariani Di Felippo[1,2], Fabrício E.S. Tosta[1],
and Thiago A.S. Pardo[1,3(✉)]

[1] Interinstitutional Center for Computational Linguistics (NILC),
São Carlos, SP, Brazil
arianidf@gmail.com, fabricio3341@hotmail.com
[2] Language and Literature Department (DL),
Federal University of São Carlos (UFSCar), Rodovia Washington Luís,
km 235 - SP 310, São Carlos 13565-905, Brazil
[3] Institute of Mathematical and Computer Sciences (ICMC),
University of São Paulo (USP), Avenida Trabalhador São-carlense,
400, São Carlos 13566-590, Brazil
taspardo@icmc.usp.br

Abstract. We define Multilingual Multi-Document Summarization (MMDS) as the process of identifying the main information of a cluster with (at least) two texts, one in the user's language and one in a foreign language, and presenting it as a summary in the user's language. Although it is a relevant task due to the increasing amount of on-line information in different languages, there are only baselines for (Brazilian) Portuguese, which apply machine-translation to obtain a monolingual input and superficial features for sentence extraction. We report our investigation on the application of *conceptual frequency* measure to build a summary in Portuguese from a bilingual cluster (Portuguese and English). The methods tackle two additional challenges: using Princeton WordNet for nouns annotation and applying MT to translate selected sentences in English to Portuguese. The experiments were performed using a *corpus* of 20 clusters, and show that lexical-conceptual knowledge improves the linguistic quality and informativeness of extracts.

Keywords: Multilingual · Multi-document · Summarization · Concept · Extract

1 Introduction

As the amount of on-line news texts in different languages is growing at an exponential pace, Multilingual Multi-Document Summarization (MMDS) is a quite desirable task. It aims at identifying the main information in a cluster of (at least) two texts, one in the user's language and one in a foreign language, and presenting it as a coherent/cohesive summary in the user's languages. However, MMDS is a highly challenging task, since

J. Silva et al. (Eds.): PROPOR 2016, LNAI 9727, pp. 38–49, 2016.
DOI: 10.1007/978-3-319-41552-9_4

it requires merging content in different languages as well as dealing with the classical multi-document issues, such as capturing the most relevant content, and maintaining the coherence/cohesion of summary by treating redundancy.

The few previous methods usually consist of two steps: translation of the foreign texts and summarization [1–4]. The first step is performed by some machine-translation (MT) engine, producing a monolingual multi-document cluster. Then, an extractive[1] multi-document summarization (MDS) method is used to build the summaries, which sometimes treats redundancy. About the input, Roark and Fisher [2] extract sentences from the machine-translated and the original texts in the user's language. Consequently, the summaries present ungrammatical sentences and disfluencies resulting from MT. Evans et al. [1, 3] only extract sentences from the translated texts, and replace them with similar ones from the text in the user's language. This method avoids the MT problems, but the content selection does not take into account the information from the text in the user's language. As an attempt to address both problems, Tosta et al. [4] extract sentences from machine-translated and original texts, and only replace selected sentences with MT problems by similar ranked ones from the text in the user's language. The research of Tosta et al. [4] was the first on MMDS involving the (Brazilian) Portuguese language. About the summarization step, the extractive methods are predominantly superficial, based on features such as *word frequency, sentence position,* etc., which usually have lower cost and are more robust, but produce poor results.

We turn to the use of *conceptual knowledge* in MMDS, which has already been used in other summarization tasks in order to achieve a better content selection (e.g., [6–9]). This work makes the assumption that such knowledge allows to take into account information from all source texts in their original language to perform content selection, producing better summaries both in terms of informativeness, since the selection is based on salient concepts, and linguistic quality, because only summary sentences in a foreign language require to be translated.

Particularly, we report our investigation on 2 methods for summarizing a bilingual cluster (Portuguese and English) to produce an extract in Portuguese. Both methods use the frequency of occurrence of the nominal concepts in the cluster to score the sentences. The scoring yields a ranking in which the sentences with the most frequent or redundant concepts are in the top positions. Given the sentence ranking, one content selection strategy is taking the top-ranked sentences in the user's language, avoiding redundancy. The other one only consists of selecting the top-ranked sentences, independently of language, also avoiding redundancy. If sentences in the foreign language are selected, they are automatically translated to the user's language.

Our experiments were performed using the CM2News *corpus*[2] [10], with 40 news texts grouped by topic in 20 clusters. Each cluster has 1 text in Portuguese and 1 in English.

---

[1] Summarization technique that involves ranking sentences using some scoring mechanism, picking the top scoring sentences, and concatenating them in a certain order to build the summary [5].

[2] http://www.nilc.icmc.usp.br/nilc/index.php/team?id=23#resource.

The concepts of CM2News were derived from Princeton WordNet[3] (WN.Pr) [11] in a semi-automatic annotation process, including (i) translation of each noun in Portuguese to English (since the *synsets* are in English), and (ii) selection of the *synset* that represents the underlying concept/sense of each noun in Portuguese and English. The experiments show that the conceptual knowledge improves summaries in terms of linguistic quality and informativeness, confirming our hypotheses.

This main contributions of this work are: being the first investigation that proposes semantic methods for MMDS of (Brazilian) Portuguese texts, outperforming a *first-sentence* baseline method [4]; providing a semantic layer of annotation to the CM2News *corpus*, and adaptation of an editor for multilingual sense annotation.

In Sect. 2, we describe some related works. In Sect. 3, we describe the lexical-conceptual methods. In Sect. 4, the *corpus* annotation is described. The evaluation will be discussed in Sect. 5. In Sect. 6, some final remarks will be given.

2 Related Work

The closest works to ours are [1–4]. Roark and Fisher [1] take as input a cluster of some translated texts to English, some English spoken language texts, and some English texts. The method ranks the sentences from all the texts based on 9 superficial features and sets a high preference for English sentences when selecting them from the ranking to compose the English extract. Of the nine features, 8 are different versions of *tf-idf, log-likelihood ratio,* and *log-odds ratio* lexical measures, and the ninth is the position of the sentence in the text. The method was trained on a subset of 80 clusters from DUC 2005 using the SVMlight machine-learning algorithm, but the authors do not provide details about evaluation.

Evans et al. [3] aim at generating an English extract from a cluster of English texts and machine translations of Arabic texts into English. The machine-translated sentences are ranked by DEMS [6], a summarizer which apply 3 main criteria of relevance: identifying importance-signaling words through an analysis of lead sentences in a large *corpus* of news, identifying high-content verbs through a separate analysis of subject-verb pairs news corpus, and finding the dominant concepts[4] in the input clusters of texts. Additionally, the sentence relevance also relies on some of the most widely superficial features, such as *position*, which increases the weight of sentences near the beginning of texts, and *length*, which penalizes sentences that are shorter or longer than a threshold, etc. The sentences selected from the rank are replaced with similar sentences from the English texts. The similarity is computed at clause or phrase level,

---

[3] A semantic network of English in which the meanings of word forms and expressions of noun, verb, adjective, and adverb classes are organized into "sets of synonyms" (*synsets*). Each *synset* expresses a distinct concept/sense and the *synsets* are interlinked through conceptual-semantic (i.e., hyponymy, meronymy, entailment, and cause) and lexical (i.e., antonymy) relations [11].

[4] The nouns are grouped into concept sets using WN.Pr *synsets*, and hyponymy relation. To build a set, the highly polysemous nouns are not disambiguated, but replaced by others that are strongly related with the same verb (e.g., "officer" is replaced by "policeman" due to the relation with "arrest"). Having the sets, the sentence ranking is based on the concepts frequency [6].

which requires the syntactic simplification of the English sentences. Next, the similarity is performed by Simfinder [12], which uses lexical and syntactic features. For evaluation, the authors have used the DUC 2004 *corpus*, which contains 24 topics with English texts, Arabic texts, Arabic-to-English machine translations, and 4 human summaries. Using ROUGE[5] [13], the automatic evaluation shows that the similarity-based summarization approach outperforms a *first-sentence* baseline[6]. In an early work, Evans et al. [1] have developed a multilingual version of the English-based summarizer Columbia Newsblaster[7]. This version starts with machine-translated texts, and also replaces the extracted sentences with similar ones in English. However, the similarity is computed at the sentence level, not requiring any syntactic simplification of the non-English sentences.

Tosta et al. [4] have proposed 2 *baselines* using 10 clusters to build extracts in Portuguese. Each cluster is composed of 3 news texts, each one in a different language (English, Spanish and Portuguese). The methods are considered *baseline* because they rely on: (i) translation of the foreign texts to Portuguese using MT[8], and (ii) selection of the relevant sentences using established superficial features, i.e., *word frequency* and *sentence position* [14]. To avoid redundancy, the traditional *word overlap* measure is calculated between each candidate sentence of the rank and the summary sentences. If an ungrammatical translated-sentence is selected, *word overlap* is also used to find a similar sentence from the Portuguese text. The methods were intrinsically evaluated according to the linguistic quality of their summaries. The authors have used the 5 criteria of DUC [15]: (i) grammaticality (i.e., no occurrence of datelines, capitalization errors or ungrammatical sentences), (ii) non-redundancy (i.e., no unnecessary repetition), (iii) referential clarity (i.e., easy identification of the pronouns and noun phrases references), (iv) focus (i.e., it should only contain information that is related to the rest of the summary), and (v) structure and coherence (i.e., it should be well-structured, not just be a heap of related information). In such evaluation, the *sentence position* method had better results.

All methods overviewed in this section first apply MT to translate the foreign texts, obtaining a monolingual cluster. However, when the content is extracted exclusively from machine-translated texts, the summary might contain sentences that are ungrammatical and difficult to understand, since MT is far from perfect. And, when the approaches use texts that were automatically translated to guide selection from the texts in the preferred language, relevant information that exclusively occurs in the preferred language is not selected to compose the summary. Thus, it would be more appropriate to take all the texts in their original language, since the goal is to detect the most

---

[5] ROUGE (*Recall-Oriented Understudy for Gisting Evaluation*) computes the number of common n-grams among the automatic and reference/human summaries, being able to rank automatic summaries as well as humans would do, as its author has shown [14].

[6] In this method, the first-sentence from each text in the cluster is selected until a maximum of words/bytes is reached, and, if the first sentence was already included from each text in the set, the second sentence from each text is included in the summary, and so on [3].

[7] http://newsblaster.cs.columbia.edu/.

[8] http://translate.google.com/.

relevant information of the "cluster". Moreover, the approaches are mainly based on flat text features. Thus, in this paper, we exploit deep linguistic information for MMDS, particularly the conceptual knowledge. Some works have already focused on concepts and their relationships for different summarization tasks under the assumption that they provide a richer representation of the source. For example, Wu and Liu [8] detect the main subtopics of texts by indexing the words to the concepts of a domain-related ontology[9]. The second-level concepts with higher counts codify the main subtopics. Paragraphs that are "closest" to the subtopics are selected. A similar idea but with additional structural features was proposed by Hennig et al. [9] for sentence scoring. The features they used were *tag overlap, subtree depth* and *subtree count*. Next, we describe our extractive MMDS strategies.

3 Lexical-Conceptual Strategies for MMDS

For describing the extractive MMDS strategies, we take into account the traditional summarization phrases: *analysis, transformation* and *synthesis* [17]. The analysis corresponds to the texts understanding, producing an internal representation of their content. The transformation performs summarization operations on the internal representation, producing the summary internal representation. In the synthesis, the summary internal representation is linguistically realized into the final summary. In our methods, the analysis consists of identifying the concepts expressed by (common) nouns (words, expressions, and abbreviations), which are the most frequent word class, covering part of the main content of the texts. To identify the nominal concepts, we use WN.Pr as the conceptual repository. We acknowledge that the granularity of the concepts inventory in WN.Pr is often too fine-grained, resulting in difficulties for finding the *synset* that best represents an underlying concept. Even though, the decision of using WN.Pr was due to (i) its widespread use in the area for summarization and also for other applications, (ii) it has been manually produced, and (iii) the current partial development state of most of the similar resources for Portuguese. Since a concept in WN.Pr is codified by a set of synonyms word forms in English (i.e., a *synset*), the annotation of the nouns from texts in Portuguese has an additional challenge: the translation of nouns to English. Here, we have performed an automatic annotation with subsequent manual or human revision. In Sect. 4, we describe the *corpus* as well as the semi-automatic annotation procedure.

The transformation corresponds to the content selection. To select the sentences, our methods perform 4 steps: (i) computing the compression rate (i.e., the desired summary size), (ii) calculation of the frequency of each nominal concept in the cluster, (iii) scoring all the sentences according to the frequency of occurrence of their nominal concepts in the cluster, and (iv) ranking the sentences by their score. Particularly about the step (ii), the *concept frequency* measure captures the content of the multilingual cluster by counting the occurrence of the concepts underlying synonyms (i.e., different words that express the same concept) and equivalences (i.e., expressions of a concept in

---

[9] This "ontology" consists of a generalization/specialization hierarchy of concepts (i.e., a taxonomy).

different languages). For example, the 2 sentences in Table 1 are from the same cluster and the concepts expressed by nouns were annotated. The numbers encoded by the symbols "< >" indicate the *synset* ID of the noun concept, and the numbers in parenthesis codify the frequency of each concept/*synset* in the cluster. The nouns "manifestante" (Portuguese) and "protester" (English), for instance, express the same concept (i.e., "*a person who dissents from some established policy*"), which is codified by the ID <10002760> ({dissenter, dissident, protester, objector, contestant}). The frequency of the concept in the cluster is 16, and this value is associated to every occurrence of a noun that lexicalizes the referred concept. Once the measure is specified for all concepts, sentences are ranked according to the sum of the frequency of their constitutive concepts. The score of the sentence in Portuguese is 51 and it occupies the first position of the rank, while the sentence in English, with a score = 28, occupies the 12th position. Being composed of the most frequent concepts, the top-ranked sentences are descriptive of the main topic of the cluster. Thus, highly ranked sentences are very suitable for the summary.

Table 1. Example of sentence scoring and ranking based on concept frequency measure.

Sentences	Score	Rank
Um grupo<31264>(6) de **manifestantes<10002760>(16)** conseguiu furar o bloqueio<8376948>(2) da Polícia Militar e chegar ao estádio<4295881> (14) Mané Guarrincha neste sábado<15164570>(4), horas<15227846> (2) antes do jogo<7470671(5) de abertura<7452699>(2) da Copa das Confederações[a]	51	1^{rst}
Brazil's<9379111>(4) opening<74522699>(2) Confederations Cup match<7470671>(5) was affected by **protesters<10002760>(16)** that left 39 people<7942152>(1) injured	28	12th

[a]"*A group of protesters broke through the military police line and got to the Mané Guarrincha stadium on Saturday, hours before the Confederations Cup's opening match*"

Given the rank, one of our selection strategies, called CF (*concept frequency*), performs the sentence selection exclusively based on the rank, independently of the source language. Specifically, CF starts selecting the best-ranked sentence to compose the summary (in Portuguese), and, if it happens that this sentence (as any other along the content selection) is in English, it is automatically translated to Portuguese. After the first selection, if the compression rate is not reached, the 2nd best-ranked sentence is a candidate to compose the summary. Since the input is a multi-document cluster, checking for redundancy between the candidate sentence and the previously selected one is necessary, because the summary should reflect the diverse topics of the cluster without redundancy. In order to avoid redundancy, we assume a threshold (i.e., a pre-established limit) that the new selected sentence may have in relation to any of the previously selected sentences. Thus, if this limit is reached, the new sentence is considered redundant and ignored, and the summarization process goes to the next candidate sentence; otherwise, the sentence is included in the summary. In case of ties (i.e., sentences with the same relevance score in the rank) between a machine-translated sentence and an original sentence in Portuguese, the CF method picks the shortest one. This whole process is repeated until the desired summary length is achieved. The CF method was proposed under the assumption: the application of a late-translation strategy, in which the MT is only used to translate the selected sentences in English to Portuguese, minimizes the problems in the summaries that are caused by the full MT of the source texts.

The other strategy, called CFUL (*concept frequency* + *user language*), is driven by the user's language. It exclusively selects the top-ranked sentences from the text written in Portuguese language to compose the summary, also avoiding redundancy. In case of ties between two original sentences in Portuguese, the CFUL method uses the same criterion applied by CF, i.e., picking the shortest one. Consequently, the final summary only contains sentences in such preferred language. This approach relies on the assumption that a summary built exclusively with original sentences in Portuguese reflects the most relevant information of the cluster, since the concepts that occur in the English text are also taken into account for sentence ranking.

Finally, in the synthesis stage, the methods produce the extracts, as the vast majority of the works in automatic summarization today. So, the CF and CFUL methods simply juxtapose the sentences selected from the rank, ordering them according to their position in their corresponding source texts.

4   The CM2News *Corpus*

For testing the MMDS methods, we have used the CM2News *corpus* [10]. It has 40 original news texts (in a total of 19,984 words) grouped by topic in 20 clusters. Each cluster is composed of 2 news texts, 1 in English and 1 in (Brazilian) Portuguese, both on the same topic, and 1 human summary in Portuguese (abstract[10]), which corresponds to the 30 % of the size of the biggest text of the cluster (i.e., 70 % compression rate). The clusters cover different domains: world, politics, health, science, entertainment, and environment. Since the *corpus* was not semantically annotated, we have carried out the annotation of the nominal concepts as follows.

Each cluster was semi-automatically annotated by groups of 2 or 3 experts with the support of an easy-to-use annotation tool adapted for this task. For each new cluster under analysis, the groups were mixed, trying to avoid any annotation bias. The task was carried out by 12 computational linguists in daily meetings of 90 or 120 min, during 15 consecutive days. The annotation training took 1 day.

The mentioned annotation tool/editor is called MulSen[11] (*Multilingual Sense Estimator*), an adaptation of NASP[12] [19]. Given a cluster, the editor firstly performs an automatic pre-processing task over the source texts, which is the morphosyntactic annotation. To address this task, it incorporates two part-of-speech (POS) taggers, one for each language [16, 18]. Once the nouns are tagged, MulSen translates the nouns from the text in Portuguese to English, which is necessary considering that WN.Pr is our conceptual repository. The translation is done using the online bilingual dictionary

---

[10] Summaries that contain some degree of paraphrase of the input.

[11] http://www.icmc.usp.br/pessoas/taspardo/sucinto/resources.html.

[12] We thank Fernando A. A. Nóbrega for helping adapting the tool.

WordReference®[13,14]. When the text in English is under annotation, MulSen just skips the MT stage. Finally, the editor suggests the *synsets* that better represent the concepts. The suggestions result from the application of *word sense desambiguation* (WSD) algorithms for English and Portuguese languages [19]. Thus, the WSD methods generate a pre-annotation of the nouns, which should be validated (or not) by the experts to complete the process. The tool allows the manual revision of the POS tagging, MT, and conceptual annotation (or *synset* selection) outputs.

To annotate the nouns, the experts have followed 4 generic and 4 specific rules.

The 4 generic instructions are: (i) firstly annotate the text in English of a cluster, since its vocabulary can provide appropriate translations for the annotation of the nouns in Portuguese, (ii) annotate the POS silence, i.e., nouns that were not automatically detected, (iii) ignore the POS noise, i.e., words that were wrongly annotated as nouns, and (vi) annotate all the different occurrences of a concept (i.e., synonyms and equivalences) in the cluster with the same (and more adequate) *synset*.

The first specific rule establishes the annotation of the multiword expressions head with a *synset* that codify the concept of the whole expression, since the taggers do not detect multiword expressions. For instance, in the Portuguese sentence "*Um dos manifestantes levou gás de pimenta no rosto*" ("*One of the protesters was hit in the face by pepper spray*"), "gás" was annotated with the *synset* {pepper spray} ("*a nonlethal aerosol spray made with the pepper derivative oleoresin capiscum*") because it is part of the expression "gás de pimenta", which is not detected by the taggers. The second rule determines that the annotators should analyze all the possible translations provided by MulSen as well as their respective *synsets* before completing the process. It is important because the adequate translation may not be the first in the list of alternatives provided by the editor. The third rule is for the cases where translations have to be manually inserted in the editor, because the editor could not (i) find any translation in WordReference or (ii) provide an appropriate one among the suggested list. For inserting a translation, the third rule establishes that the annotators should test all the possible equivalences found in others resources before finally adding the more appropriate in MulSen. The forth rule determine that, if there is not a proper *synset* to codify a concept of a noun, it should be selected a more generic one. This means that, if any of the *synsets* activated by the chosen translation is not adequate, the annotators should look for a satisfactory hypernym *synset*.

In the next section we report our experiments and the results that we obtained.

5   Evaluation and Results

The evaluation was carried out over the CM2News *corpus*. For each cluster, we manually built 1 extract based on CF and 1 based on CFUL. We have applied a 70 % compression rate (in relation to the longest text), and *word overlap* to avoid redundancy, such as [4]. Regarding the CF method, we used Microsoft Bing® for translating

---

[13] http://www.wordreference.com/.

[14] We have excluded others resources (e.g., *Google Translation*) because of use/license limitations.

the summary sentences in English to Portuguese. The strategies were analyzed based on the informativeness and linguistic quality of the extracts. Our methods were compared to the best *baseline* of Tosta et al. [4], i.e., *sentence position* method with redundancy treatment.

To analyze the quality of the extracts, we used the 5 criteria of DUC [15]. The criteria were manually analyzed by 15 computational linguists. The 20 clusters of CM2News were divided in 5 groups of 4 clusters. Each group was composed of the summaries generated by CF and CFUL, totalizing 8 extracts. The analysis of each group was performed by 3 different judges. Given a summary, the judges scored each of the 5 textual properties through an online form. For all properties, judges had a scale from 1 to 5 points, being 1 = very poor, 2 = poor, 3 = barely acceptable, 4 = good, and 5 = very good. Looking to the average values (Table 2), one may see that the CFUL method outperforms the CF strategy and the *baseline* in all the criteria, indicating that the content selection based on the combination of conceptual knowledge and user's language is better at dealing with textually factors in the summaries. This performance is not surprising, since the sentences come exclusively from one of the source texts. It is interesting to comment that this simulates a usual behavior in human summarization, which is choosing a source as basis for MDS [5]. One may also see that CF outperforms the *baseline* in 4 (except for "structure and coherence") from the 5 criteria, which confirms the hypothesis that the late-translation approach produces fewer textual problems. Even for structure and coherence, the *baseline* performance was not significantly higher than CF (2.8 and 2.6, respectively).

Regarding informativeness evaluation, we used the traditional automatic ROUGE measure [13], which is mandatory in the area. Particularly, we used ROUGE-1, which measures the amount of unigram overlap between reference summaries and automatic summaries, and ROUGE-2, which measures the amount of bigram overlap. We have chosen these two measures because unigrams and bigrams are the most frequent *n*-grams in language. The average results for ROUGE-1 and ROUGE-2 in terms of *recall*, *precision* and *f-measure* are shows in Table 3. Basically, *recall* computes the amount of common n-grams in relation to the number of n-grams in the reference summaries, *precision* computes the number of common n-grams in relation to the n-grams in the automatic summary, and the *f-measure* is the harmonic mean of the previous 2 measures, being a unique indicator of the system performance. According to Table 3, one may see that CFUL method outperforms the CF strategy and the *baseline* in the 2 measures. To statistically determine if the differences in performance were significant, we have performed a Wilcoxon signed-rank test with 95 % confidence, which confirmed the difference. These results indicate that our hypothesis – that the summaries built exclusively with original sentences in Portuguese reflects the most relevant information of the cluster, since the concepts of the English text are also taken into account for sentence ranking – hold. It is important to say, however, that such results are only indicative of what we may expect from the CF and CFUL methods, since our *corpus* for quality and ROUGE evaluation was small (20 clusters). For a more reliable result, we would need to apply the methods for a bigger *corpus*, which remains as future work.

Table 2. Linguistic quality evaluation of summaries with DUC criteria.

Criteria	CF	CFUL	Baseline
Grammaticality	3.5	**4.3**	3
Non-redundancy	3.4	**4.3**	3
Referential clarity	3.3	**3.7**	3.2
Focus	3.5	**4.1**	4
Structure and coherence	2.6	**3.4**	2.8

Table 3. Informativeness evaluation of summaries with ROUGE.

Method	Avg. ROUGE-1			Avg. ROUGE-2		
	Recall	Precision	F-measure	Recall	Precision	F-measure
CF	0.355	0.328	0.341	0.155	0.144	0.149
CFUL	**0.373**	**0.369**	**0.371**	**0.174**	**0.175**	**0.174**
Baseline	0.313	0.271	0.285	0.038	0.032	0.034

6   Final Remarks

As far as we know, this is the first investigation on deep methods for MMDS involving Portuguese as the user's language. We showed that concept-based methods tend to produce extracts with better informativeness and linguistic quality level than a *sentence position baseline*. Other contributions of this work are the annotation of a corpus with noun concepts and the adaptation of an annotation tool, which are freely available for use. However, it is important to recognize that the methods suffer from well-known drawbacks, which are the dependence of linguistic knowledge and the effective lack of scalability. In this line, it is possible to consider to use, for instance, automatic tools for WSD. For Portuguese, one might consider the use of the general purpose methods of Nóbrega and Pardo [20]; for English, several tools are available, as the one of Pedersen and Kolhatkar [21]. The overall performance of the MMDS methods will certainly drop, but their benefits would still be valuable. Some other future works include (i) exploring the construction of automatic and reference summaries with different compression rates, under the assumption that smaller extracts have fewer language problems, and (ii) investigating the impact on redundancy treatment of using a *concept overlap* strategy for redundancy identification (instead of word overlap, as we have done in this paper).

Acknowledgments. We thank the Brazilian National Council for Scientific and Technological Development (CNPq) (#483231/2012-6), the State of São Paulo Research Foundation (FAPESP) (#2012/13246-5, #2015/17841-3), and Coordination for the Improvement of Higher Level or Education Personnel (CAPES) for the financial support.

References

1. Evans, D.K., Klavans, J.L., Mckeown, K.R.: Columbia NewsBlaster: multilingual news summarization on the web. In: North American Chapter of The Association for Computational Linguistics: Human Language Technologies, p. 1–4, Boston (2004)
2. Roark, B., Fisher, S.: OGI OHSU baseline multilingual multi-document summarization system. In: Multilingual Summarization Evaluation (MSE), Michigan, USA (2005)
3. Evans, D.K., Klavans, J.L. Mckeown, K.R.: Similarity-based multilingual multi-document summarization. Technical report CUCS-014-05. Columbia University, New York (2005)
4. Tosta, F.E.S., Di-Felippo, A., Pardo, T.A.S.: Estudo de métodos clássicos de sumarização automática no cenário multidocumento multilíngue. In: 4th Workshop de IC em Tecnologia da Informação e da Linguagem Humana, pp. 34–36, Fortaleza, Brazil (2013)
5. Mani, I.: Automatic Summarization. John Benjamins Publishing Co., Amsterdam (2004)
6. Schiffman, B., Nenkova, A., Mckeown, A.: Experiments in multi-document summarization. In: 2nd International Conference on HLT Research, pp. 52–58, San Francisco (2002)
7. Ye, S., Chua, T.-S., Kan, M.-Y., Qiu, L.: Document concept lattice for text understanding and summarization. Inf. Process. Manag. **43**(6), 1643–1662 (2007)
8. Wu, C.-W., Liu, C.-L.: Ontology-based text summarization for business news articles. Comput. Appl. **2003**, 389–392 (2003)
9. Hennig, L., Umbrath, W., Wetzker, R.: An ontology-based approach to text summarization. In: 3th Workshop on Natural Language Processing and Ontology Engineering (NLPOE), pp. 291–294, Toronto, Canada (2008)
10. Tosta, F.E.S.: Aplicação de conhecimento léxico-conceitual na Sumarização Multidocumento Multilíngue. 2013. Dissertação (Mestrado em Linguística)–Departamento de Letras, Universidade Federal de São Carlos (2014)
11. Fellbaum, C. (ed.): Wordnet: an Electronic Lexical Database (Language, Speech and Communication). MIT Press, Massachusetts (1998)
12. Hatzivassiloglou, J.L., Klavans J.L., Holcombe, M.: Simfinder: a flexible clustering tool for summarization. In: NAACL Automatic Summarization Workshop, p. 9, Pittsburgh (2001)
13. Lin, C-Y.: ROUGE: a package for automatic evaluation of summaries. In: Workshop on Text Summarization Branches Out (2004)
14. Kumar, Y.J., Salim, N.: Automatic multi-document summarization approaches. J. Comput. Sci. **8**(1), 133–140 (2012). ISSN 1549-3636
15. Dang, H.T.: Overview of DUC 2005. In: Document Understanding Conference (2005)
16. Ratnaparkhi, A.: A maximum entropy model for part-of-speech tagging. In: Conference on Empirical Methods in Natural Language Processing, Philadelphia, PA (1996)
17. Sparck-Jones, K.: Automatic summarizing: factors and directions. In: Mani, I., Maybury, M.T. (eds.) Advances in Automatic Text Summarization, pp. 1–14. MIT Press, MA (1999)
18. Schmid, H.: Probabilistic part-of-speech tagging using decision trees. In: International Conference on New Methods in Language Processing, pp. 44–49, Manchester, UK (1994)
19. Nóbrega, F.A.A.: Desambiguação lexical de sentidos para o português por meio de uma abordagem multilíngue mono e multidocumento. Dissertação (Mestrado em Ciências de Computação e Matemática Computacional) - ICMC, USP, São Carlos (2013)

20. Nóbrega, F.A.A., Pardo, T.A.S.: General purpose word sense disambiguation methods for nouns in Portuguese. In: Baptista, J., Mamede, N., Candeias, S., Paraboni, I., Pardo, T.A., Volpe Nunes, Md.G. (eds.) PROPOR 2014. LNCS, vol. 8775, pp. 94–101. Springer, Heidelberg (2014)

21. Pedersen, T., Kolhatkar, V.: WordNet::SenseRelate::AllWords - a broad coverage word sense tagger that maximizes semantic relatedness. In: North American Chapter of the Association for Computational Linguistics/Human Language Technologies Conference, pp. 17–20, Boulder, Colorado (2009)

Domain-Specific Hybrid Machine Translation from English to Portuguese

João Rodrigues$^{(\boxtimes)}$, Luís Gomes, Steven Neale, Andreia Querido,
Nuno Rendeiro, Sanja Štajner, João Silva, and António Branco

Department of Informatics, Faculty of Sciences, University of Lisbon,
Lisbon, Portugal
{joao.rodrigues,luis.gomes,steven.neale,andreia.querido,
nuno.rendeiro,sanja.stajner,jsilva,antonio.branco}@di.fc.ul.pt

Abstract. Machine translation (MT) from English to Portuguese has
not typically received much attention in existing research. In this paper,
we focus on MT from English to Portuguese for the specific domain of
information technology (IT), building a small in-domain parallel corpus
to address the lack of IT-specific and publicly-available parallel corpora
and then adapted an existing hybrid MT system to the new language
pair (English to Portuguese). We further improved the initial version of
the EN-PT hybrid system by adding various modules to address the most
frequently occurring errors in the initial system. In order to assess the
improvements achieved by each of these dedicated modules, we compared
all versions of our MT system automatically. In addition, we conduct and
report on a detailed error analysis of the initial and final versions of our
system.

Keywords: Hybrid machine translation · TectoMT · Lexical semantics · IT domain · Portuguese

1 Introduction

Phrase-based statistical machine translation (PBSMT) models are generally considered to be the state-of-the-art for any language pair and domain for which
large enough parallel corpora exist. For many language pairs, however, training corpora of sufficient size are limited to only a few domains. For English to
Portuguese machine translation (MT), for example, large parallel corpora are
available for just two particular domains – legal documents (the JRC-Acquis
corpus [10]), and parliamentary discussions (the Europarl corpus [9]).

In this paper, we address the problem of English to Portuguese machine
translation for the IT domain, focusing on the conversations of real users with
technical support. In this scenario, users first ask a question in Portuguese which
is machine translated into English, and then the answer is searched for in an
English database, automatically translated back to Portuguese and presented
back to the user. As there are no publicly available parallel corpora for the IT

J. Silva et al. (Eds.): PROPOR 2016, LNAI 9727, pp. 50–61, 2016.
DOI: 10.1007/978-3-319-41552-9_5

domain, we compiled a small in-domain corpus, the QTLeap Corpus[1], consisting of 4,000 utterance pairs (2,000 questions and 2,000 answers) from the IT domain [8], under the QTLeap project[2].

Following the widespread assumption that rule-based and hybrid MT systems give better results for domains and language pairs for which limited parallel data is available – a result of their capacity to make generalisations and thus better overcome data sparsity – we opted for building a hybrid MT system. Our starting point is the TectoMT system [20] which we have adapted from English-Czech to the English-Portuguese language pair. Guided by a detailed human evaluation and error analysis of our initial English-Portuguese TectoMT system, we then added four new modules to handle the most frequently occurring mistakes produced by the initial system.

2   Related Work

Our summary of related work is divided into two sections – firstly, we summarize previous studies on MT from English to Portuguese (Sect. 2.1), and secondly we introduce the hybrid MT system (TectoMT) from which our system for English-Portuguese was built (Sect. 2.2).

2.1   English to Portuguese Machine Translation

Previous studies of MT from English to Portuguese are very scarce, with most reporting on the results of phrase-based statistical MT (PBSMT) systems. Examples of this include results reported on the JRC-Acquis corpus [10] (BLEU = 55) and on the substantially smaller FAPESP corpus of scientific news texts [2] (BLEU = 46). Scores for domain-specific PBSMT systems [6] are substantially lower – trained on Europarl and tested on TED talks and the magazine of Portuguese airline *TAP*, they report BLEU scores 20 and 19 respectively. Scores achieved using Google Translate were better (although still low) for the same task – 28 and 26, respectively.

Recently, two studies were released that report the performance of a baseline hybrid MT system from English to Portuguese for the IT domain compared with a baseline PBSMT system on the same domain [17,19]. In this paper we go one step further, enhancing the baseline TectoMT system from English to Portuguese with specific modules dedicated to reducing the recurrent errors in the baseline system. Furthermore, an extensive human evaluation is performed and reported.

2.2   TectoMT - A Hybrid Machine Translation System

TectoMT is a hybrid system, incorporating elements of statistical and rule-based MT into a modular framework that can be adapted to include various NLP

---

[1] Available from: http://www.meta-share.org/.
[2] http://www.qtleap.eu.

tasks in a single pipeline [20]. The system handles translation over three phases: analysis (of the source language), transfer (of information from source to target language), and synthesis (into the target language). The analysis and synthesis phases are primarily modular – allowing for independent, statistical and/or rule-based NLP tools and processes to be wrapped as 'blocks' and combined to form scenarios (combinations of blocks) specific to required tasks – while the transfer phase that links the two is primarily statistical.

TectoMT is based on two levels of structural representation – a shallow analytical layer (a-layer) and a deep tectogrammatical layer (t-layer) that describes the linguistic meaning of a sentence according to functional generative description (FGD) theory [16]. The translation process goes thorough these two levels of representation, both of which represent input sentences as labeled dependency trees of varying complexity:

- *a-trees*, with each token in the sentence being represented as an *a-node* constructed from:
  - original word forms
  - lemmas
  - part-of-speech (POS) tags
  - morphological information
- *t-trees*, with each token in the sentence being represented as a *t-node* constructed from:
  - deep lemmas (usually identical to the surface lemma)
  - functors (FGD theory-based semantic role labels)
  - grammatemes (person, number, tense, modality etc.)
  - formemes (morphosyntactic information such as `v:to+inf` for infinitive verbs or `n:into+X` for a prepositional phrase).

In a typical example, the analysis phase will involve input sentences being parsed and processed by different scenarios of blocks to construct a-layer trees, which are then propagated upwards to construct t-layer trees. The transfer phase then carries on, whereby t-lemmas (lemmas from the t-layer) are translated and formemes and grammatemes converted from source to target language [3, 20] – this phase is mostly statistical, and based on maximum entropy (MaxEnt) models enriched with specific translation dictionaries and a small number of hand-crafted rules for handling out-of-vocabulary words. Finally, primarily rule-based scenarios in the synthesis perform the reverse of the analysis phase, transforming translated t-trees into a-trees and then linearizing these into output sentences in surface form. For Portuguese, many of the modules in the analysis and synthesis phases are language-specific and handle problems such as word order, agreement (e.g. subject-predicate agreement or noun-adjective agreement), insertion of grammatical words (such as prepositions, articles, particles, etc.), inflections, and capitalization.

3   English-Portuguese TectoMT Systems

In this section, we describe our initial, baseline EN-PT TectoMT system (Sect. 3.1), its improved version (Sect. 3.2), and four modifications to the improved version (Sects. 3.3, 3.4, 3.5 and 3.6) that each focus on addressing the different problems highlighted in a detailed human evaluation of the initial system.

3.1   First EN-PT TectoMT System (System 1)

Building on the original English-Czech TectoMT system to produce our initial English-Portuguese version was primarily focused on adapting the rule-based modules used in the synthesis phase scenario. In the analysis phase, the conversion of source sentences in English to a-trees was already handled by various blocks of NLP tools that perform sentence splitting, tokenization, morphological tagging and dependency parsing. We followed the existing English-Czech annotation pipeline developed for the CzEng 1.0 parallel corpus [4] – using the Morče tagger [18] and the Maximum Spanning tree parser [11] trained on the CoNLL-2007 conversion of the Penn Treebank [13] – and kept the same rule-based blocks for creating a-trees and then t-trees as were used in the original English-Czech version of TectoMT [20].

When translating the English t-trees into Portuguese t-trees in the transfer phase, the transfer of t-lemmas and formemes is handled simultaneously by producing an n-best list of translation variants using t-lemma and formeme translation models (TM). For each t-lemma or formeme for a given source (English) t-tree, the translation model estimates the probability of different translation variants given the source t-lemma or formeme and any additional context. This probability is calculated as a linear combination of:

- *Discriminative Translation Models* – a prediction based on features extracted from the source tree using a MaxEnt model.
- *Dictionary Translation Models* – a dictionary of possible translations with relative frequencies (these models, which do not take contextual features into account, are called *static* models in TectoMT's source code).

After English t-trees have been translated into Portuguese t-trees the synthesis phase begins, for which Portuguese-specific rule-based blocks were written (in Perl) to handle tasks such as word ordering, insertion of negations, prepositions, conjunctions, agreement, formation of compound verbs, and so on. Where possible, existing tools for Portuguese [5] have been used to construct the scenario for synthesis, owing to their greater level of accuracy over the tools available in the original TectoMT system, with new rule-based blocks being created in order to integrate these tools into the TectoMT pipeline [15].

This initial, baseline version of the TectoMT system for English-Portuguese was trained on the whole Europarl corpus [9]. The synthesis scenario was improved iteratively, controlling for both the MT output (as BLEU) and a human error analysis of 1,000 sentences from a small in-domain corpus in each step. This set of 1,000 sentences was obtained from the same corpus as the training and test sets, without any overlapping between them. After each iteration – usually involving the addition of new blocks in the synthesis scenario – the MT output (as BLEU) was checked and a human error analysis performed by two linguistic experts. These experts – both native speakers of Portuguese – analyzed the most frequently missing n-grams (up to 3-grams) and the t-trees at the starting point of the synthesis phase, using their analysis to suggest rules for enforcing better synthesis – the transformation of t-trees to output sentences in Portuguese.

3.2   Second EN-PT TectoMT System (System 2)

The second version of the EN-PT TectoMT system saw the introduction of some improvements over the initial, baseline system. Building on the first version of the system, tokenization, lemmatization, morphological analysis, part-of-speech (PoS) tagging and dependency parsing were improved. For this second version of the EN-PT TectoMT system, improvements were also made to the analysis phase firstly by adding missing lemmas for use with the POS-tagger and by adding extra rules for tokenization.

Improvements were made in the synthesis phase of the second version of the EN-PT TectoMT system, namely by adding missing lemmas to the LX-Inflector component of the LX-Suite in order to handle nominal expressions and to the LX-Conjugator component in order to handle verbal expressions. An additional block for handling the insertion of quotation marks in quoted expressions was also added to the synthesis scenario. Next follows an example of the resulting translation using system 1 (a) and system 2 (b) with this block:

(a) *No separador de Slides em [...]*
(b) *No separador de 'Slides' em [...]*

Over the next few subsections, we describe the implementation of additional modules built to improve the second version of the EN-PT TectoMT system to address various problems discovered in the human evaluation of error analysis on the first system.

3.3   EN-PT TectoMT with Word Sense Disambiguation (System 2 + WSD)

The transfer phase in TectoMT is based on lemma-to-lemma translation models, but lemmas themselves are often ambiguous, and can be represented by multiple meanings. We thus experimented with using additional information from source language (English) word sense disambiguation (WSD) – the computational task of determining the correct meaning of a word in a particular context – in the

TectoMT transfer. For each a-layer node created in the analysis phase, we add additional contextual features containing word sense information from both the current node and its parent node to the Discriminative (MaxEnt-based) translation model. This work has been described in greater detail in previous work [12]. Next follows an example of the resulting translation using system 1 (a) and system 2 (b) with the WSD embedded information:

(a) *No domínio de notificação de Windows há o ícone de Panda.*
(b) *Na **área** de notificação de Windows há o ícone de Panda.*

English word senses were obtained using the UKB system [1], a collection of tools and algorithms for performing graph-based WSD over a pre-existing knowledge base. For a given word, UKB is able to query a graph-based representation of WordNet [7] and return the appropriate synset identifier that represents the meaning of the given word, using its surrounding words as context. In addition to synset identifiers, we also provide supersenses to the translation model as features – supersenses are the 45 semantic files by which synset identifiers are organized in WordNet, allowing senses to be generalized across semantic classes like PEOPLE, GROUP or ARTIFACT.

3.4   EN-PT TectoMT with Hidden Entities (System 2 + HideIT)

The error analysis of the initial, baseline version of the EN-PT TectoMT system suggested that a substantial number of translation errors originate from the incorrect handling of named entities (NEs), especially those that are domain-specific (IT) and thus cannot be successfully captured by named entity recognition and classification (NERC) tools. To address this, we experimented with the implementation of a rule-based component called *HideIT* to account for domain-specific entities that do not require translation such as URLs, shell commands, and code snippets. Next follows an example of the resulting translation using system 1 (a) and system 2 (b) with the HideIT block:

(a) *Envie um correio qualidade@pcmedic. PT.*
(b) *Envie um correio a **qualidade@pcmedic.pt.***

The HideIT component consists of two blocks. The first block is applied at the very start of the translation pipeline – just after the tokenization of the source text and before any meaningful linguistic processing takes place – and attempts to recognize such entities using manually gathered heuristics from 2,000 sentences from the in-domain development corpus. Recognized entities are then replaced with an appropriate placeholder (e.g. xxxCMDxxx or xxxURLxxx for shell command and URL, respectively), while the original values are stored as metadata. The second block is applied at the very end of the translation pipeline, and extracts the values that were recognized earlier and stored as metadata and swaps them with the placeholders that were introduced by the first block to hide the entities from the core processing components of the translation pipeline.

3.5   EN-PT TectoMT with Added Gazetteer (System 2 + Gazetteer)

We also focused on trying to obtain correct translations and localizations of NEs in the IT domain – such as menu items, button names, sequences and messages – that are expected to appear in a fixed inflectional form. The fact that such NEs are fixed allowed us to match expressions from a specialized lexicon (gazetteer) in the source text and replace them with their equivalent expressions in the target language. The English-Portuguese gazetteer was collected from four sources: localization files of VLC,[3] LibreOffice,[4] KDE,[5] and IT-related Wikipedia articles.

Following the tokenization of text in the analysis phase, expressions in the gazetteer are searched for in the source sentence. Matched expressions – which can span several neighbouring tokens – are then replaced by a single-word place-holder. Then, in the transfer phase, these placeholders are replaced in the t-trees by the corresponding expressions stored in the gazeteer from the target language. Note that this step is performed before translating any of the other words of the source sentence.

3.6   EN-PT TectoMT with Domain Adaptation (System 2 + DomAdapt)

The error analysis of the first version of the EN-PT TectoMT system also high-lighted many incorrectly translated domain-specific words and phrases that still could not be addressed using the new HideIT and Gazetteer implementations. To address this problem, we experimented with domain adaptation during the trans-fer phase by interpolating translation models from a general domain (Europarl) and the IT domain (2,000 utterances from the QTLeap corpus described in the Introduction to the paper enriched with parallel terminology from both the Microsoft Terminology Collection,[6] and LibreOffice localization data[7]). This interpolation helps to account for some of the errors in the output of the initial system originating from a lack of in-domain training data.

The interpolation was not applied only to the lexical transfer (of lemmas, as in the experiments with HideIT and Gazetteer), but also to the transfer of formemes. It had been noticed that the IT domain formeme Translation Models (TMs) had a different distribution of probabilities to the general domain (Europarl) TMs, and so it was ventured that the interpolation of formeme TMs could also be benefi-cial. The EN-PT TectoMT system trains four standard TMs from parallel train-ing data – a Dictionary formeme TM, a Discriminative formeme TM, a Dictionary t-lemma TM, and a Discriminative t-lemma TM. For the interpolation of these

---

[3] http://downloads.videolan.org/pub/videolan/vlc/2.1.5/vlc-2.1.5.tar.xz.
[4] http://download.documentfoundation.org/libreoffice/src/4.4.0/ libreoffice-translations-4.4.0.3.tar.xz.
[5] svn://anonsvn.kde.org/home/kde/branches/stable/l10n-kde4/pt/messages.
[6] Available from: http://www.microsoft.com/Language/en-US/Terminology.aspx.
[7] Available from: https://www.libreoffice.org/community/localization/.

TMs, each of the four TMs was assigned an interpolation weight (1.0 for the Dictionary formeme and Discriminative t-lemma TMs, and 0.5 for the Discriminative formeme and Dictionary t-lemma TMs). Next follows an example of the resulting translation using system 1 (a) and system 2 (b) with domain adaptation:

(a) *No menu de desempenhar escolhe voltar a celeridade normal.*
(b) *No menu de* **Reproduzir** *escolhe voltar a* **velocidade** *normal.*

4 Evaluation

Our evaluation of all of the systems described in the previous Section has consisted of two methods – an automatic evaluation of MT output and a manual evaluation of error analysis performed by linguistic experts.

4.1 Automatic Evaluation

We performed an automatic evaluation of MT output (as BLEU [14]) for all of the described EN-PT TectoMT systems: System 1, System 2, System 2 + WSD, System 2 + HideIT, System 2 + Gazetteer, System 2 + DomAdapt, and System 2+ (System 2 enriched with all four additional modules – WSD, HideIT, Gazetteer, and DomAdapt). The results are presented in Table 1.

Table 1. BLEU scores for all systems.

Experiment	BLEU	BLEU-BLEU(System 2)
System 1	19.34	−0.48
System 2	19.82	0.00
System 2 + WSD	20.07	+0.25
System 2 + HideIT	20.16	+0.34
System 2 + Gazetteer	20.76	+0.94
System 2 + DomAdapt	21.80	+1.98
System 2+	22.42	+2.60

Table 2 shows the number of errors found by the linguists in each system multiplied by four (an estimate of the likely number of errors that would occur in 100 sentences, this was due to the interest in the "density" of each error type rather than the total number, notice that the values are a mean value of errors found by two annotators), as well as the absolute difference and the relative difference between number of errors found in System 2+ and System 1.

The results in Table 1 show that the largest improvements to the system are achieved by making use of Gazetteers (specialized lexicons) and by interpolating general and IT-domain TMs, while the addition of WSD and HideIT modules also yield slight improves of the system. Changes in the analysis and synthesis phases from System 1 to System 2 also led to substantive improvements. The full system (System 2+) – which incorporates all of the previously described improvements and additional modules – achieves good results (BLEU = 22.42).

4.2 Error Analysis

To gain better insight into the translation quality achieved by System 1 and by System 2+, we asked two linguistic experts (both native speakers of Portuguese) to analyze the specific errors made by each system on a subset of 25 sentences. The errors they discovered were then classified according to the Multidimensional Quality Metrics (MQM) framework[8] (with some slight modifications):

1. Accuracy
   (a) Addition
   (b) Mistranslation
   (c) Omission
   (d) Overly literal
   (e) Untranslated
2. Fluency
   (a) Grammatical register
   (b) Spelling
   (c) Typography
   (d) Grammar
       i. Word form

   A. Part of speech
   B. Agreement
   C. Tense/aspect/mood
   ii Word order
   iii Function words
       A. Extraneous
       B. Incorrect
       C. Missing
   (e) Unintelligible
3. Locale convention
4. Terminology

The results shown in the Table 2 demonstrate that there are less Accuracy errors (−43 %) in the output of System 2+, particularly errors classified as overly literal translation or mistranslation. In terms of Fluency, the output of System 2+ showed fewer spelling errors, agreement errors, word order problems and incorrect translations of function words than were present in the output of System 1. However, the number of missing function words and tense, aspect and mood errors increased from System 1 to System 2+. Taken as a whole and in context, these results suggest that translation of terminology in particular has indeed been improved in System 2+.

---

[8] http://www.qt21.eu/launchpad/content/multidimensional-quality-metrics.

Table 2. Number of errors in each system (System 1 and System 2+), and their relative difference.

Error type	System 1	System 2+	%
Accuracy	0	0	0 %
-Addition	8	6	−25 %
-Mistranslation	178	82	−54 %
-Omission	46	36	−22 %
-Overly literal	30	16	−47 %
-Untranslated	22	22	0 %
Accuracy subtotal	284	162	−43 %
Fluency	2	2	0 %
-Grammatical register	0	0	0 %
-Spelling	48	40	−17 %
-Typography	46	54	17 %
-Grammar	0	0	0 %
−Word form	0	0	0 %
—Part of speech	34	34	0 %
—Agreement	56	52	−7 %
—Tense/aspect/mood	56	100	79 %
−Word form subtotal	146	186	27 %
−Word order	74	66	−11 %
−Function words	0	0	0 %
—Extraneous	112	110	−2 %
—Incorrect	52	32	−38 %
—Missing	210	244	16 %
−Function words subtotal	374	386	3 %
-Unintelligible	0	0	0 %
Fluency subtotal	690	734	6 %
Locale convention	0	0	0 %
Terminology	12	10	−17 %

5   Conclusions

Previous research addressing MT from English to Portuguese has been scarce thus far, with the few studies that do describe this language pair generally focusing on phrase-based SMT systems. In this paper, we have described our implementation of an MT pipeline from English to Portuguese for a specific domain (IT), also creating a small, in-domain corpus to account for the lack of publicly-available parallel corpora for the domain in question. Part of this corpus was used for development of our hybrid EN-PT MT system, and the other part used for testing.

We first built an initial, baseline EN-PT hybrid MT system by adapting the existing hybrid TectoMT system from English-Czech to English-Portuguese. After performing an initial error analysis, we further improved the analysis and synthesis phases of the system and added four new modules to address most common mistakes of the initial system. Automatic evaluation of the output of the revised system using each of the newly-created module showed that each of them helps to improve the overall performance of the system, suggesting that the addition of a gazetteer (specialized lexicon) and the interpolation of general and domain-specific translation models as the most promising strategies for improving MT output. Finally, a detailed human error analysis of the initial and the final systems confirmed that the additional modules and improvements of analysis and synthesis phases implemented in the second version of the EN-PT hybrid MT system do contribute to improved MT output.

Acknowledgements. The results reported in this paper were partially supported by the Portuguese Government's P2020 program under the grant 08/SI/2015/3279: ASSET-Intelligent Assistance for Everyone Everywhere, and by the EC's FP7 program under the grant number 610516: QTLeap-Quality Translation by Deep Language Engineering Approaches.

References

1. Agirre, E., Soroa, A.: Personalizing PageRank for word sense disambiguation. In: Proceedings of the 12th Conference of the European Chapter of the Association for Computational Linguistics, EACL 2009, pp. 33–41. Association for Computational Linguistics, Athens (2009)
2. Aziz, W., Specia, L.: Fully automatic compilation of a Portuguese-english parallel corpus for statistical machine translation. In: Proceedings of the 8th Brazilian Symposium in Information and Human Language Technology. Cuiabá, MT, October 2011
3. Bojar, O., Týnovský, M.: Evaluation of tree transfer system. Technical report, Charles University in Prague (2009)
4. Bojar, O., Žabokrtský, Z., Dušek, O., Galuščáková, P., Majliš, M., Mareček, D., Maršík, J., Novák, M., Popel, M., Tamchyna, A.: The joy of parallelism with CzEng 1.0. In: Proceedings of the 8th International Conference on Language Resources and Evaluation (LREC), pp. 3921–3928 (2012)
5. Branco, A., Silva, J.R.: A suite of shallow processing tools for Portuguese: LX-suite. In: Proceedings of the 11th Conference of the European Chapter of the Association for Computational Linguistics (EACL) (2006)
6. Costa, A., Luís, T., Coheur, L.: Translation errors from english to portuguese: an annotated corpus. In: Proceedings of the 9th International Conference on Language Resources and Evaluation (LREC) (2014)
7. Fellbaum, C.: WordNet: An Electronic Lexical Database. MIT Press, Cambridge (1998)
8. Gaudio, R.D., Burchardt, A., Branco, A.: Evaluating machine translation in a usage scenario. In: Proceedings of LREC (2016). (to appear in print)
9. Koehn, P.: Europarl: a parallel corpus for statistical machine translation. In: Proceedings of the Tenth Machine Translation Summit, pp. 79–86 (2005)

10. Koehn, P., Birch, A., Steinberger, R.: 462 machine translation systems for Europe. In: Proceedings of the MT Summit XII (2009)
11. McDonald, R., Pereira, F., Ribarov, K., Hajič, J.: Non-projective dependency parsing using spanning tree algorithms. In: Proceedings of the Conference on Human Language Technology and Empirical Methods in Natural Language Processing (EMNLP), pp. 523–530 (2005)
12. Neale, S., Gomes, L., Branco, A.: First steps in using word senses as contextual features in maxent models for machine translation. In: Proceedings of the First Workshop on Deep Machine Translation, DMTW-2015, pp. 64–72 (2015)
13. Nilsson, J., Riedel, S., Yuret, D.: The CoNLL 2007 shared task on dependency parsing. In: Proceedings of the CoNLL shared task session of EMNLP-CoNLL, pp. 915–932 (2007)
14. Papineni, K., Roukos, S., Ward, T., Zhu, W.J.: BLEU: a method for automatic evaluation of machine translation. In: Proceedings of ACL (2002)
15. Rodrigues, J., Rendeiro, N., Querido, A., Štajner, S., Branco, A.: Bootstrapping a hybrid MT system to a new language pair. In: Proceedings of LREC (2016). (to appear in print)
16. Sgall, P., Hajicová, E., Panevová, J.: The Meaning of the Sentence in its Semantic and Pragmatic Aspects. Springer Science & Business Media (1986)
17. Silva, J., Rodrigues, J., Gomes, L., Branco, A.: Bootstrapping a hybrid deep MT system. In: Proceedings of the Fourth Workshop on Hybrid Approaches to Translation (HyTra), pp. 1–5. ACL (2015)
18. Spoustová, D., Hajič, J., Votrubec, J., Krbec, P., Květoň, P.: The best of two worlds: cooperation of statistical and rule-based taggers for czech. In: Proceedings of the Workshop on Balto-Slavonic Natural Language Processing, pp. 67–74 (2007)
19. Štajner, S., Rodrigues, J., Gomes, L., Branco, A.: Machine translation for multilingual troubleshooting in the IT domain: a comparison of different strategies. In: Proceedings of the Deep Machine Translation Workshop (DMTW), pp. 106–115 (2015)
20. Žabokrtský, Z., Ptáček, J., Pajas, P.: TectoMT: highly modular MT system with tectogrammatics used as transfer layer. In: Proceedings of the Third Workshop on Statistical Machine Translation, pp. 167–170 (2008)

Analysis of Temporal Adverbial Phrases for Portuguese–Chinese Machine Translation

Siyou Liu[✉] and Ana Luisa Varani Leal

Faculty of Arts and Humanities, University of Macau, Macau S.A.R., China
helen.liu103@gmail.com, analeal@umac.mo

Abstract. Adverbial phrase (AdP) contains rich and indispensable information, however, translating them properly is one of big challenges for machine translation (MT) systems. In this paper, we systematically present a contrastive analysis of MT of AdPs from Chinese to Portuguese. Our study is conducted on *The International Chinese Newsweekly* corpus which consists of 46 Chinese texts (ST) with their respective Portuguese translations given by a state-of-the-art Portuguese–Chinese MT system and human beings (HT). By comparing the syntactic structures of these texts in ST, PCT, and HT sides, we found that nearly 90 % MT outputs suffer structural inconsistency with poor translation qualities. Therefore, we discuss and finally propose a series of grammar rules to address the problem. We believe that this work could inspire both MT researchers and industries to boost the performance of Portuguese–Chinese MT systems.

Keywords: Temporal adverbial phrase · Machine translation · Portuguese–Chinese · Contrastive analysis · Syntactic structure · Syntax rule

1 Introduction

An adverbial is an adverb, adverbial phrase or adverbial clause, which gives us additional information about a verb or a sentence in order to modify or describe it (e.g. time, place, or manner of the action in the sentence). In other words, adverbials can answer a series of questions such as *where, when, how, why, how often, how long, how much* as shown in Table 1 [1]. Adverbial is an indispensable component in a language, because it contains additional but important information for communication. However, for the state-of-the-art MT system, no matter from semantic perspective or from grammatical perspective, it is still a big challenge to adequately translate it. Therefore, the study on the equivalence of adverbials between languages is significant to improve the performance of MT.

In the field of Linguistics, there are some studies exploring and demonstrating differences between languages [11,13,15,17]. For example, Xia [15] investigates Chinese Adverbial systemically and tries to explain the differences between Chinese and Portuguese from aspects such as meanings of words, construction of words, ways of expressing grammar meanings, words grammar functions, words order, languages culture mentality, etc.

© Springer International Publishing Switzerland 2016
J. Silva et al. (Eds.): PROPOR 2016, LNAI 9727, pp. 62–73, 2016.
DOI: 10.1007/978-3-319-41552-9_6

Table 1. Questions that can be answered by the Adjunct adverbials

Question word	Type	Examples of Adjunct adverbial
How?	Manner	e.g. carefully, with enthusiasm
When?	Time	e.g. yesterday, on Tuesday, after I left
Where?	Place	e.g. there, in the kitchen, where I was
Why?	Reason	e.g. for no reason, since I am poor
How often?	Frequency	e.g. twice, monthly
How long?	Duration	e.g. for two years

In the field of Machine Translation, there are too much work to be realized in order to deal with divergences between languages as well. Wang et al. [11,13] try to deal with dropped pronoun problem in translating automatically from pro-drop language to non-pro-drop language and proposes new approach to improve the translation quality of the dropped pronoun for dialogue MT. However, it's not usual to verify studies that combine these two fields in the same approach and try to resolve inconsistencies in MT through a detailed contrastive analysis of a specific linguistic phenomena between two languages.

The Machine Translation system used in this work is PCT3.0 *Portuguese – Chinese MT system*[1], which has been developing by the Natural Language Processing & Portuguese – Chinese Machine Translation Laboratory (NLP2CT) at University of Macau. The development objective of this Machine Translation system is to satisfy the increasing demand for translating tools to serve the dual official language needs of the Territory of Macau. It is a rule-based system and applies the scheme of annotation in syntax trees in the representation of bilingual examples and the Constraint Synchronous Grammar [14].

In this paper, we mainly study temporal adverbial phrase ("when" type of adverbials as shown in Table 1) to find solutions in terms of rules and syntax heuristics in order to solve inconsistencies found in Chinese–Portuguese MT. Generally, the contributions of this work are observed in three aspects: First, we manually build tree-based and sentential corpora for the following syntactic analysis between Chinese and Portuguese, which contains Chinese source sentences (ST), Portuguese human translations (HT) and MT system outputs (PCT). Second, we systematically compare the differences on syntactic structures among ST, HT and PCT, and also evaluate their translation qualities. Through the contrastive analysis, we find that nearly 90 % MT outputs suffer structural inconsistencies with poor translation quality. Finally, we propose some grammar rules to solve the inconsistencies found in the PCT outputs and improve the machine translation quality. We believe that our work would inspire both MT researchers and industries to boost the performance of Portuguese–Chinese MT system.

The paper is organized as follows. To make the reader understand our work more smoothly, we introduce the theoretical underpinnings particularly on temporal adverbial phrase in standard Chinese and Portuguese in Sect. 2. Then Sect. 3

---

[1] Available at http://nlp2ct.cis.umac.mo/views/pct/pct.html.

is composed of three parts: in the first two parts, we briefly describe our research, including the corpus and methodology adopted, while in the third part, we report our analysis and case study results, pointing out the problems and shortcomings of the translation program PCT that emerged from the descriptive and contrastive analysis, particularly in the aspect of temporal adverbial phrase translation. In Sect. 4, we discuss the problems and propose a set of simple but effective grammar rules for current MT systems. And we also report an oracle experimental result to prove the feasibility of our method. Finally, we draw the conclusion in Sect. 5.

2   Theoretical Background

In generative grammar, an adverbial is an element that provides additional information about a verb or a sentence modifying or describing it (e.g. time, place, or manner of action in a sentence). According to [1], "three different types of adverbial functions are distinguished: (1) adverbial adjunct, (2) adverbial disjunct, and (3) adverbial conjunct." The adverbial adjunct is optional to modify the part of sentence, the subject or the principal verb. The adverbial disjunct modifies the entire sentence, not just the verb. The main function of the adverbial conjunct is expressing textual relations such as linking clauses. For example,

(1) I finished the article last Friday. (Adjunct adverbial)
(2) Fortunately, I finished the article. (Disjunctive adverbial)
(3) Next, I'll finish the article. (Conjunct adverbial)

In (1), "last Friday" is the modifier of the verb "finished", which indicates when the action happens. In (2), "Fortunately" conveys the speaker's attitude toward the content of the sentence; the adverbial modifies the entire sentence, rather than the verb "finished". In the (3), the adverbial "next" provides explicit information about how the sentence relates to a previous one: the action "finish the article" happens after or upon the completion of another action.

2.1   Adverbial Phrases in Portuguese

The aforementioned definition of adverbials is considered in the generative grammar, which means that they are also suitable for Portuguese. Based on the perspective proposed by Cunha and Cintra [4], "as the name indicates, the adverbial adjunct is a term with adverbial value, which denotes some circumstance of the fact expressed by the verb, or intensifies the meaning of the verb, or an adjective, or an adverb". This means that adverbial adjunct can modify verbs, adjectives, or adverbs in a sentence, answering questions about manner, time, place, reason, frequency and duration. For example,

(4) Mas, atualmente, cerca de 16 mil de tripulantes também declararam que
     se as duas partes não pudessem chegar ao consenso sobre o problema dos

salários, iriam iniciar <u>de imediato</u> uma grande greve, a situação poderia ficar <u>ainda mais</u> grave para a companhia aérea. (Text11 Autema-Syntree[2])

But, <u>currently</u>, about 16 thousand flight attendants have <u>also</u> stated that if both sides could not reach a consensus on the issue of wages, they would <u>immediately</u> start a big strike, and the situation could get <u>too much</u> serious for the airline company.

In example (4), all the underlined parts are adverbial adjuncts: the adverb *atualmente* [currently] adds some temporal information for the whole sentence; the adverb *também* [also] can be treated as a manner modifier of the verb *declararam* [stated]. In addition, the adverbial phrases *de imediato* [immediately] and *ainda mais* [too much] modify, respectively, the verb *iniciar* [start] and the adjective *grave* [serious], answering questions concerned with "*when*" and "*how*".

In terms of the definition of adverbial phrase in Portuguese, Cunha and Cintra [4] propose that adverbial phrase refers to the "conjunct of two or more words which has the same function as adverbs", and belongs to adverbial adjunct.

The present study focuses on the "*when*" type of questions, i.e., temporal adverbial adjunct phrases(AdP). For instance in example (4), the temporal adjunct AdP is *de imediato* [immediately], which, in according to the Portuguese standard grammar, is a prepositional phrase, for it begins with the preposition *de* [of]. Furthermore, all the examples of temporal adverbial provided by [4,7,8], indicate that the prepositional phrase composes the temporal adverbial phrase in Portuguese language.

2.2   Adverbial Phrases in Chinese

In Chinese Grammar, the adverbial may occur either at the beginning of the sentence or after the topic or subject, and modifies the entire sentence [2]. According to [15], there are 13 kinds of adverbials, such as adverbial of time, place, manner, and reason. Temporal adverbials can be a temporal noun (5), a temporal adverb (6), a prepositional phrase (7), a directional phrase (8), and a quantitative phrase (9). For instance,

The temporal adverbial in example (5) is the temporal noun *muqian* [currently/at the moment]. In example (6) it is the temporal adverb *zhengzai* [be + V-ing]. In (7) it is the prepositional phrase *zai ershiwu ri* [on 25th]. In example (8) it is the directional phrase *ershisi xiaoshi yihou* [24 h later], in which the last word *yihou* [later] in Chinese is a directional noun. In example (9), according to the Chinese standard grammar, the phrase *yijiubajiu nian* [1989 year] is a quantitative phrase, for it begins with the number "1989", which means this temporal adverbial is a quantitative phrase.

---

[2] Annotation and Analysis of Bilingual Syntactic Trees for Chinese/Portuguese (Autema-Syntree), University of Macau Research Grant MYRG102 (Y2-L2)-FSH11-ALL, a research project aimed at solving structural inconsistency of translated texts. This work is a part of the research project.

(5) 目前/n , 约/adj 一万六千/num 名/n 空乘人员/n ...
Currently, about sixteen thousand flight attendants ...

(6) 中国/n 目前/n 正在/adv 建设/v ...
China at the moment is constructing ...

(7) ...将/adv 在/p , 二十五/num 日/n 召集/v 两/num 党/n 议员/n ...
... will **on the 25th day** convene both parties deputy...

(8) 二十四/num 小时/n 以后/n , 双方/n 发表/v 声明/n ...
24 hours later, both sides issued a statement ...

(9) 一九八九/num 年/n , 杨宪益/n 公开/adv 指责/v 中国/n 政府/n ...
In 1989 year, Yang Xianyi publically accused the Chinese government ...

The above contrastive analysis shows that the constituents of Chinese temporal adverbial phrase are more diverse than in Portuguese, which admits only prepositional adverbial phrases. The following analysis (in Sect. 4) will show more details about the contrast of temporal AdPs in Chinese and Portuguese.

3 Methodology

Our work is based on Autema–SynTree project – *Annotation and Analysis of Bilingual Syntactic Trees for Chinese/Portuguese*, and focuses on the phrasal organization in Chinese and Portuguese, which is necessary to develop syntactical structures and improve PCT system. The next subsection presents details about our corpus. And then, in Subsect. 3.2, we demonstrate the uniform process of our artificial analysis.

3.1 Corpus

All of the source texts (in Chinese) of our corpus are extracted from *The International Chinese Newsweekly, 2009/12–2010/03*. In order to make the data representative, we randomly select 46 texts for analysis in this paper. Besides, the two type of translations (in Portuguese) are generated respectively by human translators and MT system. To compare the syntactic structures, we build the syntactic trees for source sentences, human translations as well as MT outputs by using SyntaxTree Builder[3].

The statistics of our corpus are detailed in Table 2. In total, the source texts have 222 sentences and there are around 200 words in each text. In the totality of these sentences, 51 of them contain temporal Adverbial Phrases (23 %), while the number of sentences in HT with temporal AdPs is the same.

---

[3] Avaliable at http://nlp2ct.cis.umac.mo/syntaxToolkit/builder.html.

Table 2. Statistics data of our corpus

Data	Text Num.	Sent. Num.	%
ST with temporal adverbials	41	51	23 %
HT with temporal adverbials	41	51	23 %
Corpus of the project	46	222	100 %

3.2 Steps

Generally, the approach of analysis is straightforward, and its implementation is almost entirely realized by human beings, which means that the steps are performed sequentially and hierarchically. The analysis process includes 7 stages as follows,

(1) The sentences containing temporal AdPs were identified both in Chinese and in Portuguese;
(2) The Chinese temporal AdPs in the STs were analyzed and classified according to the traditional grammar;
(3) The organization of the Chinese temporal AdPs' constituents was described in order to determine which phrases can work as temporal adverbials;
(4) The Portuguese temporal AdPs provided in the PCTs and in the HTs were respectively analyzed and classified according to the lexico-grammar;
(5) The organization of the Portuguese temporal AdPs' constituents was described, both in the PCTs and HTs;
(6) The contrast of phrasal organizations in the PCTs and HTs, as well as in the HTs and STs, consisted of mirroring the syntactic trees of each pair of texts to one another to identify inconsistency in the PCTs;
(7) The contrasts and differences were discussed and summarized, with a view to figuring out how to improve PCT in what concerns the temporal AdPs.

4 Analysis and Solutions

4.1 Comparative Analysis

Based on the Generative Grammar, we contrasted the syntactic structures of the temporal adverbial in Chinese ST with those provided by PCT translations and human translations.

On the basis of the Portuguese syntax rules and a contrastive analysis, as shown in Table 3, 44 sentences were identified out 51 (86 %) because of the inadequate renditions provided by PCT. More specifically, 19 AdPs (43 %) were rendered without preposition, 3 AdPs (7 %) correspond to the order sequence of the Part-to-Whole-Principle (PWP), and 9 (20 %) were provided with prepositions placed in incorrect positions. Other cases include 13 occurrences of different nature (30 %).

Table 3. Statistic Inadequacies Data

Case	Inadequate Sent. Num.	Inadequate Sent. %
Lack of preposition	19	43%
Problem of the PWP	3	7%
Incorrect position of preposition	9	20%
Others	13	30%
Total	44	100%

Having found the nature of the inadequate machine translation renditions, we first focused on the instances of lack of preposition in the Portuguese AdPs.

(10)

ST ... 文献/n 出版社/n 于/p 二月/n 二十八/num 日/n 发布/v
... Academic Press **on February 28 day** published

HT A ACSC **em 28 de fevereiro** lançou ...
The CSSAP **on 28 of February** published ...

PCT Editorial ACSC **em 28 de fevereiro** lançou ...
Editorial CSSAP **on 28 of February** published ...

(11)

ST ... 将/adv 在/p 二十五/num 日/n , 召集/v ...
... will **on the 25 day** convene ...

HT ... vai convocar, **em 25 deste** mês ...
... will convene, **on 25th this month** ...

PCT ... em breve **em vinte e cinco dias** convocam ...
... in soon **on 25 days** convene ...

(12)

ST 三月/n 一/num 日/n , 为/p 迎接/v 世博会/n
March one day, to welcome the Expo ...

HT **Em um de março**, para a Expo Mundial ...
On one of March, for the Expo International ...

PCT **Um de março**, para acolher Expo Mundial ...
one of March, to welcome Expo International ...

(13)

ST 日本/n 媒体/n 二月/n 二十二/num 日/n 报导/v ...
Japanese media **February 22 day** reported ...

HT A imprensa japonesa informou que **em 22 de Fevereiro** ...
The press Japanese reported that **on 22 of February** ...

PCT A mídia do Japão **vinte e dois de fevereiro** conversar ...
The media of Japan **22 of February** talk ...

Comparing examples (10), (11) with examples (12), (13), we observe that not only examples (10) and (11) show instances including prepositions in both HT and PCT translations, but also the Chinese prepositions *yu* [on] and *zai* [on] in (10) and (11) are adequately translated into Portuguese as *em* [on/in], both in human and machine translations, while examples (12) and (13) show instances including prepositions only in the human translations, but not in ST or in the PCT renditions. In other words, those Chinese AdPs which begin with a preposition are adequately translated by PCT with a preposition. However, according to Chinese standard grammar, the presence of preposition as head in the AdPs of time is optional, and the Chinese AdPs in (12) and (13) are quantitative phrases because they start with numbers, but when we translate them into Portuguese, we have to transform them to prepositional phrases, which are not done by PCT system.

Therefore, we observe that what the machine translation does is a literal translation – translating word-by-word. When the Chinese ST features a prepositional phrase, the translation system translates it as a prepositional phrase, shown in Examples (10) and (11). Nevertheless, when the Chinese ST has a quantitative phrase, and the machine rendition is a noun phrases, as in examples (12) and (13).

In addition, we also observe the difference of their locations, demonstrated in examples (12) and (13). In Chinese grammar, when a Noun Phrase serves as a temporal AdP, it is located at the beginning or in the middle of the sentence. However, in Portuguese, the temporal AdP likely occur in the end of the sentence, but if using comma to separate it from other elements in the sentence, it can be placed anywhere as well.

Looking to the analysis of the order of the time noun phrase as shown in Example (14).

(14)

ST   ... 于/p 二月/n 二十七/num 日/n 晚上/n ...
      ... on February 27th day night ...

HT   Na noite de vinte e sete de fevereiro, CNN televisionou ...
      In the night of 27 of February, CNN broadcasted ...

PCT   CNN em vinte e sete de fevereiro a noite produzir
      CNN on 27 of February the night produce ...

In with [10], the time order in Chinese should be year–month–day–hour, which is known as Whole-to-Part Principle, while in Portuguese the structure order is the opposite, i.e., hour–day–month–year. In ST (14) above, the Chinese structure order is *eryue ershiqi ri wanshang* [February 27th night], whose description of time is from month to day, then to a specific time of the day. However in the HT, the temporal adverbial phrase is *Na noite de vinte e sete de fevereiro* [at the night of 27th of February], which shows the inverse order of the Portuguese structure, from the specific time of the day to day, to month. However, in the PCT, the output *em vinte e sete de fevereiro a noite* corresponds partially to the standard temporal word order in Portuguese standard grammar, that is, the order is more close to that found in Chines standard grammar (from general to specific).

Example (15) below shows an instance of difference in prepositional position.

ST 是/v 一九四六/num 年/n 以来/n 跌幅/n 最/d 高/a 一/num 年/n
shì　　yī jiǔ sì liù　　nián　　yǐ lái　diē fú　zuì　gāo　　yī　　nián
... is **1946 year since** drop range highest one year.

(15) HT <u>desde</u> **1946**, é o ano de maior queda.
     <u>Since</u> **1946**, it is the year of bigger drop.

PCT ser o ano **1946** <u>desde</u> Die Fu são no máximo o ano um.
    ... to be **the year 1946** <u>since</u> Die Fu are in the biggest the year

The difference in Example (15) can be demonstrated more clearly via the syntax trees. Figures 1 and 2 show syntactic structure trees contrasts between ST and HT (because of space limitation, only the AdP is shown below).

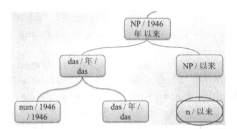

Fig. 1. Syntactic Tree of Chinese ST (*yilai* [since] is a temporal noun, and it is typically placed at the end of the temporal AdP in Chinese.)

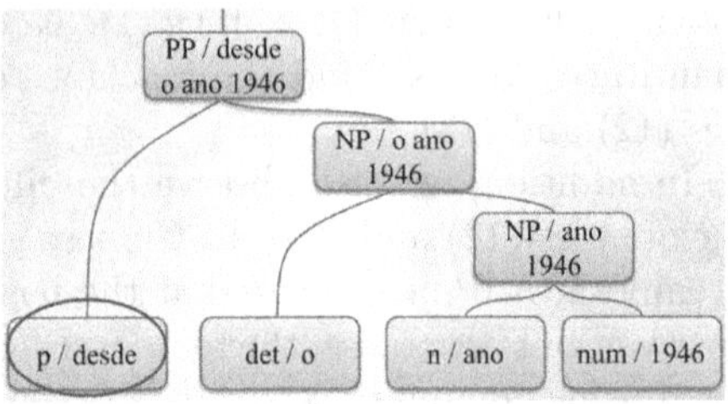

Fig. 2. Syntactic Tree of Portuguese HT (*desde* [since] is a preposition, and it is typically placed at the beginning of the temporal AdP in Portuguese.)

In (15), the noun *yilai* [since] occurs in the final position of time phrase *1946 nian* [year], while the correspondent in Portuguese is the preposition *desde* [since], typically in first position. In Fig. 1, we see that when the Noun Phrase *yilai* [since], in the end of the temporal AdP, is translated in to Portuguese, its location is shifted to the beginning of the temporal AdP. However, in the PCT translation system, the preposition *desde* [since] is placed after the temporal Noun Phrase *o ano 1946* [the year 1946], which means that the system has performed a literal translation.

4.2 Grammar Rules

Drawing on the description and contrast of the syntactic structures of temporal AdPs in Chinese ST and Portuguese translations, our research teams discussed the grammar rules to address the temporal AdP problems in rule-based MT systems, which are listed as follows, (Table 4)

(1) When PCT translates a Chinese structure of temporal Noun Phrase (month + date) that realizes a temporal AdP, it should be rendered as a structure composed of preposition and nominal phrase, that is, [preposition "*em*" (in/on) + Noun Phrase].

Table 4. Proposed grammar rules

Inadequacies	Proposed rules	Examples
Lack of preposition	**ZH**: [AdP(P *zài/yú* + temp. NP)]/[AdP(temp. NP)]	**ZH**: (*zài/yú*) *sānyuè wŭ rì* '(em) março 5 dia'
	PT: [AdP(P *em* + temp.NP)]	**PT**: *em cinco de março*
Problem of the PWP	**ZH**: [AdP (P + NP *year-month-day-hour*)]	**ZH**: *liùyuè shíqī rì zăoshàng* 'Junho 17 dia manhã'
	PT: [AdP (P + NP *hour-day-month-year*)]	**PT**: *manhã de dezassete de junho*
Incorrect position of preposition	**ZH**: [temp.AdP (temp.NP + dir.N *yĭlái/yĭhòu*)]	**ZH**: *nàtiān yĭhòu* 'aquele dia após'
	PT: [temp.AdP (P *desde/após* + NP)]	**PT**: *após aquele dia*

(2) When PCT translates a structure of temporal AdP from Chinese into Portuguese, the word order should follow the typical Portuguese pattern, namely: hour/part of day–day–month–year.

(3) The system should have a rule for the mandatory position of the preposition in the phrases in Portuguese realizing adverbials of time. In Chinese, temporal AdPs can consist of temporal noun, temporal adverb, prepositional phrase, directional phrase, and quantitative phrase, in which the preposition is not obligatory. However, all types of Chinese temporal AdPs must be translated into Portuguese with a preposition at the beginning of the phrase.

(4) When PCT translates a Chinese structure of temporal AdP with *yihou* [after], *yilai* [since], and temporal Noun Phrase, it should be noted as temporal nouns or temporal AdPs, while whole combination should be noted as the internal rule for Portuguese AdP [Adverbial Phrase (Preposition Phrase + Noun)], and the position of the preposition should be adjusted to initial position.

There are a number of works can be employed to intergrate our grammar rules, syntax structure or additional information into MT system such as factored translation [5,6], model adaptation [6,12], hierarchical models [3], syntax-based model [16] etc. As implementation is out of our main work in this paper, we will not discuss more about how to insert them into MT system.

5   Conclusion and Future Work

Motivated by the differences in temporal AdP, particularly in its syntactical structures, between Portuguese and Chinese, we investigate the machine translation of temporal AdP in Portuguese – Chinese language pair. Through the deep contrast and analysis, we identified the problems and defects of MT outputs, especially in temporal AdP. Then we propose four grammar rules to address these imperfections for rule-based MT system. With regard to the future, we intend to explore integrating rule-based MT systems with statistical MT systems, after inserting our proposed rule into the rule-based MT systems. We will use more evaluation method such as BLEU [9] to test the feasibility of our proposed method.

References

1. Brinton, L.J., Brinton, D.: The Linguistic Structure of Modern English. John Benjamins Publishing, Amsterdam (2010)
2. Charles, L., Thompson, S.: Mandarin Chinese: A Functional Reference Grammar. University of California Press, Berkeley (1981)
3. Chiang, D.: A hierarchical phrase-based model for statistical machine translation. In: Proceedings of the 43rd Annual Meeting on Association for Computational Linguistics, Ann Arbor, Michigan, USA, pp. 263–270 (2005)
4. Cunha, C., Cintra, L.F.L.: Nova gramática do português contemporâneo [New Grammar of Contemporary Portuguese]. Edição João Sá da Costa, Lisboa (1995)
5. Koehn, P., Hoang, H.: Factored translation models. In: Proceedings of the 2007 Joint Conference on Empirical Methods in Natural Language Processing and Computational Natural Language Learning, Prague, Czech Republic, pp. 868–876 (2007)
6. Koehn, P., Schroeder, J.: Experiments in domain adaptation for statistical machine translation. In: Proceedings of the 2nd Workshop on Statistical Machine Translation, Prague, Czech Republic, pp. 224–227 (2007)
7. Mateus, M.H.M.: Gramática da língua portuguesa [Grammar of Portuguese Language]. Editorial Caminho, Lisboa (1989)
8. Montenegro, H.: Os adverbiais na estrutura verbal: estudo sintáctico-semântico-pragmático (1999)
9. Papineni, K., Roukos, S., Ward, T., Zhu, W.J.: Bleu: A method for automatic evaluation of machine translation. In: Proceedings of the 40th Annual Meeting on Association for Computational Linguistics, Philadelphia, Pennsylvania, pp. 311–318 (2002)
10. Tai, J.H.: Hanyu Yufa Gainian Xitong Chutan [First Systematical Exploration of Conceptions in Chinese Grammar]. Dangdai Yuyanxue [Contemporary Linguistics], Beijing, China (2001)
11. Wang, L., Tu, Z., Zhang, X., Li, H., Way, A., Liu, Q.: A novel approach for dropped pronoun translation. In: Proceedings of the North American Chapter of the Association for Computational Linguistics, San Diego, California, USA (2016)
12. Wang, L., Wong, D.F., Chao, L.S., Lu, Y., Xing, J.: A systematic comparison of data selection criteria for smt domain adaptation. Sci. World J. **2014**, 1–10 (2014)
13. Wang, L., Zhang, X., Tu, Z., Li, H., Liu, Q.: Dropped pronoun generation for dialogue machine translation. In: Proceedings of the IEEE International Conference of Acoustics, Speech and Signal Processing, Shanghai, China (2016)
14. Wong, F., Chao, S.: Pct: Portuguese-Chinese machine translation systems. J. Transl. Stud. **13**(1–2), 181–196 (2010)
15. Xia, Y.: The Study of Chinese adverbial and its translation into Portuguese. Ph.D. thesis, University of Jinan, Guangzhou (2001)

16. Yamada, K., Knight, K.: A syntax-based statistical translation model. In: Proceedings of the 39th Annual Meeting on Association for Computational Linguistics, Toulouse, France, pp. 523–530 (2001)
17. Yu, X.: Contributo para Análise dos Adverbiais em Português e em Chinês [Contribution to Analysis of Adverbials in Portuguese and em Chinese]. Master's thesis, Instituto Politécnico de Macau, Macau (2008)

Determining the Level of Clients' Dissatisfaction from Their Commentaries

Ana Catarina Forte[1(✉)] and Pavel B. Brazdil[1,2]

[1] LIAAD – INESC Tec, Porto, Portugal
fortecatarina@gmail.com
[2] FEP, University of Porto, Porto, Portugal

Abstract. We present a study in the area of sentiment analysis of clients' commentaries transcribed by assistants of a help-desk service of one Portuguese telecommunications company. We have adopted a lexicon-based approach to determine the polarity of the sentiment of each commentary, based on the so called *opinion words*. This task was by no means easy, as not many tools are available for the Portuguese language. The initial results with the off-the-shelf solutions were rather poor. This has motivated us to carry out a number of enhancements, including, for instance, enriching the given lexicon with *domain specific* terms, formulating specific rules for negation and amplifiers. Automatic pruning of some of the lexicon terms has led to a significant improvement in performance. As our final system achieved a very good performance, our work should be of interest to others working on domain specific solutions for languages where ready-made solutions are not available.

Keywords: Sentiment analysis in Portuguese · Lexicon-based approaches · Lexicon enrichment · Negation of sentiment words · Amplifiers and attenuators

1 Introduction

In recent years characterized by emergence of social networks, blogs and forums, individuals and organizations used these means to express their opinions. As a result, there has been a continuous increase in the daily volume of information generated by users of social platforms, which created new challenges. First, it is not humanly possible to read all this information. Another problem is that organizations may be unaware of what the customers' opinions are. The customers may publish their opinions, sometimes quite negative ones, which are disseminated quickly reaching a large number of people and the company may be unaware of this. Thus organizations strive to continuously update their information, so as to understand the customers' behavior better, enabling them to predict their next steps.

According to some sources [3], 80 % of customer data captured within enterprises is in unstructured format, in the form of call center transcriptions or e-mails from customers and employees. It is relevant to study how customers manifest themselves over time, enabling to detect changes in their preferences and identify patterns of their behavior or detect that customers may effectively stop paying for certain services or products.

© Springer International Publishing Switzerland 2016
J. Silva et al. (Eds.): PROPOR 2016, LNAI 9727, pp. 74–85, 2016.
DOI: 10.1007/978-3-319-41552-9_7

This situation has motivated us to undertake this study. The data comes from one customer service department of a major Portuguese telecommunications company. This service involves questions concerning customer satisfaction, promotional campaigns, password change and description of problems that need to be resolved. It provides means for the customers to express their opinions and ask questions. This service is coordinated by technical assistants who answer phone calls from customers, while transcribing the contents of each call in a form of *commentaries*.

Our aim was to analyze this data and determine in an automatic way the polarity of each commentary. We have adopted a lexicon-based approach and employed two lexicons (*Afinn, Sentilex*) initially, but only *Sentilex* could be used directly with Portuguese texts. The terms of *Afin* had to be translated from English into Portuguese. Unfortunately, the results were rather disappointing. This has motivated us to carry out various enhancements whose aim was to improve the results. One involved the inclusion of *domain specific* terms in the lexicon which lead to a marked improvement in performance. We have also formulated specific rules determining how negation should be treated in Portuguese, and similarly, investigated also *intensifiers* and *attenuators*. Each of these led again to improved results. Perhaps the most surprising finding was that the method of automatic pruning of some of the lexicon terms proved beneficial for performance. The performance of the final system was indeed very good in terms of predictive accuracy of (81.9 %) and also, in terms of cost-sensitive analysis described later.

The structure of the rest of this paper is as follows: Sect. 2 presents a review of the literature in the area of sentiment analysis and special attention is paid to works focusing on Portuguese texts. Section 3 starts with an overview of a typical sentiment analysis system and then continues with the description of different enhancements of the system that led to improved results. Section 4 discusses the results obtained and Sect. 5 presents conclusions and suggestions for future work.

2   Related Work

Sentiment analysis (SA) is a field of study that examines opinions, feelings, evaluations, attitudes and emotions in relation to entities such as products, services, organizations, people, events and topics [10]. Most work in this area is carried out for the English language [12, 13]. For the Portuguese language the available resources are rather scarce. As it has rather specific characteristics (e.g. multiple verbal forms, specific rules for negation), developing a SA system can be considered a challenge.

Several methods have been proposed for the automatic determination of a sentiment value in text documents [10, 13, 17]. The most common approach involves considering this a text classification task, where the aim is to determine whether a document expresses a *negative, neutral* or *positive* opinion. The two main approaches are a lexicon-based approach and a machine learning approach [22].

The machine learning approach uses a set of features to learn a classifier. The features represent *terms* or in general *unigrams*, *bigrams* or *multi-word units*, which may in addition be characterized by part-of-speech tags. Although this approach may obtain good results, the transfer of the model - trained classifier - to another domain may not lead to good results [2].

The lexicon-based system can be developed manually [21, 23] or automatically [6, 24, 25]. This approach uses a *sentiment lexicon*, which includes words or expressions together with their polarities. It is important that the lexicons have a good coverage of words that appear in the texts that need to be processed.

As for the number of polarity levels, some researchers have considered just three (e.g. *negative, neutral and positive or −1, 0, 1* if we use numbers), while other used more. In the work reported here we use 7 levels *(−3, −2, −1, 0, 1, 2, 3)*, where the value *−3 (3)* represents the most negative (positive) sentiment value. We have adopted 7 levels because our client – the telecommunications company – has requested this.

Some authors claim that lexicon-based methods are more robust than the machine learning approaches (ML) [22]. However, our initial experiments with the lexicon-based approach did not result in good performance in our application domain. This indicates that this approach may also suffer from fragilities attributed by Taboada et al. [22] to ML approaches.

One study in Portuguese [18] determines the sentiment of texts from Twitter messages relative to five political leaders involved in the 2011 parliamentary elections. The task was first to identify the political leaders and then determine the polarity of each Twitter message. The sentiment analysis was based on lexical polarity, lexical-syntactic patterns and inference rules.

Some people have also studied the so called *polar expressions*. Polanyi et al. [15] have proposed the following rule to deal with such expressions: *"If a given polar expression includes negation, the sentiment value of the associated expression should be reversed"*. This rule is also applicable to Portuguese texts, although as we will see further on, in some situations it is useful to use a somewhat different rule.

Another line of work involves considering *amplifiers* and *attenuators* of the sentiment value. The most common way to identify these terms is by a recourse to a list of *adverbs* and *adjectives*. Polanyi et al. [15] modified the sentiment value of the subsequent word by adding or subtracting 1. For example, suppose the amplifier *very* has been used within the expression *very beautiful*. As *beautiful* has a sentiment value of +2, the resulting value of the expression is $2 + 1 = 3$.

Brooke [1] has proposed to treat *amplifiers* in a somewhat different way. Instead of adding (or subtracting) a fixed value, the author proposed to increase (or decrease) the sentiment value by a certain percentage. However, the author did not provide details on how the percentage is determined.

3   Adapting the Approach of Sentiment Analysis for Portuguese

As some readers may not be familiar with sentiment analysis (SA), the next subsection provides an overview of a typical lexicon-based approach. This serves also as a basis for describing the enhancements carried out for Portuguese (Subsect. 3.2).

3.1 Overview of a Typical Lexicon-Based Approach to Sentiment Analysis

The typical lexicon-based sentiment analysis method involves basically two major steps, pre-processing and calculation of the sentiment value of texts using sentiment lexicons. Both are described in more detail below.

Pre-processing. Many pre-processing steps used in SA tasks are commonly used also in classification of texts or clustering. Therefore our description here is rather brief, as readers may consult other sources [4, 7, 20].

One important step involves transformation of capital letters to lower case enabling to identify words that are similar, but not written in exactly the same way. The following steps normally include removal of numbers, punctuation symbols, short words (e.g. consisting of 1–2 characters), blanks and *stopwords*. All these items are normally not relevant and their elimination does not usually effect the outcome.

Many systems apply also *stemming* which consists of eliminating the suffixes, while retaining only the root of the word. Stemming algorithms depend strongly on word formation rules. So, direct application of an English *stemmer* may result in an undesirable outcome when applied to other languages including Portuguese [8]. This is why some researchers have adopted *lemmatization* instead, which selects a representative form (e.g. infinitive) from a given set of words with the same root.

Usage of Lexicons to Calculate the Sentiment Value. After the preprocessing has been carried out, the chosen sentiment lexicon is used to derive the sentiment value of a given text. This is done by processing all words. If a particular word appears in the sentiment lexicon, its sentiment value is recorded. The sentiment value of the given text is determined by summing up all the sentiment values encountered. This process is illustrated by the following example:

cliente quer (2) desativar (-2) serviços pois vai sair sem período fidelização
(the customer wants to stop the services, because he will leave without the loyalty period)

The only words that appear in a sentiment lexicon are *quer* and *desativar* with sentiment values of +2 and −2. So the resulting sentiment value of this phrase is 0.

3.2 Enhancements of the Basic Approach for Sentiment Analysis in Portuguese

The results of the basic system just described, when applied to our document set in Portuguese were rather disappointing, because the predictive accuracy was rather low.

This is the reason why we have set out to carry out different enhancements with the objective to improve of performance, acceptable to the telecommunications company.

3.2.1   Our Dataset and the Lexicons Used

The document collection use here consists of commentaries of clients that were gathered in a period of about 3 months in 2014, totaling 1724 cases. Each document includes one commentary, consisting of about 17 words on average.

Lexicon *Sentilex:* This sentiment lexicon was elaborated for the Portuguese language and contains more than 82 k entries. It was developed specifically for data mining applications and classification of sentiments in Portuguese [19]. The entries can be *inflected forms of words,* which can be included within expressions. Expressions can contain a one or more words (*compound expressions*) and typical idioms. The sentiment values are on an integer scale ranging from −1 to 1.

Lexicon *Afinn* and *AffinP:* Lexicon *Affin* includes 2477 words and phrases in English, whose sentiment value range from −5 to 5, where −5 (5) is associated with very negative (positive) words [11]. This lexicon was translated into Portuguese manually which was quite time consuming task, lasting about 3 weeks. The Portuguese version, *AffinP,* includes 1997 words.

Domain Specific Lexicon *AffinPD:* We have noted that the lexicons above did not include many of the terms that were used in clients' commentaries. So, for instance, the words *comprar* and *fidelização* (*buy* and *loyalty agreement*) which do not exist in *Sentilex,* nor in *AffinP,* are indicative of a positive action on the part of client. The word *reclama* (*complains*) exists only in *AffinP,* but not in *Sentilex.* It transmits clearly a negative sentiment.

This analysis has motivated us to explore the hypothesis that elaborating a domain specific lexicon may lead to a marked improvement of performance. We have decided to use *AffinP* as a starting point, as this lexicon attained better performance than *Sentilex.* Some of sentiment values in *AffinP* were altered, taking into account the specificities of the domain of telecommunications. The domain specific lexicon is referred to as *AffinPD* and includes 6915 words. It uses the sentiment scale of *Afinn* ranging from −5 to 5. As this lexicon was elaborated manually, this task was rather time consuming and required about 2 months to complete.

The percentage of words covered by *AffinPD* of our text documents was around 88.3 %, whereas in the other lexicons this rate was quite low (reaching about 20 %). Our experimental results presented later show that the addition of domain specific terms leads to a marked improvement of performance.

Optimizing the Lexicon by Pruning: It is conceivable that not all terms in the given lexicon are useful and that maintaining them there may in fact be detrimental to performance. We have decided to investigate this issue. This study was conducted only for *AffinPD* lexicon. The method used for constructing the optimized lexicon is similar to the feature selection method called *Backward Elimination* [4]. The method processes all terms in turn. In each step the system determines the effect of eliminating a particular term on the overall performance of the system. If the performance decreased, the term is maintained, but otherwise it is be dropped. In this phase we have used a validation set (different from the test set) which consisted of about 600 commentaries accompanied by a sentiment value.

3.2.2 Different Enhancements Carried Out

Enhanced Pre-processing. The pre-processing steps described in Sect. 3.1 were re-used in the system described in this section, which includes several enhancements described next.

Restricted Stemming: It may happen that some words that have different sentiment value before stemming become indistinguishable after stemming. Some examples are shown in Table 1.

Table 1. Undesirable effects of stemming for some Portuguese words

Lexicon	Word	Sentiment value	Root
AffinP	*afetado* (affected)	−2	afet
AffinP	*afeto* (feeling)	3	afet
Sentilex	*fala* (speaks)	0	fal
Sentilex	*falido* (bankrupt)	−1	fal

For this reason, in this work stemming is ***applied selectively*** only to words that do not appear in the sentiment lexicon. Usage of lemmatization would have been another possible alternative.

Rewrite Rules: The texts written by operators often include several different variants of the same word (e.g. *desativar, desactivar*). This is undesirable, as some of these variants may not appear in the lexicon. Therefore, we have decided to formulate a set of rules (635 in total) that substitute certain abbreviations and misspelled variants by their standardized form. Below we give two examples of such rules:

$$clt \rightarrow cliente$$
$$desactivar \rightarrow desativar$$

The second rule transforms the word *desactivar* from the old orthographic convention to the new one.

Removal of Short Texts: Our enhanced preprocessing comprises also the removal of commentaries which include only 1 word, as these transmit hardly any information.

Domain-Specific *Stopwords:* This stage focuses on building a list of domain-specific *stopwords*, or predefined phrases recorded by operators that do not transmit any unusual sentiment value. This list includes 345 items. Consider, for instance:

***Boa** tarde, bem-vindo ao **apoio** ao cliente se pretende **ajuda** sobre um tema (...)*
***Good** afternoon, welcome to customer **support** if you want **help** on a topic (...)*

The words *boa (good), apoio (support), ajuda (help)* would be attributed a positive sentiment value and hence influence the final scores. However, this does not make sense, as these words do not convey any sentiment value on the part of the client.

3.2.3 Dealing with Polar Expressions

Our system deals also with expressions that include *negation* which are sometimes referred to as *polar expressions*. These may include, for instance, a negation word, such as *não* (*no* or *not* in English) followed by some other word or expression, such as *divertido* (*funny* in English). If we did not introduce any special treatment, we could obtain a counterintuitive result. Suppose the sentiment value of *não* is −1 (as in *AffinP*) and the sentiment value of *divertido* is 3, the resulting value would be positive.

The rule proposed by Polanyi et al. [15]: *"If a given polar expression includes negation, the sentiment value of the associated expression should be reversed"*, suggests a solution, which is also applicable to Portuguese texts. So, in our example with *não* (−1) *divertido* (3), the resulting value would be −1 × 3 = −3. In the following this rule is referred to as *inversion rule* and is used in all cases where the word appearing after negation has a positive sentiment.

However, there are cases when the word appearing after negation has a negative sentiment. The application of the above rule would lead to an incorrect value. Consider, for instance, the phrase "cliente *não* (−1) quer ser mais *incomodado* (−2)" (customer *doesn't* want to be more *disturbed*). It does not seem right to attribute it a positive sentiment value resulting from −1 × (−2) = 2. In our system this phrase is attributed the value of −2 (the sentiment value of *incomodado*). We will call this rule *corrected inversion rule* in the following. The results presented in Sect. 4 indicate that the *corrected inversion rule* is preferable to the original rule.

Our system permits to process various other words besides the negation word *não*. The list assembled contains 33 different words that modify the meaning of the subsequent text. The list includes, for instance *não, nunca, jamais, nem, ninguém, nenhum, negativamente, tampouco,* among others.

3.2.4 Amplifiers and Attenuators

Polanyi et al. [14], define *contextual valence shifters* as contextual phenomena that modify the polarity of another term. According to the authors these can be divided into three groups: *inverters* invert the polarity of a polarized item, *intensifiers* (or *amplifiers*) intensify it and *attenuators* diminish it [9, 16]. Sentiment modifiers are mostly *adverbs* (*attractively, annoyingly*), *adjectives* (*better, bad*), *sentence connectors* (*although, however, but*), *modal verbs* (*might, could, possibly*) or even *nouns* (*freedom, bankruptcy*).

We have followed this line of work, but adapted it to Portuguese, while focusing mainly on sentiment modifiers that are *adverbs* and *adjectives*. Words such as *demasiado (too much), bastante (in sufficient number), pouco (a little)* and *menos (less)* do not exist in the sentiment lexicons analyzed. To resolve this issue, we have assembled a list of 166 amplifiers and 35 attenuators. The process for calculating the sentiment value of sentences is as follows:

> If an *amplifier* is encountered:
> > If the value of the subsequent text is positive (negative), add (subtract) 1.
> If an *attenuator* is encountered:
> > If the value of the subsequent text is positive (negative), subtract (add) 1.

3.2.5   Normalizing the Sentiment Values

As the ranges of the sentiment values in the lexicons used are different, the resulting sentiment value of the given text (commentary) attributed by our system could depend on which lexicon has been used. This is obviously undesirable. We have therefore incorporated a normalization step, whose aim to transform the sentiment values generated into a normalized range spanning from −3 till 3. This range was suggested by the company who require that we not only identify the negative (positive) cases, but distinguish how grave the situation is.

Normalization is relatively simple to carry out. Basically, it is necessary to divide all values by normalization constant $N_{AFFINPD}$ or $N_{SENTILEX}$, depending on which lexicon is being used. After some analysis we have determined that the setting $N_{AFFINPD} = 4$ and $N_{SENTILEX} = 1$ is the most appropriate.

The sentiment values are illustrated by the following example that was already used in Sect. 3.1. Here we show the effects of using the enriched lexicon, negation rules and normalization.

cliente quer desativar *(-5)* **serviços pois vai** sair *(-3)* **sem** *(-1)* **período** fidelização *(3)*

Let us examine the effects of negation in this example. If we did not use it, the final sentiment value (before normalization) would be $-5-3-1 + 3 = -4$. The inversion rule for negation transforms the last 3 words into *sem período fidelização (−3)* and hence the score is $-5-3-3 = -11$. This value needs to be normalized, which leads to $-11/N_{AFFINPD} = -11/4 = -2.75$. Finally, the sentiment value is mapped to the nearest integer value −3.

3.3   Evaluation of the Sentiment Classifier

The data for the evaluation consisted of 200 texts accompanied by their sentiment value ranging between −3 and 3. The evaluation was carried out by comparing the sentiment values generated with the ones that were attributed by a person (ground truth). Two different performance measures were used - *success rate* and *mean cost per case*. The use of cost-sensitive analysis was motivated by the fact that the sentiment values −3, −2, .. 3 are on ordinal scale. It is therefore useful to consider how far a particular prediction is from the correct value. Assuming that −3 and 3 are one unit apart, the distance between −3 and −2, for instance, is a fraction of that distance, i.e. $1/6 = 0.17$ (after rounding). So this error has the cost of 0.17. The mean cost per case is calculated by calculating the mean of the costs of individual documents.

4   Results

The results presented here are relative to our test dataset consisting of about 200 cases accompanied by a sentiment value. The evaluation involved the set of 13 different variants of our sentiment analysis system described next. We have conducted experiments to determine to what extent the different facets incorporated in these variants would improve the performance. Figure 1 shows the results.

1. Lexicon *AffinP*: This variant represents the very basic SA system with *AffinP*. It achieved a modest success rate of 33.33 % and mean cost of 16.85 %.
2. Lexicon *AffinPD*: Similar to variant 1, but the system uses an enriched lexicon containing domain-specific words. The success rate improves by 5.9 %.
3. Pre-processing: This variant extends the previous variant and uses the basic pre-pre-processing method described in Sect. 3.1.
4. Rewrite rules, as described in Sect. 3.2.2. This led to quite significant reduction of the mean cost by 1.88 %.
5. Removal of texts with a single word (see Sect. 3.2.2).
6. Elimination of domain-specific stopwords (see Sect. 3.2.2).
7. Elimination of stopwords: We have reused the list of 203 Portuguese stopwords available in R with the command *stopwords('portuguese')*.
8. Restricted stemming (see Sect. 3.2.2).
9. Negation 1: This variant incorporates the *inversion rule* (see Sect. 3.2.3).
10. Negation 2: This variant extends the variant 8 by incorporating the *corrected inversion rule* (see Sect. 3.2.3). This led to a significant cost reduction of 3.87 %.
11. Amplifiers (see Sect. 3.2.3): This variant extends the variant 10. This led to a cost reduction of 1.25 %.
12. Attenuators (see Sect. 3.2.3).
13. Optimized Lexicon *AffinPDO:* The lexicon is pruned following the method described in Sect. 3.2.1. The final version had a very good mean cost of 4.54 %.

Each of the following versions extends its predecessor unless stated otherwise.

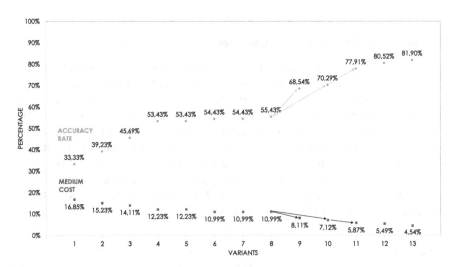

Fig. 1. Results of different variants of our system

We have carried out also experiments with *Sentilex* lexicon which is available for Portuguese texts. Unfortunately, this system had poor results. The basic solution achieved the success rate of 31.34 % and the mean cost of 18.1 % (i.e. worse than our version 1 with *AffinP*). This lexicon had rather poor coverage of the words (only about 5 %) that appear in our commentaries. Also, the range of sentiment values range from −1 to 1, which is probably insufficient to capture the nuances of sentiment.

5 Conclusions

Our aim was to develop a sentiment analysis system which could be applied to Portuguese texts. This appeared a challenge, as not many tools were available. Applying off-the-shelf solutions in conjunction with lexicon Sentilex, available for Portuguese, gave poor results. We conjecture that this is due mainly to the fact that its terms do not cover the domain of telecommunications well, and also, the range of just three sentiment values is not able to capture the varied degrees of sentiments.

We have therefore analyzed a set of techniques with the objective of determining whether these could lead to improved performance. Translating terms of the chosen lexicon (*Affin*) into Portuguese and enriching it with domain-specific terms was rather time consuming, but the effort paid off. It led to a marked improvement of performance. There are many specific terms in the telecommunications that were not present in the original lexicon. Defining rules enabling to translate words into its standardized form was also useful.

We have also incorporated rules that deal with negation and adapted the inversion rule described in literature to Portuguese. We did not limit ourselves just to the negation word *não*, but defined an extended set of 33 *negation words*. We have investigated situations when the word following the *negation word* has a negative sentiment. The experiments on pre-classified data indicated that the *corrected negation rule* is preferable to the usual *inversion method* (as in [15]). Finally, we have incorporated rules that modify the sentiment value of expressions that include *amplifiers* and *attenuators*, which required that a list of words be built for the Portuguese language. The final system reached an excellent performance both in terms of accuracy and mean cost.

Future work may involve the incorporation of an automatic spell-checker for Portuguese. Elaborating rules manually to correct abbreviated or misspelled words is time-consuming. We could also investigate the possibility of generating the enriched lexicon through automated methods, rather than through a manual approach. As this task is rather time-consuming, a great deal of time could be saved. Also, it would make our system adaptable to other domains.

Acknowledgements. We wish to thank Nuno Paiva for defining this very interesting problem in the first place and to João Cordeiro for his useful comments on this work. We acknowledge the support of Project NORTE-01-0145-FEDER-000020, financed by the *North Portugal Regional Operational Programme* (NORTE 2020), under the PORTUGAL 2020 Partnership Agreement and support from ERDF – *European Regional Development Fund through the Operational Programme for Competitiveness and Internationalisation* - COMPETE 2020 Programme within project «POCI-01-0145-FEDER-006961», and by National Funds through the FCT – *Fundação para a Ciência e a Tecnologia* (Portuguese Foundation for Science and Technology) as part of project UID/EEA/50014/2013.

References

1. Brooke, J.: A Semantic Approach to Automatic Text Sentiment Analysis. M.A. thesis, Simon Fraser University, Burnaby, B.C., Canada (2009)
2. Gamon, M., Aue, A.: Automatic identification of sentiment vocabulary: exploiting low association with known sentiment terms. In: Proceedings of the ACL WS on Feature Engineering for ML in Natural Language Processing, Ann Arbor, Michigan, pp. 57–64. ACL (2005)
3. Forte Consultancy *Group* Company: Text Mining - Going Way beyond Just Listening to the Voice of the Customer. Forte Consultancy (2010)
4. Guyon, I., Gunn, S., Nikravesh, M., Zadeh, L.A.: Feature Extraction: Foundations and Applications. Studies in Fuzziness and Soft Computing. Springer, NY (2006)
5. Haddi, E., Liu, X., Shi, Y.: The role of text pre-processing in sentiment analysis. Procedia Comput. Sci. **17**, 26–32 (2013)
6. Hatzivassiloglou, V., McKeown, K.R.: Predicting the semantic orientation of adjectives. In: Proceedings of the 35th Annual Meeting of the ACL and 8th Conference of the European Chapter of the ACL, Madrid, pp. 174–181 (1997)
7. Hemalatha, I., Saradhi-Varma, G.P., Govardhan, A.: Preprocessing the informal text for efficient sentiment analysis. Int. J. Emerg. Trends Technol. Comput. Sci. **1**(2), 58–61 (2012)
8. Honored, A., Leon, R., O'Donnel, R., Sinclair, D.: A word stemming algorithm for the Spanish language. In: Proceedings of the Seventh International Symposium on String Processing Information Retrieval (SPIRE 2000), p. 139. IEEE Computer Society, Washington (2000)
9. Kennedy, A., Inkpen, D.: Sentiment classification of movie reviews using contextual valence shifters. Comput. Intell. **22**(2), 110–125 (2006)
10. Liu, B.: Sentiment Analysis and Opinion Mining: Synthesis Lectures on Human Language Technologies. Morgan and Claypool Publishers, San Rafael (2012)
11. Nielsen, F.: A new ANEW: evaluation of a word list for sentiment analysis in microblogs. In: Proceedings of the ESWC 2011 Workshop on Making Sense of Microposts: Big things come in small packages 718, CEUR Proceedings, pp. 93–98 (2011)
12. Pang, B., Lee, L.: Opinion mining and sentiment analysis. Found. Trends Inf. Retr. **2**(1–2), 1–135 (2008)
13. Pang, B., Lee, L., Vaithyanathan, S.: Thumbs up?: sentiment classification using machine learning techniques. In: Proceedings of the ACL 2002 Conference on Empirical Methods in Natural Language Processing (EMNLP 2002), vol. 10, pp. 79–86. ACL (2002)
14. Polanyi, L., Zaenen, A.: Contextual valence shifters. In: Proceedings of AAAI Spring Symposium on Exploring Attitude and Affect in Text, pp. 106–111 (2004)
15. Polanyi, L., Zaenen, A.: Contextual valence shifters. In: Shanahan, J.G., Qu, Y., Wiebe, J. (eds.) Computing Attitude and Affect in Text, vol. 20, pp. 1–10. Springer, Dordrecht (2006)
16. Quirk, R., Greenbaum, S., Leech, G., Svartvik, J.: A Comprehensive Grammar of the English Language. Longman, London (1985)
17. Saraswat, M., Patel, R.: A survey on sentiment analysis. Int. J. Res. Eng. Technol. Manag. **2**(6) (2014)
18. Silva, M.J., Team, R.: Notas sobre a realização e qualidade do twitómetro. Technical report, University of Lisbon, Faculty of Sciences, LASIGE, Lisbon, Portugal (2011)
19. Silva, M.J., Carvalho, P., Sarmento, L.: Building a sentiment lexicon for social judgement mining. In: Caseli, H., Villavicencio, A., Teixeira, A., Perdigão, F. (eds.) PROPOR 2012. LNCS, vol. 7243, pp. 218–228. Springer, Heidelberg (2012)

20. Soman, K.P., Diwakar, S., Ajay, V.: Data Mining: Theory and Practice. PHI Learning (2006)
21. Stone, P.J., Dunphy, D.C., Smith, M.S., Ogilvie, D.M.: The General Inquirer: A Computer Approach to Content Analysis. MIT Press, Cambridge (1966)
22. Taboada, M., Brooke, J., Tofiloski, M., Voll, K., Stede, M.: Lexicon-based methods for sentiment analysis. Comput. Linguist. **37**(2), 267–307 (2011)
23. Tong, R.M.: An operational system for detecting and tracking opinions in on-line discussions. In: Working Notes of the ACM SIGIR 2001 Workshop on Operational Text Classification, NY, pp. 1–6 (2001)
24. Turney, P.: Thumbs up or thumbs down? Semantic orientation applied to unsupervised classification of reviews. In: Proceedings of 40th Meeting of the Association for Computational Linguistics, Philadelphia, PA, pp. 417–424 (2002)
25. Turney, P., Littman, M.: Measuring praise and criticism: inference of semantic orientation from association. ACM Trans. Inf. Syst. **21**(4), 315–346 (2003)

Comparing Approaches to Subjectivity Classification: A Study on Portuguese Tweets

Silvia M.W. Moraes, André L.L. Santos[(✉)], Matheus Redecker,
Rackel M. Machado, and Felipe R. Meneguzzi

Faculdade de Informática, PUCRS, Avenida Ipiranga, 6681, Prédio 32,
90619-900 Porto Alegre, RS, Brazil
{silvia.moraes,felipe.meneguzzi}@pucrs.br,
andre.leonhardt.santos@gmail.com,
{matheus.redecker,rackel.machado}@acad.pucrs.br

Abstract. In this paper, we compare lexicon-based and machine learning-based approaches to define the subjectivity of tweets in Portuguese. We tested SentiLex and WordAffectBR lexicons, and Sequential Machine Optimization and Naive Bayes algorithms for this task. In our study, we used the Computer-BR corpus that contains messages about the technology area. We obtained better results using the Comprehensive Measurement Feature Selection method and the Sequential Machine Optimization algorithm as the classifier. We achieved considerable accuracy when we included the polarities of words in the vector space model of tweets.

Keywords: Subjectivity classification · Sentiment analysis · Natural language processing

1 Introduction

In the past decade, people have used social web to express and share their 'sentiments' about products and services. Texts published in social media (e.g., Twitter, Facebook, forums, blogs, and user forums) have become important sources of information for organizations. The analysis of these snippets of text is a way of monitoring the opinion and response from the clients of these organizations [1]. The area of research that automatically performs this processing is known as Sentiment Analysis or Opinion Mining. In this area, textual information can be categorized into two main types: facts and opinions. Opinions, unlike facts, describe people's sentiments, appraisals, or feelings toward entities, events, and their properties. The task of defining whether a sentence expresses an opinion or a fact can be treated as a classification problem. This task is called subjectivity classification [2]. The subjectivity classification is a stage that precedes the Opinion Mining. When used, it improves Opinion Mining performance by preventing noisy and irrelevant extraction [3, 4]. In approaches that use machine learning algorithms for polarity classification, the improved results can be attributed to the balancing of training sets. Some authors mention that the imbalance of such approaches is caused by the class of objective sentences, which usually has a larger number of samples [5].

© Springer International Publishing Switzerland 2016
J. Silva et al. (Eds.): PROPOR 2016, LNAI 9727, pp. 86–94, 2016.
DOI: 10.1007/978-3-319-41552-9_8

In this paper, we compare two traditional approaches to subjectivity classification: based on lexicon, and machine learning algorithms [10]. We tested SentiLex and WordAffectBR lexicons, and Sequential Machine Optimization (SMO) and Naive Bayes algorithms to determine the subjectivity of the tweets from Computer-BR corpus, we that built for this study. This corpus is composed of messages in Portuguese language about the technology area. We categorized the tweets according to their sentiment orientations (polarities). We considered as subjective those sentences with positive or negative polarity, and as objective sentences the remaining ones. In the approach using machine learning, we tested a new method for feature selection: the Comprehensive Measurement Feature Selection (CMFS), indicated for unbalanced corpora [15, 16]. Our best results were obtained using this method and the SMO algorithm as classifier. We achieved an accuracy on average of 78.51 % when we included the polarities of words in the vector space model of tweets. Besides the results obtained from the research described in this paper, we also consider Computer-BR corpus as one of our contributions to the sentiment analysis area.

This paper is organized into 7 sections. In Sect. 2, we present some works related to ours. In Sect. 3, we introduce the subjectivity classification task and approaches used to treat it. In Sect. 4, we describe the corpus we created and the pre-processing realized. In Sects. 5 and 6, we detail the approaches used to define the subjectivity of the sentences. In Sect. 7, we present our conclusions. Finally, the acknowledgment and references are presented.

2 Related Work

The opinion mining from web texts is a non-trivial Natural Language Processing task, for this reason it has received much attention. Most literature on sentiment analysis for Portuguese language addresses polarity classification at sentence and aspect (feature) level. In applications at the sentence-level, in which sentences are classified as positive, negative, or neutral, the accuracy ranges from 55 % to 71,79 % [6–8, 17]. In these applications, the best results were obtained from the Sequential Minimal Optimization (SMO) [6, 7] and Naive Bayes [3] algorithms. In lexicon-based approaches, the accuracy for Portuguese language is around 57.3 % [9]. It is worth mentioning that linguistic resources for sentiment analysis in Portuguese language are still developing [13, 14]. The lack of benchmarks corpora, for example, makes more challenging the comparison of results.

According to researchers in the area, the performance of these tasks improves when we perform the subjectivity classification in a previous step [3, 4]. Initially, Kamal [3] classifies sentences in English as subjective and objective, and later he performs features-based sentiment analysis for the sentences defined as subjective. In the subjectivity classification stage, the author has reached an accuracy of 91.6 % with the Naive Bayes algorithm. Lambov et al. [18] says that the classification of subjectivity is a specific domain problem, showing that the results fall around 20 % for across domains.

3   Subjectivity Classification

We understand the subjectivity classification as a task to define whether a sentence is objective or subjective. Objective sentences express facts, while subjective sentences express opinions. Opinions, unlike facts, describe people's sentiments, appraisals, or feelings toward entities, events, and their properties [2]. In our study, we determined the classification of sentences according to their polarities (positive, negative, or neutral), that indicated the sentiment orientation. The algorithm assigns the polarity by the presence of certain adjectives, verbs, and nouns in the sentences. For example, words such as *fast* (adjective), *to love* (verb) and *joy* (noun) expressed sentiments with positive polarity, whereas words such as *slow* (adjective), *to hate* (verb) and *sadness* (noun) indicated negative polarity to the sentences. The polarity was neutral when the word was neither positive nor negative. We considered as subjective sentences those with positive or negative polarity. The remaining sentences (with neutral polarity) were considered objective.

According to Madhat et al. [10], in the sentiment analysis area, the most common techniques use approaches based on lexicons or machine learning algorithms. We used the lexicons SentiLex-PT [11] and WordNet-Affect [12], and the SMO and Naive Bayes algorithms in our study. We chose these resources and techniques, because they are widely known [3, 6, 7] and achieves good results. Figure 1 shows the pipeline that we implemented to define the subjectivity of the sentences. In the following sections this pipeline is detailed.

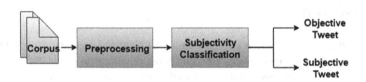

Fig. 1. Subjectivity classification

4   Computer-BR Corpus and Preprocessing

To perform this study, we built a corpus in Portuguese, the Computer-BR. This corpus consists of 2,317 tweets that was extracted in the period from January 1st to September 18th, 2015. The Corpus has 34,437 tokens and 4,653 types. To build it, we used keywords related to computers: notebook, analysis, testing, and so on. We relied on 4 human annotators who defined the polarity of the tweets, 3 of them participated in the whole process of annotation and the fourth decided the final polarity only in cases of disagreement. Although our study is only on subjectivity classification, the annotation considered 4 classes: irony (−2), negative (−1), neutral (0) and positive (1). We built the Computer-BR also for its application in future works. The final kappa index was 0.69. It is worth mentioning that three annotators were from the Computer Science area and one from the Linguistics area. The Table 1 shows the polarity distribution in the corpus.

As the treatment of irony is not included in this work, tweets classified into the irony class became negative. Thus, 443 was the total amount of tweets classified as

Table 1. Sentiments distribuition in the Computer-BR corpus.

Classes	#Tweets (%)
Irony	39 (1,7 %)
Negative	404 (17,4 %)
Neutral	1,677 (72,4 %)
Positive	197 (8,5 %)

Table 2. Examples of tweets from Computer-BR corpus.

Tweet	Polarity
TO IRADOO! (I'M ANGRYY!)	negative
Bateria do meu notebook já era... (Battery of my notebook is gone...)	neutral
Apaixonada pelo meu Notebook☺♥□ (I am in love with my Note-book☺♥□)	positive
Aiii que maravilha, meu notebook parou de ligar! (Ahhh, wonderful, my notebook no longer switches on!)	irony, negative

negative in our study. We intend to investigate the polarity classification including irony in a future work. Table 2 shows some examples of tweets from the corpus.

Web texts, especially those posted on microblogs, have a lot of noisy and uninformative pieces (HTML tags, scripts and advertisements). In these texts, it is common the repetition of vowels, punctuation problems, misspelling, emoticons, colloquialism, unconventional use of upper and lower cases, and out-of-vocabulary words (abbreviations, acronyms, and slang). Portuguese texts from the technology domain still use technical terms in English. All these factors reduce the efficiency of automatic classifiers. To minimize this problem, it was necessary to normalize the tweets. So, in the texts processing stage, we removed (or treated) special characters and hashtags, transformed emoticons and hyperlinks into text, and replaced abbreviations and slang with usual expressions, such as "*vc*" into "*você*" (you) and "*novis*" into "*novidade*" (news).

After this stage, we used the parser VISL[1] to supply the morphosyntactic annotation to the texts. Although this tool provides a richer annotation on linguistic information, in our study, to simplify, we used only lemmas and Part-of-Speech (PoS) tags of the words.

In the following sections, we describe studies on subjectivity classification.

5 Lexicon-Based Subjectivity Classification

The lexicon-based approach depends on finding the opinion lexicon that is used to analyze the text. This approach is divided into dictionary-based approach and corpus-based approach that use statistical or semantic methods to find sentiment polarity [10]. In our study, we used the dictionary-based approach. We did tests with

---

[1] http://beta.visl.sdu.dk/visl/pt/.

two lexicons: SentiLex-PT and WordNetAffect BR. SentiLex-PT [11] is a lexicon for Portuguese language that has 7,014 lemmas (4,779 adjectives, 1,081 nouns, 489 verbs, and 666 expressions). Each lemma shows the grammatical category, the target in a sentence; for each target, the polarity associate with it (positive, neutral, or negative), and the last information is about the method of assignment (if it was manual or with a tool named Judgment Analysis Lexicon Classifier - JALC). WordNetAffect BR [13] was built from WordNet terms translated from English into Portuguese, with terms that connote different emotions. The 289 words, including adjectives and nouns, were manually translated.

Initially, we did some tests with both lexicons separately, however, when we decided to use them together, we obtained a small improvement. Therefore, in our study, we are looking for the polarity of the words of tweets in both lexicons.

In the next section, we present the strategies we use to define the subjectivity of tweets.

5.1 Strategies to Identify Subjective Tweets

We tested three different heuristics to define the class of the tweets. All heuristics consider the polarity of the words (lemma) from the lexicon. If the word did not exist in the lexicon, its polarity was considered neutral.

- Heuristic 1 – The sum of polarities: This heuristic consists of adding the polarity of each tweet token [10]. If the result of this sum is non-zero, the tweet has some sentiment. The formula is represented in (1), where $|sent|$ is the number of tweet tokens and $term_i$ represents each token present in the sentence of tweet.

$$subjectivity_{sent} = \sum\nolimits_{i=1}^{|sent|} polarity(term_i) \qquad (1)$$

- Heuristic 2 – The number of words with polarity: This heuristic[2] assumes that if n tweet words have polarity, then the tweet is subjective. The equations for this heuristic are in (2) and (3), where $n \in [1; 4]$.

$$subjectivity_{sent} = \sum\nolimits_{i=1}^{|sent|} countPolarity(n, term_i) \qquad (2)$$

$$countPolarity(n, term_i) = \begin{cases} 1, & if\ tweet\ has\ n\ terms\ with\ polarity \\ 0, & otherwise \end{cases} \qquad (3)$$

- Heuristic 3 – The proportion of words with polarity: This heuristic is similar to the previous one, however, it considers the number of tokens with polarity in relation to the total of tweet tokens [8]. The equation for this heuristic is in (4). The decision about the subjectivity of the tweets was based on a threshold. We tested the threshold value in the range $[0.5; 0.35]$.

---

[2] This heuristic is a small variation of the strategy proposed in [9].

$$subjectivity_{sent} = \sum_{i=1}^{|sent|} \frac{countPolarity(term_i)}{|sent|} \qquad (4)$$

5.2   Results

To evaluate the results we adopted the usual measures in information retrieval: precision, recall, F-measure, and accuracy. Table 3 shows the best results obtained with the studied strategies.

Table 3. Lexicon-based approach results.

Heuristic	Precision	Recall	F-measure	Accuracy
1	0.64	0.64	**0.64**	0.70
2 ($n = 2$)	0.57	0.63	0.62	**0.74**
3 (threshold >=0.25)	0.50	0.64	0.57	0.72

By employing the most basic strategy (heuristic 1), we obtained the best F-measure. However, the number of words with polarity (heuristic 2, for n = 2) achieved the best accuracy. It is important to mention that we used the polarity of nouns, verbs, and adjectives. We also checked if the use of only adjectives in the heuristic 2 would be enough, but the results were worse.

The lexicon-based approach had limitations. Sometimes the meaning of a word in the lexicon did not correspond to its meaning in the sentence. In our study, the high number of errors is also due to the number of advertisements in the corpus. The tweets with advertisements are objective, but if they have words that promote products and these words have polarities, the tweet was classified as subjective. This is the case of the tweet: *"VENDO NOTEBOOK MARCA DELL Excelente condição!"* (SELL NOTEBOOK BRAND DELL Excellent condition!).

6   Machine Learning-Based Subjectivity Classification

In this approach, we used the classification algorithms SMO and Naive Bayes, both from Weka[3] tool. We chose these algorithms for being frequently referenced in related works [3, 6, 7]. The feature selection is a fundamental stage of this approach. For this reason, we tested a new method at this stage: the CMFS [15, 16]. We chose this method because it is new, suitable for class imbalance, and showed good results in [16]. We compared this new method to the usual method based on frequency. These methods are described in the next section.

---

[3] http://www.cs.waikato.ac.nz/ml/weka/.

6.1   Methods for the Features Selection

Regardless the method used, initially we made a list of words for both tweet classes: objective and subjective. For each class, we used the features selection methods to define the relevance of the words. Based on this relevance, we ranked n most important words (for n ranging from 10 to 100). To generate the Bag-of-Words (BoW) we chose two strategies which we called "union" and "exclusion". The strategy "union" merges n most relevant words for each class. The strategy "exclusion" merges, but also to excludes the words common to both classes. We also tested two vector space model to the tweets: binary and based on polarity. In the latter, we replaced the binary values by the polarities of words. The polarity was defined from the lexicons used in Sect. 4. We used three methods to select the most relevant words:

- Absolute Frequency (*fa*): indicates the number of occurrences of a word w_k in a class c_j.
- Relative Frequency (*fr*): corresponds to the relation between the number of occurrences of a word and the total number of words of the class. The equation for this frequency is in (5).

$$fr(w_k, c_j) = \frac{fa(w_k, c_j)}{|W|} \qquad (5)$$

- Comprehensive Measurement Feature Selection (CMFS): indicates the significance of a word w_k in one class c_j, against the occurrences of the same word in the corpus. According to Yang et al. [16], this significance can be reached by multiplying the probability that the word w_k occurs in the category c_j, $P(w_k|cj)$, and the probability that the word w_k belongs to the category c_j, when the word w_k occurs, $P(c_j|w_k)$. The equation for this frequency is in (6).

$$CMFS(w_k, c_j) = \frac{P(w_k|c_j)P(C_j|W_k)}{P(W_k)} \qquad (6)$$

6.2   Results

As in Sect. 5.2, the usual measure were used to evaluate the results. Tables 4 and 5 show the best results obtained with the studied methods for the features selection applying SMO and Naive Bayes classification algorithms, respectively. The *BoW* column represents the number of attributes used, following the strategy of the most relevant words for each class.

When we used SMO, the best method for the feature selection was CMFS based on polarity. Combining this method with the strategy "exclusion", we had improvements in both F-measure and accuracy. Nevertheless, with the Naive Bayes approach, we had a good accuracy with the relative frequency method and the strategy "exclusion", but this combination had the worst results for F-measure.

Table 4. Approach results using SMO

Method	Strategy	BoW	Precision	Recall	F-measure	Accuracy
Fa	Union	133	75,30	77,00	74,90	77,04
Fr	Union	133	75,30	77,00	74,90	77,04
CMFS	Exclusion	125	76,80	78,00	75,30	78,03
CMFS + pol.	Exclusion	125	77,60	78,50	**75,70**	**78,51**

Table 5. Approach results using Naive Bayes

Method	Strategy	BoW	Precision	Recall	F-measure	Accuracy
Fa	Exclusion	21	71,80	73,20	63,70	73,20
Fr	Exclusion	21	71,80	73,20	63,70	**73,20**
CMFS	Union	171	71,40	71,30	**71,30**	71,26
CMFS + pol.	Union	16	70,20	71,80	70,80	71,77

7 Conclusion

This paper describes two different approaches to subjective classification, one uses machine learning and the other lexicon based. We propose different heuristics to highlight the differences between them and compare the results. The evaluation of the results showed that machine learning obtain better results than lexicon based. For the lexicon based approach the best result was 74 % of accuracy, while in machine learning approach the best result was 78 %. For future studies we want to classify the subjective tweets into positive or negative polarities.

Acknowledgments. Our thanks to Dell for the financial support of this work.

References

1. Feldman, R.: Techniques and applications for sentiment analysis. Commun. ACM **56**(4), 82–89 (2013)
2. Dale, R., Moisl, H., Somers, H. (eds.): Handbook of Natural Language Processing. CRC Press, Boca Raton (2000)
3. Kamal, A.: Subjectivity Classification using Machine Learning Techniques for Mining Feature-Opinion Pairs from Web Opinion Sources (2013). arXiv preprint arXiv:1312.6962
4. Fersini, E., Messina, E., Pozzi, F.A.: Subjectivity, polarity and irony detection: a multi-layer approach. In: Proceedings of the First Italian Conference on Computational Linguistics CLiC-it 2014 & the Fourth International Workshop EVALITA (2014)
5. Drury, B., de Andrade Lopes, A.: A comparison of the effect of feature selection and balancing strategies upon the sentiment classification of Portuguese news stories. In: Proceedings of ENIAC (2014)
6. Santos, A.P., Ramos, C., Marques, N.C.: Sentiment classification of Portuguese news headlines. Int. J. Softw. Eng. Appl. **9**(9), 9–18 (2015)

7. Rosa, R.L., Rodríguez, D.Z., Bressan, G.: SentiMeter-Br: a social web analysis tool to discover consumers' sentiment. In: 2013 IEEE 14th International Conference on Mobile Data Management (MDM), vol. 2, pp. 122–124. IEEE (2013)
8. Morgado, I.C.: Classification of sentiment polarity of Portuguese on-line news. In: Proceedings of the 7th Doctoral Symposium in Informatics Engineering, pp. 139–150 (2012)
9. Filho, P.P.B., Pardo, T.A., Aluısio, S.M.: An evaluation of the Brazilian Portuguese liwc dictionary for sentiment analysis. In: 9th Brazilian Symposium in Information and Human Language Technology, Fortaleza, Ceara (2013)
10. Medhat, W., Hassan, A., Korashy, H.: Sentiment analysis algorithms and applications: a survey. Ain Shams Eng. J. 5(4), 1093–1113 (2014)
11. Carvalho, P., Silva, M.J.: SentiLex-PT: principais características e potencialidades. Oslo Stud. Lang. 7(1), 425–438 (2015)
12. Pasqualotti, P.R., Vieira, R.: WordnetAffectBR: uma base lexical de palavras de emoções para a língua Portuguesa. RENOTE 6, 1–10 (2008)
13. Généreux, M., Martinez, W.: Contrasting objective and subjective Portuguese texts from heterogeneous sources. In: Proceedings of the Workshop on Innovative Hybrid Approaches to the Processing of Textual Data, pp. 46–51. Association for Computational Linguistics (2012)
14. Moraes, S., Silveira, M., Manssour, I.: 7x1-PT: um Corpus extraído do Twitter para Análise de Sentimentos em Língua Portuguesa. BRACIS, STIL (2015)
15. Yang, J., Liu, Y., Zhu, X., Liu, Z., Zhang, X.: A new feature selection based on comprehensive measurement both in inter-category and intra-category for text categorization. Inf. Process. Manage. 48(4), 741–754 (2012)
16. Yang, J., Qu, Z., Liu, Z.: Improved feature-selection method considering the imbalance problem in text categorization. Sci. World J. (2014)
17. Souza, M., Vieira, R.: Sentiment analysis on twitter data for Portuguese language. In: Caseli, H., Villavicencio, A., Teixeira, A., Perdigão, F. (eds.) PROPOR 2012. LNCS, vol. 7243, pp. 241–247. Springer, Heidelberg (2012)
18. Lambov, D., Dias, G., Noncheva, V.: High-level features for learning subjective language across domains. In: Proceedings of International AAAI Conference on Weblogs and Social Media ICWSM (2009)

TOPIE: An Open-Source Opinion Mining Pipeline to Analyze Consumers' Sentiment in Brazilian Portuguese

Ellen Souza[1,2(✉)], Tiago Alves[1], Ingryd Teles[1], Adriano L.I. Oliveira[2],
and Cristine Gusmão[3]

[1] MiningBR Research Group, Federal Rural University of Pernambuco (UFRPE),
Serra Talhada, PE, Brazil
`eprs@uast.ufrpe.br, tiagocalumbi@gmail.com,`
`ingryd_vanessa_@hotmail.com`
[2] Centro de Informática, Federal University of Pernambuco (CIn-UFPE), Recife, PE, Brazil
`{eprs,alio}@cin.ufpe.br`
[3] Programa de Pós-graduação em Engenharia Biomédica, Centro de Tecnologia e Geociências,
Federal University of Pernambuco (CTG-UFPE), Recife, PE, Brazil
`cristinegusmao@gmail.com`

Abstract. The growth of social media and user-generated content (UGC) on the Internet provides a huge quantity of information that allows discovering the experiences, opinions, and feelings of users or customers. These electronic *Word of Mouth* statements expressed on the web are prevalent in business and service industry to enable a customer to share his/her point of view. However, it is impossible for humans to fully understand it in a reasonable amount of time. Opinion mining (also known as Sentiment Analysis) is a sub-field of text mining in which the main task is to extract opinions from UGC. Thus, this work presents an open source pipeline to analyze the costumer's opinion or sentiment in Twitter about products and services offered by Brazilian companies. The pipeline is based on General Architecture for Text Engineering (GATE) framework and the proposed hybrid method combines lexicon-based, supervised learning, and rule-based approaches. Case studies performed on Twitter real data achieved precision of almost 70 %.

Keywords: Text mining · Text classification · Opinion mining · Sentiment analysis · Portuguese language · GATE

1 Introduction

The growth of social media and user-generated content (UGC) on the Internet provides a huge quantity of information that allows discovering the experiences, opinions, and feelings of users or customers. The volume of data generated in social media has grown from terabytes to petabytes [1].

These electronic *Word of Mouth* statements expressed on the web are much prevalent in business and service industry to enable customer to share his/her point of view [2]. Companies, on the other hand, have used this information to improve the quality of its

© Springer International Publishing Switzerland 2016
J. Silva et al. (Eds.): PROPOR 2016, LNAI 9727, pp. 95–105, 2016.
DOI: 10.1007/978-3-319-41552-9_9

products and services. Studies informed that 88 % of Brazilian companies use at least one social network [3].

Twitter has 320 million[1] monthly active users and it oversees 1 billion 'tweets' every day. Since it is a rich source of real-time information, many entities (companies, politicians, government, etc.) have demonstrated interest in knowing the opinions of people about services and products.

However, it is not possible for humans to fully understand UGC in a reasonable amount of time, which explains the growing interest by the scientific community to create systems capable of extracting information from it [4]. Specific processing methods, techniques and algorithms are required in order to extract knowledge from unstructured texts [5].

According to Liu [6], sentiment analysis, also known in the literature as opinion mining, is the field of study that analyzes people's sentiments, evaluations, attitudes, and emotions about entities such as products, services, organizations, individuals, issues, events, topics, and their attributes expressed in textual input. This is accomplished through the opinion classification of a document, sentence or feature into categories: 'positive', 'negative', and 'neutral'. This kind of classification is referred to in the literature as sentiment polarity or polarity classification.

Given that the Portuguese language is one of the most common spoken languages in the world - with almost 270 million people[2] speaking some variant of the language, and it is also the second most frequent language on Twitter [7], the goal of this work is to provide an open-source opinion mining pipeline for 'tweets' written in Brazilian Portuguese. The pipeline is based on the GATE [8] framework and the proposed hybrid method combines lexicon-based, supervised learning, and rule-based approaches in order to achieve better results and reduce context dependence [9].

The contributions of our paper are as follows:

1. An open-source application (see Footnote 3) to evaluate the customers' opinion about products and services offered from Brazilian companies;
2. Experimental execution on a set of real micro-blogging posts to evaluate the proposed pipeline;
3. A dataset[3] containing 33306 *tweets* of the most complaints companies according to Consumer Protection and Defense Program agency (PROCON);
4. Finally, we include a gold standard (See footnote 3) containing 543 *tweets* manually labeled into three classes (polarity): positive, negative and neutral;

The remainder of this paper is structured as follows: Sect. 2 provides the related work. Section 3 presents the opinion mining process. Section 4 details the pipeline, called TOPIE. Section 5 presents an experimental evaluation with a real case study. Finally, Sect. 6 contains the conclusions and directions for future work.

---

[1] www.about.twitter.com.

[2] Brazil (202,7 M), Mozambique (24,7 M), Angola (24,3 M), Portugal (10,9 M), Guinea-Bissau (1,7 M), East Timor (1,2 M), Equatorial Guinea (722,254), Macau (587,914), Cabo Verde (538,535) and São Tomé e Príncipe (190,428). From US/CIA - World Factbook (July, 2014).

[3] https://sites.google.com/site/miningbrgroup/.

2   Related Work

Works dealing with Portuguese language are scarce, and most efforts are directed towards the English language [2, 6, 10]. After performing a systematic mapping review on several scientific databases up to the year 2014, twenty-five papers relating the application of opinion mining for Portuguese (Brazilian and European variants) were identified, from which eight papers [3, 11–18] have analyzed the consumers' sentiment like the study proposed in this paper.

The work of Nascimento *et al.* [11], and Santos and Ladeira [17] make use exclusively of machine learning as opinion mining approaches. On the other hand, Avanço and Nunes [12], Freitas and Vieira [13], Santos *et al.* [18], and Chaves *et al.* [14] built their sentiment analysis using lexicon-based approaches. The works of [3, 15, 16] are more similar to our proposal as they adopted hybrid approaches, that is lexicon-based, supervised learning, and rule-based approaches.

The work of Evangelista and Padilha [3] propose a tool to classify web publications made on social networks such as Facebook and Twitter as positive, negative and neutral. For the classification task, this tool uses SentiWordNet lexical resource and the Naive Bayes algorithm. In the realized experiments, the collected publications to classification are from three Brazilian e-commerce companies. The results were very relevant for publications from Twitter using Naive Bayes algorithm for the e-commerce domain, with a correct classification rate of between 71 % and 95 %.

The works of Rosa *et al.* [15, 16] analyzes Brazilian Consumers' Sentiments on Twitter for the hair cosmetics domain using SentiMeter-Br. For the Portuguese dictionary performance validation, the results were compared with the SentiStrength algorithm and evaluated by three Specialists in the field of study. The polarity of the short texts was also tested through machine learning, with correctly classified instances of 71.79 % by Sequential Minimal Optimization (SMO) algorithm.

Related papers presented in the last two paragraphs differ from our proposal in some aspects: (1) our opinion mining system was evaluated for multiple domains, such as telecommunication, telephony, and banking; (2) the sentiment analysis of our approach is executed in feature level through entities recognition; (3) the open-source pipeline as well as the dataset used to build and evaluate the opinion mining system are available for the use of other researchers; and (4) we have used only one tool (GATE) [8] to perform all the opinion mining steps.

3   Opinion Mining Steps

Figure 1 presents the process used to build the proposed opinion mining application. It is based on the work of Aranha and Passos [19] and divided into four steps which are detailed in the following subsections. Resources and dataset are publicly available (See footnote 3).

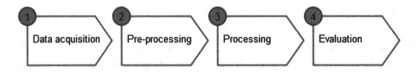

Fig. 1. Opinion mining steps based on [19]

3.1 Data Acquisition

This step included the dataset collection and gold collection construction. We built a web crawler to collect data from Topsy[4] database. The crawler received as input the companies' identification on Twitter and returned a list of tweets in JSON file. We have collected more than 30000 *tweets* of companies that received the most complaints in 2013 according to PROCON (Table 1). The reason why those companies were selected was that we would like to compare results from Twitter and PROCON.

Table 1. List of companies that received the most complaints in 2013 (PROCON)

Business Segment	Companies	Collected *tweets*
Telecommunication/Telephony	Oi	3442
Telecommunication/Telephony	Claro	4652
Telecommunication/Telephony	TIM	4037
Telecommunication/Telephony	Vivo	3463
Banking	Itaú,	4411
Banking	Bradesco	3753
Banking	Banco de Brasil	1954
Banking	Caixa Econômica Federal	2993
Telephony	Samsung	1538
Telecommunication	Sky	3063
TOTAL		33306

For the gold collection, a total of 543 'tweets' were manually labeled between three classes: 'positive', 'negative', or 'neutral' by at least two researchers.

3.2 Pre-processing

We performed the following pre-processing techniques on input data: tokenization, sentence split, part-of-speech (POS) and named entity recognition (NER). They are detailed in Sect. 4.

---

[4] http://topsy.com/.

3.3 Processing

The opinion mining pipeline was built on the top of GATE [8] and it is based on [20], as detailed in Sect. 4. Our classification system performs the sentiment analysis at the feature level of granularity. The polarity is determined using a hybrid approach that combines machine learning as well as lexicon and rule-based.

A lexicon dictionary (*Gazetteers*) with more than 10000 words was built manually from the analysis of 2000 *tweets* extracted from Corpora. The words are distributed in several categories as shown on the *left side* in Fig. 2. The *conditional* balloon brings together all terms used to describe a condition, for example, 'if' and 'case'. *Organization* lists the names of companies and some variations since it is common to find the use of informal language in social networks. *Feeling* contains terms related to an affirmation or denial, for example: 'for sure'. *Relationship* has words focused on descriptions of some relationships between the company and the person who made the post. An example might be words like 'bought' and 'waiting'. *Emotion* lists words that are related to the emotional state of the user, such as 'fear', 'sadness' and 'anger'. The *Emoticon* balloon contains the emotions icons used to describe a feeling.

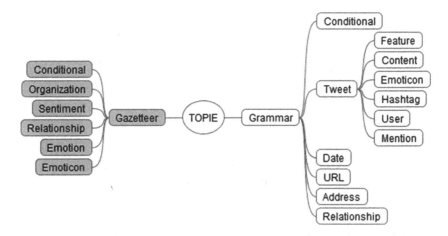

Fig. 2. Dictionary categories and grammatical rules

The balloons on the *right side* in Fig. 2 contains the built grammar rules to extract conditional, tweet general information, date, URL, address, and relationship data. The grammar rules were developed using JAPE, which is an annotation patterns engine based on regular expression [8]. Figure 3 shows an example of a rule in Jape. Supposing that the token 'Smile' is found in an opinion, the rule from Fig. 3 will annotate 'affect' to the *category* attribute and 'joy' to the *kind* attribute. Figure 4 shows the annotation result for the token "ruim" (bad), which was classified as a negative feeling (sadness) and was catch by the "ruleNegationOption" rule.

The Support Vector Machine (SVM) was selected as it is among the most frequently used classifiers and has presented good overall result for the opinion mining tasks [2, 10].

```
1  Phase: LookupAffect
2  Input: Lookup
3  Options: control = appelt
4  Rule: AffectLookup ( {Lookup.majorType == affect} ):tag
5  --> :tag.AffectLookup = {category = :tag.Lookup.category,
   kind = :tag.Lookup.minorType }
```

Fig. 3. Affect lookup rules using JAPE annotation pattern

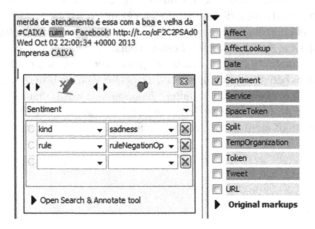

Fig. 4. Sentiment feature attributes for token 'ruim' (bad)

3.4 Evaluation

We performed three experiments with different sizes and classes balanced and selected the precision, recall, and f-measure to evaluate the results as they very frequent opinion mining measures [10].

4 TOPIE: Twitter Opinion Mining Information Extraction

Figure 5 presents the application architecture built in this study based on the work of Bontcheva *et al.* [20]. The authors proposed the TwitIE pipeline, which is a modification of the GATE ANNIE [8] open-source pipeline for news text. The TwitIE pipeline was customized to micro-blog texts for five languages (English, Dutch, German, French and Spanish) at every stage and includes Twitter-specific data import and metadata handling. The pipeline main components are represented in seven boxes and are detailed below. *Tweets* are read in JSON format and converted into GATE document content. The pipeline output is a XML annotated file, which can be used outside GATE.

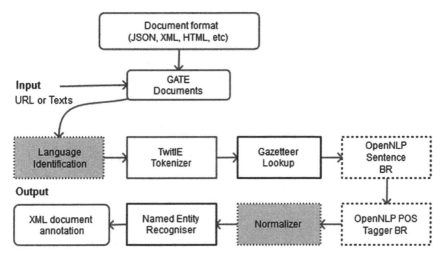

Fig. 5. TOPIE pipeline based on [20]

The gray boxes (Language Identification and Normalizer) are the ones that we did not use in our pipeline. The boxes with dashed border (Sentence Splitter and POS) were replaced by OpenNLP resources already available for the Portuguese language. The boxes with double border (Gazetteer and NER) were developed for this study. The box with simple border (Tokenizer) was reused without modification.

4.1 Language Identification

The TwitIE system uses a *TextCat* algorithm adaptation to Twitter. The *TextCat* original implementation recognizes more than sixty languages, including Portuguese. As our pipeline was built and evaluated only for a single language (Brazilian Portuguese), we did not make use of this component.

4.2 TwitIE Tokenizer

The TwitIE tokenizer is an adaptation of ANNIE's English tokenizer [8]. This component was used in our pipeline without modification. We have tried the Portuguese OpenNLP Tokenizer, but it performed worse and the reason was that it was not adapted for micro-blog texts, such as 'tweets'. Such general purpose tokenizers need to be adapted to work correctly on social media, in order to handle specific tokens like URLs, *hashtags*, user mentions, special abbreviations, and emoticons [20]. Figure 6 shows the component results for the token 'bradesco'.

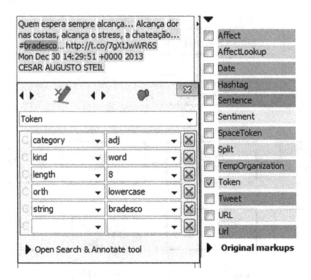

Fig. 6. The tokenizer results on GATE annotation set view

4.3   Gazetteer Lookup

The gazetteer consists of lists such as cities, organizations, days of the week, etc. It also contains the names of useful indicators, such as typical company designators (e.g. 'Ltd.'), titles, etc. The gazetteer lists are compiled into finite state machines, which can match text tokens. As shown in Fig. 2 we constructed a list with more than 10000 words extracted from the built corpora. Those lists are publicly available (See footnote 3) and can be reused in other opinion mining systems for Portuguese.

4.4   OpenNLP Sentence PT

The sentence splitter is required for the POS tagger component. The OpenNLP Sentence[5] - developed specifically to split sentences written in Portuguese, was reused without modification.

4.5   OpenNLP POS Tagger PT

POS Tagging is used to detect subjective messages by identifying the grammatical classes of the words used in the text [21]. Usually adjectives, adverbs and some substantives and verbs have polarity values [12]. The OpenNLP POS tagger PT (See footnote 5) developed specifically to Portuguese was reused in our pipeline without modification.

---

[5] http://opennlp.sourceforge.net/models-1.5/.

4.6 Normalizer

This component performs the NLP normalization task, as a solution for overcoming or reducing linguistic noise, very common in micro-blog texts. As this component is language-dependent and it does not give support to Portuguese, it was removed from our pipeline. However, we consider as future work, to implement a normalizer component specifically to micro-blogs texts written in Brazilian Portuguese.

4.7 Named Entity Recognizer

This component was developed specifically for this study. Grammar rules were built using JAPE [7] to extract information about entities and features as shown in Fig. 2 (right side). Entities such as products, services, organizations, individuals, issues, and their attributes expressed in textual input are mapped in this step Fig. 7 presents the overall results after the pipeline execution on GATE annotation set view.

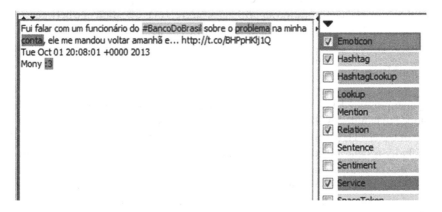

Fig. 7. An example of the pipeline results on GATE annotation set view

5 Case Study

In order to evaluate the proposed pipeline, using the GATE *plug-in Learning*, three experiments were performed using different configurations. Table 2 reports the classification results in precision, recall, and the f-measure on all datasets for the multiple kernel-based SVM classifier according to gold collection described in Sect. 3. The datasets were built randomly and classes are unbalanced.

The dataset 3 from experiment 1 reaches the best precision (66,5 %) which has little significance when compared with the work of Bontcheva *et al.* [20] that reaches precision of 77 %. The main reason is that we did not use specific components to handle short texts from micro-blogs, such as POS tagging which plays an important role identifying word classes that suggest sentiment polarity.

We observed that the used Tokenizer treats hashtags and user mentions as two tokens with a separate annotation HashTag covering both which interferes in final results.

Table 2. Experiments's results for multiple kernel-based SVM classifier.

		Size	Precision	Recall	F-measure
Experiment 1	Dataset 1	≈500	62,0	23,8	34,4
	Dataset 2	≈500	63,8	27,5	38,5
	Dataset 3	≈500	66,5	28,9	40,3
Experiment 2	Dataset 1	≈1000	62,3	26,4	37,0
	Dataset 2	≈1000	59,9	24,2	34,5
	Dataset 3	≈1000	58,7	22,0	32,0
Experiment 3	Dataset 1	≈2000	65,7	26,0	37,3
	Dataset 2	≈2000	60,5	25,2	35,5
	Dataset 3	≈2000	63,1	26,3	37,1

Also, some characters that compose the date and time fields were identified as emotion icons. However, our result is significant when comparing to the ones presented for the Portuguese language presented in Sect. 2, which reaches precision of 71.79 % by Sequential Minimal Optimization algorithm and correct classification rate between 71 % and 95 % using Naive Bayes algorithm for a particular domain.

6 Conclusion

This paper presented TOPIE, a Twitter opinion mining information extraction pipeline developed to handle 'tweets' written in Brazilian Portuguese. The pipeline was built on the top of the GATE framework [8] and it is based on the work of Bontcheva *et al.* [20]. The polarity is determined using a hybrid approach that combines machine learning as well as lexicon and rule-based. The sentiment analysis is executed in the level of feature or aspect through entities recognition. Case studies performed on real data from Twitter achieved precision of almost 70 % using SVM classifier for the telecommunication, telephony, and banking domains. Issues related to micro-blog were discussed, and the requirement for domain adaptation demonstrated. There is still a significant gap due to insufficient linguistic context and the noisiness of 'tweets'.

By releasing TOPIE as open-source, and making dataset and gold collection publicly available, we hope to give researchers an easily repeatable, baseline system against which they can compare new algorithms. As future work, we plan to evaluate data using different classifiers such as Naïve Bayes and Decision Tree, and compare our corpus-based lexicon to already available lexicon resources for the Portuguese Language, such as SentiLex-PT and OpLexicon. We plan also to adapt a Portuguese POS tagger and implement the normalization task as texts from micro-blog are written in an informal way, with grammatical errors, spelling mistakes, as well as ambiguous and ironic.

References

1. Marine-Roig, E., Anton Clavé, S.: Tourism analytics with massive user-generated content: A case study of Barcelona. J. Destin. Mark. Manag., 1–11 (2015)

2. Ravi, K., Ravi, V.: A survey on opinion mining and sentiment analysis: Tasks, approaches and applications. Knowl.-Based Syst. **89**, 14–26 (2015)
3. Evangelista, T.R., Padilha, T.P.P.: Monitoramento de Posts Sobre Empresas de E-Commerce em Redes Sociais Utilizando Análise de Sentimentos (2013)
4. Balazs, J.A., Velásquez, J.D.: Opinion mining and information fusion: a survey. Inf. Fusion **27**, 95–110 (2016)
5. Hotho, A., Andreas, N., Paaß, G., Augustin, S.: A brief survey of text mining, 1–37 (2005)
6. Liu, B., Zhang, L.: Chapter 1 A survey of opinion mining and sentiment analysis, pp. 1–49
7. Poblete, B., Garcia, R., Mendoza, M., Jaimes, A.: Do all birds tweet the same? characterizing twitter around the world categories and subject descriptors. Society, 1025–1030 (2011)
8. Cunningham, H., Maynard, D., Bontcheva, K., Tablan, V., Ursu, C., Dimitrov, M., Dowman, M., Aswani, N.: Developing language processing components with GATE (a user guide). Univ. Sheff. 2006, 1–457 (2001)
9. Pereira, A.: Sentiment analysis for streams of web data: a case study of Brazilian financial markets, pp. 167–170
10. Pang, B., Lee, L.: Opinion Mining and Sentiment Analysis. Found. Trends® Inf. Retr. **2**, 1–135 (2008)
11. Nascimento, P., Aguas, R., Kong, X., Osiek, B., De Souza, J.: Análise de sentimento de tweets com foco em notícias (2009)
12. Avanco, L.V., Nunes, M.D.G.V.: Lexicon-based sentiment analysis for reviews of products in Brazilian Portuguese. In: 2014 Brazilian Conference on Intelligent Systems, pp. 277–281 (2014)
13. De Freitas, L.A., Vieira, R.: Ontology-based feature level opinion mining for Portuguese reviews. In: WWW 2013 Companion - Proceedings of the 22nd International Conference on World Wide Web, pp. 367–370 (2013)
14. Chaves, M.S., de Freitas, L.A., Souza, M., Vieira, R.: PIRPO: an algorithm to deal with polarity in Portuguese online reviews from the accommodation sector. In: Bouma, G., Ittoo, A., Métais, E., Wortmann, H. (eds.) NLDB 2012. LNCS, vol. 7337, pp. 296–301. Springer, Heidelberg (2012)
15. Rosa, R.L., Rodriguez, D.Z., Bressan, G.: SentiMeter-Br: a new social web analysis metric to discover consumers' sentiment. In: Proceedings of the International Symposium on Consumer Electronics, ISCE, pp. 153–154 (2013)
16. Rosa, R.L., Rodriguez, D.Z., Bressan, G.: SentiMeter-Br: a social web analysis tool to discover consumers' sentiment. In: Proceedings of the IEEE International Conference on Mobile Data Management, vol. 2, pp. 122–124 (2013)
17. dos Santos, F.L., Ladeira, M.: The role of text pre-processing in opinion mining on a social media language dataset (2014)
18. Graciela, A., Becker, K., Moreira, V.: Um estudo de caso de mineração de emoções em textos multilíngues (2012)
19. Aranha, C., Passos, E.: A tecnologia de mineraçao de textos. Rev. Sist. Sist. Informação **5**, 1–8 (2006)
20. Bontcheva, K., Derczynski, L., Funk, A.: TwitIE: an open-source information extraction pipeline for microblog text. In: RANLP (2013)
21. Alves, A.L.F., Grande, C., Grande, C., Firmino, A.A., De Oliveira, M.G., De Paiva, A.C.: A comparison of SVM versus naive-bayes techniques for sentiment analysis in tweets: a case study with the 2013 FIFA confederations cup categories and subject descriptors, pp. 123–130 (2014)

Language Applications – Short Papers

Evaluating Progression of Alzheimer's Disease by Regression and Classification Methods in a Narrative Language Test in Portuguese

Sandra Aluísio[1(✉)], Andre Cunha[1], and Carolina Scarton[2]

[1] Interinstitutional Center for Computational Linguistics,
University of São Paulo, São Carlos, SP 13566-590, Brazil
`sandra@icmc.usp.br`, `andre.lv.cunha@gmail.com`
[2] Department of Computer Science, University of Sheffield,
Regent Court, 211 Portobello, Sheffield S1 4DP, UK
`carol.scarton@gmail.com`

Abstract. Automated discourse analysis aiming at the diagnosis of language impairing dementias already exist for the English language, but no such work had been done for Portuguese. Here, we describe the results of creating a unified environment, entitled Coh-Metrix-Dementia, based on a previous tool to analyze discourse, named Coh-Metrix-Port. After adding 25 new metrics for measuring syntactical complexity, idea density, and text cohesion through latent semantics, Coh-Metrix-Dementia extracts 73 features from narratives of normal aging (CTL), Alzheimer's Disease (AD), and Mild Cognitive Impairment (MCI) patients. This paper presents initial experiments in automatically diagnosing CTL, AD, and MCI patients from a narrative language test based on sequenced pictures and textual analysis of the resulting transcriptions. In order to train regression and classification models, the large set of features in Coh-Metrix-Dementia must be reduced in size. Three feature selection methods are compared. In our experiments with classification, it was possible to separate CTL, AD, and MCI with 0.817 F_1 score, and separate CTL and MCI with 0.900 F_1 score. As for regression, the best results for MAE were 0.238 and 0.120 for scenarios with three and two classes, respectively.

1 Introduction

Population aging is a known social trend in developed countries that is now becoming pronounced also in developing nations. Aging, from maturity to senescence, brings changes, marked chiefly by reductions in cognitive and functional reserve. If a reduced functional reserve renders an individual vulnerable for developing illnesses, maintaining functionality constitutes an important factor to health stability in good conditions. Language abilities in elderly individuals are indicative of their general cognitive health, and important pillars to maintain functionality in a healthy aging. As improvements in health treatments increase life expectancy, needs for change emerge in public policies aimed at dealing

© Springer International Publishing Switzerland 2016
J. Silva et al. (Eds.): PROPOR 2016, LNAI 9727, pp. 109–114, 2016.
DOI: 10.1007/978-3-319-41552-9_10

with the health demands of the increasingly aged population. In this scenario, increased relevance is given to research on the identification and treatment of diseases affecting the elderly population, such as the many forms of **dementia**. Dementia is the result of a progressive and irreversible neurodegenerative disorder, which can manifest itself in several forms. The main challenge in dementia managing comes from the fact that the degeneration process can begin years – sometimes even decades – before the first cognitive effects are perceived [1]. Even though there are no available treatments that reverse the disease's progression, the consensus in the area is that, when such treatments become available, it will be imperative to begin the treatment long before clinically significant damages have occurred to the brain [2]. A prominent variation of dementia is **Alzheimer's Disease** (AD). A less-known dementia-related condition is **Mild Cognitive Impairment** (MCI), defined as a cognitive decline larger than the expected for other individuals in the same age and education group, without significant interference in day-to-day activities. MCI is the initial clinically defined stage of dementia, and patients with the variation amnestic MCI evolve to having AD at a rate of 15 % a year, *versus* 1 % to 2 % in the general population.

Language is an important source of diagnostic information that can reveal the location and extent of the brain damage, a fact that establishes language analysis as an important ally in the quest for early diagnosis and more timely and effective treatments [2]. Qualitative analyses of language production can be performed manually at reasonable cost, but manual quantitative analyses are extremely expensive and time-consuming, hindering its practical application in clinical settings. Thus, computational approaches become an appealing alternative. The work of Thomas et al. [3] proposes several lexical approaches for detecting and quantifying AD, aiming to explore whether automated techniques based on the analysis of spontaneous speech can objectively measure AD level. The authors employed eight metrics in two classification scenarios (two and four classes), where the classes were obtained from the patients' scoring on the Mini-Mental State Examination. The authors generated classification models using several feature selection configurations, and the best results obtained were 69.6 % and 50.0 % precision for the two- and four-class scenarios, respectively. Proposing a syntactical analysis-based approached, Roark and colleagues [4] analyzed speech samples produced by 74 subjects – 37 healthy controls and 37 MCI patients – in the context of the Wechsler Logical Memory I and II tests. Each speech sample was transcribed, and analyzed both manually and automatically. A total of 16 metrics was extracted from the texts, six of them related to syntactical complexity, and ten phonological ones. The authors computed the correlation between manual and automatic annotation, finding a large value (greater than 87 %). Jarold et al. [2] used lexico-semantic features and idea density to compare texts produced by patients with AD, cognitive impairment, and clinical dementia. The authors trained classification models to separate each disorder from the control group, obtaining an accuracy of 73.0 % in separating AD *versus* controls, and 82.6 % cognitive impairment *versus* controls.

Even though there are studies proposing automated approaches to discourse analysis focusing on dementia identification, they are available only for the English language. For Brazilian Portuguese, to the best of our knowledge, no such work has so far been conducted. To fulfill such gap, we report our efforts towards creating a computational tool aimed at automatically detecting linguistic and cognitive decline in dementia as early as the first symptoms, as well as identifying distinctive linguistic features for dementia diagnosis, initially focusing on AD and MCI.

2   Experimental Setup

The Coh-Metrix-Dementia system[1] receives the transcription of a narrative produced by a subject, and makes use of several Natural Language Processing resources and tools to produce a variety of textual metrics, that can be used for both manual inspection and for training Machine Learning classifiers and regressors that can predict the subject's most likely cognitive status based on the characteristics of his/her text. In its current version, our tool is capable of extracting 73 metrics from the input texts. Of these metrics, 48 were adapted from Coh-Metrix-Port [5,6] a tool intended to extract textual metrics related to cohesion, coherence, and readability. These metrics are divided into ten categories: basic counts, logic operators, content word frequencies, hypernyms, tokens, constituents, connectives, ambiguity, co-reference, and anaphoras. We added 25 new metrics, in the following categories: disfluency, latent semantic analysis, lexical diversity and syntactical complexity, and semantic density. Among these newly added metrics, the most significant one for clinical applications is **idea density** (ID). Idea density was employed in the widely known Nun Study [7], which analyzed text samples produced by nuns in their early adulthood. In the Nun Study, ID in such samples was demonstrated to correlate with the diagnosis of AD and other brain pathologies as early as 60 years before the onset of the first symptoms. There are tools capable of automatically extracting ID from text samples in the English language [8], but no such tool existed for Portuguese. Therefore, we developed a tool called IDD3[2] (*Idea Density from Dependency Trees*), which can extract ID from well-formed English and Portuguese texts. The adopted methodology and experimental evaluation results are presented in [9].

To assess the performance of Coh-Metrix-Dementia's metrics in classification and regression tasks, we analyzed transcriptions of spoken narrations of the Cinderella story produced by 60 subjects: 20 healthy controls, 20 AD patients, and 20 MCI patients. Each subject was shown 22 pictures representing the Cinderella story. The patient was then requested to narrate the story in as much detail as possible; the narrative was recorded and afterwards manually transcribed. In order to make the NLP tools invoked by Coh-Metrix-Dementia to work properly, the transcriptions had to be manually revised: they were segmented in sentences, following a semantic-structural criterion; pauses and empty sentences

---

[1] http://143.107.183.175:22380/.
[2] https://github.com/andrecunha/idd3.

(those that are unrelated to the story) were used to guide the segmentation, and standard punctuation and capitalization were added afterwards. We also replaced oral language constructs by their standard form. Each text was analyzed by Coh-Metrix-Dementia, and the corresponding metrics values were used to train classification and regression models.

2.1 Classification Methods

We tested seven classification algorithms and four feature sets in two classification scenarios. All classification algorithms used are from the WEKA toolkit: Naive Bayes (NB), Support Vector Machines (SVM) with linear kernel, Multilayer Perceptron (MLP), logistic regression (LR), JRip, J48, and Random Forest (RF). The feature sets are: only the 48 metrics already present in Coh-Metrix-Port (called CMP); all the 73 metrics (called All); only the 25 new metrics (called New); and a set of features selected by the *Correlation-based Feature Selection* algorithm [10] (called CFS), in a 10-fold cross-validation scheme. The two classification scenarios are: three classes (CTL, AD, and MCI), and two classes (CTL and MCI). CFS selected 17 features for the three-class and 16 for the two-class scenarios. The classification results were obtained in a *leave-one-out* approach. Table 1 shows the results for all models; best results appear in bold.

Table 1. F_1 score of the classification methods for different metrics sets.

Algorithm	Three classes				Two classes			
	CMP	All	New	CFS	CMP	All	New	CFS
NB	0.651	0.733	**0.767**	**0.817**	0.725	0.825	0.850	0.825
SVM	**0.669**	0.715	0.731	0.753	**0.775**	0.747	0.798	0.848
MLP	0.566	0.536	0.633	0.601	0.725	0.699	0.775	0.825
LR	0.616	0.701	0.718	0.750	0.749	0.747	0.697	0.749
JRip	0.500	0.699	0.750	0.732	0.697	0.875	0.800	0.775
J48	0.498	0.666	0.633	0.748	0.596	**0.900**	**0.900**	**0.900**
RF	0.635	**0.750**	0.733	0.752	0.750	0.799	0.850	0.850
Baseline	*0.333*	*0.333*	*0.333*	*0.333*	*0.500*	*0.500*	*0.500*	*0.500*

2.2 Regression Methods

The problem of classifying subjects according to linguistic and cognitive decay can also be modeled as a regression task; in fact, modeling with regression methods is a more realistic scenario. Individuals should present progressive decay in different degrees, and just classifying them in categorical classes is an approximation of the problem. Here, we have used values of 0, 0.5, and 1 for CTL, MCI, and AD, respectively, since these are the values of CDR-SOB

(Clinical Dementia Rating – Sum Of Boxes) for the conditions above. CDR-SOB and the values of MMSE (Mini Mental State Examination) are used to classify patients. Four regression algorithms from scikit-learn were applied to build regression models: Epsilon-Support Vector Regression (SVR) with RBF kernel, Bayesian Ridge Regressor (BR), Random Forest Regressor (RFR), and Stochastic Gradient Descent Regressor (SGD). All methods had the hyper-parameters optimized via grid search, and a 10-fold cross-validation was applied. The features used were scaled beforehand. Table 2 shows the results for all models in terms of Mean Absolute Error (MAE). RFR showed the best results on all datasets for the two classes scenario and was the best for the dataset **All** for the three classes scenario. BR was the best for all other datasets for the three classes scenario although RFR showed results very close to BR. RF is an ensemble machine learning algorithm and in many applications this algorithm produces one of the best accuracies to date. Moreover, it has important advantages over other techniques in terms of ability to handle high-dimensional problems, where the number of features is often redundant [11].

Table 2. Results for MAE of all regression models and for all metric sets.

Algorithm	Three classes				Two classes			
	CMP	All	New	CFS	CMP	All	New	CFS
SVR	0.333	0.316	0.254	0.300	0.203	0.207	0.177	0.172
BR	**0.328**	0.262	**0.250**	**0.256**	0.200	0.189	0.158	0.158
RFR	0.331	**0.238**	0.253	0.260	**0.163**	**0.120**	**0.120**	**0.136**
SGD	0.411	0.371	0.367	0.327	0.230	0.210	0.188	0.185

3 Conclusions and Future Work

We have employed seven classification and four regression methods to predict the diagnosis of a patient based on several linguistic characteristics of his text. For classification, we obtained 0.817 F_1 score for three classes, and 0.900 for two classes, both using the CFS-selected features; for regression, we obtained 0.238 MAE for three classes, and 0.120 for two classes, both using all features. However, the data set employed here is very limited in size. For the estimation of real-world, robust classification or regression models, more data samples are needed. Also, a more careful examination of the distribution of the data samples in the input space, and the subsequent design of a suitable kernel for the SVM algorithm is likely to produce more reliable results, although RF is a very competitive ensemble. Currently, our system relies heavily on manual transcription effort, since the patient's narrative is transcribed and segmented into edited sentences manually; automating these steps would considerably speed up the analysis pipeline and potentially eliminate manual annotation inconsistencies.

Acknowledgments. FAPESP supported this study (No. 2013/16182-0) and the EXPERT (EU Marie Curie ITN No. 317471) project.

References

1. Sperling, R.A., Karlawish, J., Johnson, K.A.: Preclinical alzheimer disease - the challenges ahead. Nat. Rev. Neurol. **9**(1), 54–58 (2013)
2. Jarrold, W.L., Peintner, B., Yeh, E., Krasnow, R., Javitz, H.S., Swan, G.E.: Language analytics for assessing brain health: cognitive impairment, depression and pre-symptomatic alzheimer's disease. In: Yao, Y., Sun, R., Poggio, T., Liu, J., Zhong, N., Huang, J. (eds.) BI 2010. LNCS, vol. 6334, pp. 299–307. Springer, Heidelberg (2010)
3. Thomas, C., Keselj, V., Cercone, N., Rockwood, K., Asp, E.: Automatic detection and rating of dementia of Alzheimer type through lexical analysis of spontaneous speech. In: 2005 IEEE International Conference on Mechatronics and Automation, vol. 3, pp. 1569–1574 (2005)
4. Roark, B., Mitchell, M., Hosom, J., Hollingshead, K., Kaye, J.: Spoken language derived measures for detecting mild cognitive impairment. IEEE Trans. Audio Speech Lang. Process. **19**(7), 2081–2090 (2011)
5. Scarton, C., Aluísio, S.: Análise da inteligibilidade de textos via ferramentas de processamento de língua natural: adaptando as métricas do coh-metrix para o Português. Linguamática **2**(1), 45–62 (2010)
6. Aluisio, S., Specia, L., Gasperin, C., Scarton, C.: Readability assessment for text simplification. In: Proceedings of the NAACL HLT 2010 Fifth Workshop on Innovative Use of NLP for Building Educational Applications, pp. 1–9. Association for Computational Linguistics (2010)
7. Snowdon, D., Kemper, S., Mortimer, J., Greiner, L., Wekstein, D., Markesbery, W.: Linguistic ability in early life and cognitive function and alzheimer's disease in late life: findings from the nun study. JAMA **275**(7), 528–532 (1996)
8. Brown, C., Snodgrass, T., Kemper, S., Herman, R., Covington, M.: Automatic measurement of propositional idea density from part-of-speech tagging. Behav. Res. Methods **40**(2), 540–545 (2008)
9. Cunha, A., Bender de Sousa, L., Mansur, L., Aluisio, S.: Automatic proposition extraction from dependency trees: helping early prediction of alzheimer's disease from narratives. In: 2015 IEEE 28th International Symposium on Computer-Based Medical Systems (CBMS), pp. 127–130, June 2015
10. Hall, M.A.: Correlation-based feature subset selection for machine learning. Ph.D. thesis, University of Waikato, Hamilton, New Zealand (1998)
11. Lebedev, A., Westman, E., Westen, G.V., Kramberger, M., Lundervold, A., Aarsland, D., Soininen, H., Koszewska, I., Mecocci, P., Tsolaki, M., Vellas, B., Lovestone, S., Simmons, A.: Random forest ensembles for detection and prediction of alzheimer's disease with a good between-cohort robustness. Neuroimage Clin. **6**, 115–125 (2014)

A Comparative Evaluation of QA Systems over List Questions

Patricia Nunes Gonçalves and António Horta Branco[✉]

Department of Informatics, University of Lisbon, Edifício C6,
Faculdade de Ciências Campo Grande, 1749-016 Lisbon, Portugal
{patricia.nunes,antonio.branco}@di.fc.ul.pt

Abstract. The evaluation of a Question Answering system is a challenging task. In this paper we evaluate our system, LX-ListQuestion, a Web-based QA System that focuses on answering list questions. We compare our system against other QA Systems and the results were analyzed in two ways: (i) the quantitative evaluation of answers provides recall, precision and F-measure and (ii) the question coverage that indicate the usefulness of the system to the user by counting the number of questions for which the system provides at least one correct answer. The evaluation brings interesting results that points to a certain degree of complementary between different approaches.

Keywords: QA Systems · List questions · Evaluation QA

1 Introduction

In Open-domain Question Answering the range of possible questions is not constrained, hence a much tougher challenge is placed on systems. The goal of an Open-domain QA system is to answer questions on any kind of subject domain [10]. Research in Open-domain Question Answering had a boost in 1999 with the Text REtrieval Conference (TREC)[1], which provides large-scale evaluation of QA systems thus defining the direction of research in the QA field.

List questions started being studied in the context of QA in 2001 when TREC included this type of questions in the dataset.

Finding the correct answers to List questions requires discovering a set of different answers in a single document or across several documents. An approach to answer a List question in a single document is very similar to the approach to find the correct answer to factoid questions: (i) find the most relevant document; (ii) find the most relevant excerpt and (iii) extract the answers from this relevant excerpt. On the other hand, the process to extract the answers spread over several documents raised new challenges such as grouping repeated elements, handling more information, separating the relevant information from the rest of the information, among others.

---

[1] http://trec.nist.gov/.

© Springer International Publishing Switzerland 2016
J. Silva et al. (Eds.): PROPOR 2016, LNAI 9727, pp. 115–121, 2016.
DOI: 10.1007/978-3-319-41552-9_11

Evaluation of QA Systems involves a large amount of manual effort, but it is a fundamental component to improve the systems. Traditional evaluation of QA systems use recall, precision and F-measure to measure performance of systems [8,11]. Besides the traditional evaluation, we assessed the systems by using question coverage that indicate the usefulness of the system to the user by counting the number of questions that the system provides at least one correct answer, providing another perspective of evaluation. **Paper Outline**: Sect. 2 introduces our system, LX-ListQuestion, a Web-based QA system that uses redundancy and heuristics to answer List questions. In Sect. 3 we compare the results, over the same question dataset, with other two QA systems: RapPortagico and XisQuê. Finally in Sect. 4 we present some concluding remarks.

2   LX-ListQuestion Question Answering System Architecture

The LX-ListQuestion System [6,7] is a fully-fledged Open-domain Web-based QA system for List questions. The system collect answers spread over multiple documents using the Web as a corpus. Our approach is based on redundancy of information available on the Web combined with heuristics to improve QA performance. The implementation is guided by the following design features:

- Exploits redundancy to find answers to List questions;
- Compiles and extracts the answers from multiple documents;
- Collects at run-time the documents from Web using a search engine;
- Provides answers in real time without resorting to previously stored information.

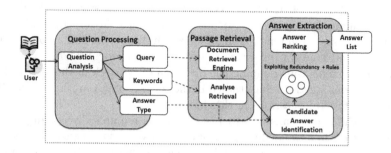

Fig. 1. LX-ListQuestion system architecture

The system architecture is composed by three main modules: Question Processing, Passage Retrieval and Answer Extraction (Fig. 1). The Question Processing module is responsible for converting a natural language question into a form that subsequent modules are capable of handling. The main sub-tasks are (i) question analysis: responsible for cleaning the questions; (ii) extraction of keywords: performed using nominal expansion and verbal expansion; (iii) transformation of the question into a query; (iv) identification of the semantic category of the expected answer; and (v) identification of the question-focus.

The Passage Retrieval Module is responsible for searching Web pages and saving their full textual content into local files for processing. After the content is retrieved, the system will select relevant sentences. The Answer Extraction Module aims at identifying and extracting relevant answers and presenting them in list form. The candidate answer identification is based on a Named Entity Recognition tool. The candidates are selected if they match the semantic category of the question. The process of building the Final List Answer and more details about implementation of LX-ListQuestion can be found on [5]. LX-ListQuestion is available online at http://lxlistquestion.di.fc.ul.pt.

3   Comparing LX-ListQuestion and Other QA Systems

Comparing LX-ListQuestion with other QA systems is crucial to providing us with an assessment of how our system is positioned relative to the state-of-the-art. In this Section we compare the results of LX-ListQuestion with two other QA systems for Portuguese: RapPortagico and XisQuê.

The evaluation has two components: the quantitative evaluation of answers and the question coverage evaluation. The quantitative analysis uses precision, recall and F-measure as metrics. Nevertheless, these metrics do not accurately reflect how effective the systems are in providing correct answers to the maximum number of questions. For that, we use the question coverage, which determine the number of questions that receive at least one correct answer.

The question dataset used in these experiments is based on Páigico Competition[2]. The whole dataset is composed by 150 questions about Lusophony extracted from the Portuguese Wikipedia [4]. For the experiments, we use a subset of 30 questions whose expected answer type is Person or Location. We pick these two types since they are the ones more accurately assigned by the underlying a Named Entity Recognition tool named LX-NER [3]. Note, however, that our approach is not intrinsically limited to only these types.

3.1   Comparing Design Features

In this Section we compare the design features of LX-ListQuestion, RapPortagico and XisQuê. As mention at Sect. 2, LX-ListQuestion is a web-based QA system that finds a list of answers retrieving documents from the web and extracting candidates answers from inside the documents. RapPortagico [9], an off-line QA system that uses Wikipedia to retrieve the answers for List questions and XisQuê [1, 2], a Web-based QA system that answers factoid questions that selects the most important paragraph of the Web pages and extracts the answer through the use of hand-built patterns. Table 1 shows the differences between the design features of the systems.

---

[2] http://www.linguateca.pt/Pagico/.

RapPortagico pre-indexes the documents using noun phrases that occur in the sentences in the corpus while LX-ListQuestion does not uses any pre-indexing of documents. RapPortagico uses the off-line Wikipedia as the source of information, while LX-ListQuestion uses the Web to find the answers. Both systems are also different in the type of answers. RapPortagico returns a List of Wikipedia pages and LX-ListQuestion returns a list of answers. The design features of LX-ListQuestion and XisQuê are to a certain extent similar. Both systems are Web-based QA systems and use the Web as the source of answers, and Google as supporting search engine. What differs between the systems is that the XisQuê answers Factoid Questions and LX-ListQuestion answers List Questions.

Table 1. Comparing Design Features of QA systems

	RapPortagico	XisQuê	LX-ListQuestion
Corpus pre-indexing	Yes. It pre-indexes the corpus using Noun Phrases	No	No
Corpus source	Off-line Wikipedia documents	Web	Web
Search engine	Lucene (indexed to documents stored into local files)	Google	Google
Type of questions	Factoid and List	Factoid	List
Type of answers	List of Wikipedia pages	Answer and Snippet	List of Answers

3.2   Quantitative Evaluation and Question Coverage

The evaluation was performed for the same set of questions for all systems. Table 2 shows the results of comparing the three systems. LX-ListQuestion obtained more correct answers than other systems, this can be verify in the recall measure with 0.120 and 0.097 and 0.055 respectively. However, it has lower precision since it returned more candidates than the other system. When comparing F-measure, LX-ListQuestion achieved slightly better results, obtaining 0.102 against 0.095 for RapPortagico and better results than XisQuê, that only obtained 0.078.

The question coverage that indicate the usefulness of the system to the user counting the number of questions for which the system provides at least one correct answer. Table 3 summarizes the number of questions answered by each system. From the 30 questions in the dataset, LX-ListQuestion provided at least one correct answer to 17 of them, against 14 of RapPortagico and 13 of XisQuê. The question coverage evaluation also allowed us to uncover an interesting behavior of these systems. For 7 questions answered by LX-ListQuestion, RapPortagico

Table 2. Evaluation of QA systems - LX-ListQuestion, RapPortagico and XisQuê

Experiments	Refer. answer list	Correct answers	All answers retrieved	Recall	Precision	F-Measure
LX-ListQuestion	340	41	460	0.120	0.089	0.102
RapPortagico		32	327	0.097	0.100	0.098
XisQuê		19	139	0.055	0.136	0.078

Table 3. Question Coverage

	LX-ListQuestion	RapPortagico	XisQuê
Number of Questions Answered	17	14	13

Table 4. Examples of answers provided by the each system

Question	Correct answers		
	LX-ListQuestion	RapPortagico	XisQuê
Cidades que fizeram parte do domínio português na India	Damao	Calecute	—
Cities that were part of the Portuguese Empire in India		Goa	—
Praias de Portugal boas para a práitica de Surf	Ericeira	—	Guincho
Good Portuguese beaches for surfing	Arrifana	—	Peniche
	Praia Vale Homens	—	
	São João Estoril	—	
Cidades Lusófonas conhecidas pelo seu Carnaval	Salvador	Mindelo Cabo Verde	Olinda
Lusophone cities known for their carnival celebrations	Recife	—	
	São Paulo	—	

Table 5. Results overview

Systems	Refer. answers list	Correct answers	All answers retrieved	Recall	Precision	F-Measure
LX-ListQuestion	340	41	460	0.120	0.089	0.102
RapPortagico		32	327	0.097	0.100	0.098
XisQuê		19	139	0.055	0.136	0.078
Combination		80	914	0.235	0.087	0.126

provided no answer. Conversely, for 5 questions answered by RapPortagico, LX-ListQuestion provided no answer. In addition, we note that when a question is answered by both systems, the answers given by each system tend to be different. Concerning XisQuê and LX-ListQuestion, we find that a large majority of correct answers given by XisQuê are different from those given by LX-ListQuestion. Namely, in 9 out of 13 questions to which XisQuê provides a correct answer, that answer is not present in the list of answers given by LX-ListQuestion. This result points towards a certain degree of complementarity between the systems. Table 4 shows some examples of questions and answers provided by each system that demonstrate the complementarity between the systems.

4   Concluding Remarks

In this paper we present an evaluation of our system, LX-ListQuestion, a Web-based QA system that uses redundancy and heuristics to answer List questions and compared the results with other two QA systems: RapPortagico and XisQuê. Our evaluation shows that our LX-ListQuestion achieved better results, with 0.102 in F-Measure, against 0.098 of RapPortagico, and 0.078 of XisQuê. The question coverage evaluation points towards a certain degree of complementarity between these systems. We observe that for a set of questions answered by LX-ListQuestion, the other systems provide no answers. Conversely, for some other questions answered by RapPortagico or XisQuê, LX-ListQuestion provided no answer. Based on our experiments, we noted that the approaches of RapPortagico, XisQuê and LX-ListQuestion may reinforce each other. To demonstrate these assumption, we built Table 5 with an overview of the results obtained in the experiments. The last row is the hypothetical combination of LX-ListQuestion, RapPortagico and XisQuê. As we can see, a QA system that combines their approaches can achieve better results and improve Recall and F-measure metrics.

References

1. Branco, A., Rodrigues, L., Silva, J., Silveira, S.: Real-time open-domain QA on the Portuguese web. In: Geffner, H., Prada, R., Machado Alexandre, I., David, N. (eds.) IBERAMIA 2008. LNCS (LNAI), vol. 5290, pp. 322–331. Springer, Heidelberg (2008)
2. Branco, A., Rodrigues, L., Silva, J., Silveira, S.: XisQuê: an online QA service for Portuguese. In: Teixeira, A., Lima, V.L.S., Oliveira, L.C., Quaresma, P. (eds.) PROPOR 2008. LNCS (LNAI), vol. 5190, pp. 232–235. Springer, Heidelberg (2008)
3. Ferreira, E., Balsa, J., Branco, A.: Combining rule-based and statistical models for named entity recognition of Portuguese. In: Proceedings of Workshop em Tecnologia da Informaçãoe de Linguagem Natural, pp. 1615–1624 (2007)
4. Freitas, C.: A lusofonia na Wikipédia em 150 topicos. Linguamatica **4**(1), 9–18 (2012)
5. Gonçalves, P.: Open-Domain Web-Based Multiple Document Question Answering forList Questions with Support for Temporal Restrictors. Ph.D. thesis, University of Lisbon, Lisbon, Portugal, 6 2015
6. Gonçalves, P., Branco, A.: Answering list questions using web as a corpus. In: Proceedings of the Demonstrations at the 14th Conference ofthe European Chapter of the Association for Computational Linguistics, pp. 81–84. Association for Computational Linguistics, Gothenburg, April 2014
7. Gonçalves, P., Branco, A.: Open-domain web-based list question answering with LX-listquestion. In: Proceedings of the 4th International Conference on WebIntelligence, Mining and Semantics, WIMS 2014, pp. 43:1–43:6. ACM, New York (2014)
8. Radev, D.R., Qi, H., Wu, H., Fan, W.: Evaluating web-based question answering systems. In: Proceedings of the Third International Conference on Language Resources and Evaluation, LREC 2002. European Language Resources Association, Las Palmas, 29-31 May 2002

 9. Rodrigues, R., Oliveira, H.: Uma abordagem ao páigico baseada no processamento e anáilise desintagmas dos tópicos. Linguamatica **4**(1), 31–39 (2012)
10. Strzalkowski, T., Harabagiu, S.: Advances in Open Domain Question Answering, 1st edn. Springer Publishing Company Incorporated, Netherlands (2007)
11. Voorhees, E.: Evaluating question answering system performance. In: Strzalkowski, T., Harabagiu, S. (eds.) Advances in OpenDomain Question Answering. Text, Speech and Language Technology, vol. 32, pp. 409–430. Springer, Netherlands (2006)

Characterizing Opinion Mining: A Systematic Mapping Study of the Portuguese Language

Ellen Souza[1,2(✉)], Douglas Vitório[1], Dayvid Castro[1], Adriano L.I. Oliveira[2], and Cristine Gusmão[3]

[1] MiningBR Research Group, Federal Rural University of Pernambuco (UFRPE), Serra Talhada, PE, Brazil
`eprs@uast.ufrpe.br, douglas.alisson17@gmail.com,`
`wellescastro@gmail.com`
[2] Centro de Informática, Federal University of Pernambuco (CIn-UFPE), Recife, PE, Brazil
`{eprs,alio}@cin.ufpe.br`
[3] Programa de Pós-graduação em Engenharia Biomédica, Centro de Tecnologia e Geociências, Federal University of Pernambuco (CTG-UFPE), Recife, PE, Brazil
`cristinegusmao@gmail.com`

Abstract. The growth of social media and user-generated content (UGC) on the Internet provides a huge quantity of information that allows discovering the experiences, opinions, and feelings of users or customers. Opinion Mining (OM) is a sub-field of text mining in which the main task is to extract opinions from UGC. Given that Portuguese is one of the most common spoken languages in the world, and it is also the second most frequent on Twitter, the goal of this work is to plot the landscape of current studies that relates the application of OM for Portuguese. A systematic mapping review (SMR) method was applied to search, select and to extract data from the included studies. Manual and automated searches retrieved 6075 studies up to year 2014, from which 25 articles were included. Almost 70 % of all approaches focus on the Brazilian Portuguese variant. Naïve Bayes and Support Vector Machine were the main classifiers and SentiLex-PT was the most used lexical resource. Portugal and Brazil are the main contributors in processing the Portuguese language.

Keywords: Text mining · Text classification · Opinion mining · Sentiment analysis · Portuguese language

1 Introduction

The growth of social media and user-generated content (UGC) on the Internet provides a huge quantity of information that allows discovering the experiences, opinions, and feelings of users or customers. The volume of this kind of data has grown to petabytes [1]. These electronic Word of Mouth (eWOM) statements are prevalent in business and service industry to enable a customer to share his/her point of view [2].

However, it is impossible for humans to fully understand UGC in a reasonable amount of time, which is why there has been a growing interest in the scientific community to create systems capable of extracting information from it [3]. Opinion

J. Silva et al. (Eds.): PROPOR 2016, LNAI 9727, pp. 122–127, 2016.
DOI: 10.1007/978-3-319-41552-9_12

mining (OM) is a sub-field of text mining in which the main task is to extract opinions from UGC [3]. OM detects, extracts, and classifies opinions concerning different topics. Common opinion classes are: positive, negative, and neutral [2].

For [2], sentiment analysis, opinion mining, and subjectivity analysis are interrelated areas of research which use various techniques taken from Natural Language Processing, Information Retrieval, structured and unstructured Data Mining. Whereas data mining is largely language independent, text mining involves a significant language component, justifying its study associated with one target language. Most text mining tasks focus on processing English documents [5], but many other languages, including Spanish and Portuguese, have also been considered.

Portuguese language is one of the most common spoken languages in the world, with almost 270 million people[1] speaking some variant of the language. Therefore Portuguese is also the second most used on Twitter, which is considered one of the main sources of UGC [6]. Thus, combining the guidelines to perform Systematic Reviews [7, 8], the goal of this article is to characterize current research that reports the use of OM for Portuguese, driven by the general research question (RQ):

- RQ: What is the current state of text mining applied to the Portuguese language?

The automated and manual search procedures retrieved 6075 papers published up to the year 2014, from which 25 were included in this study. Data extracted from primary studies were systematically structured and analyzed to answer historical, descriptive, and classificatory research questions presented below:

- RQ1: What is the evolution in the number of publications up to year 2014?
- RQ2: Which individuals, organizations, and countries are the main contributors in the research area?
- RQ3: What are the mining techniques and tools applied to Portuguese? How they were evaluated and which levels of sentiment analysis were adopted?
- RQ4: What are the characteristics of the dataset used?

Our aim is to facilitate researchers and practitioners to discover what has been achieved and where the gaps are in this field area. Section 2 provides the related work. Section 3 details the mapping protocol. Section 4 presents a set of results. Section 5 discusses the results and contains the conclusions and directions for future work.

2   Related Work

Although we have made an extensive search, we did not found any OM systematic mapping review in any language and even less specifically for Portuguese. However, we found several language independent OM surveys [2, 3, 9–12], a paper [13] describing

---

[1] Brazil (202,7 M), Mozambique (24,7 M), Angola (24,3 M), Portugal (10,9 M), Guinea-Bissau (1,7 M), East Timor (1,2 M), Equatorial Guinea (722,254), Macau (587,914), Cabo Verde (538,535) and São Tomé e Príncipe (190,428). From US/CIA - World Factbook (July, 2014).

the computational linguistics area in Brazil and a survey [14] of automatic term extraction for Brazilian Portuguese.

Balazs and Velásquez [3] present a brief overview on some of the most popular datasets used for training and validating OM systems, however no dataset for Portuguese was available. In [2], the authors present the datasets used for training and validating OM systems, but again no dataset for Portuguese was available. In our SMR, we have adopted the same organization as [2]. In [10, 11], the authors provide a brief literature survey on the problem of sentiment analysis and opinion mining. We have classified the primary studies according to the levels of sentiment analysis provided by [10]. Medhat et al. [12] give details about the language adopted but no OM system was reported for the Portuguese language.

In [13], an overview of the computational linguistics (CL) in Brazil is presented. According to the authors, research in Brazil is varied and deals mainly with Portuguese, English and Spanish processing. Research on text mining is mostly carried out by non-computational linguistics researchers. In [14], a survey of the state of the art in automatic term extraction (ATE) for the Brazilian Portuguese language is presented. According to the authors, there are still several gaps to be filled, for instance, the lack of consensus regarding formal definitions.

3   Review Method

The SMR provides a structure of the type of research reports and results that have been published by categorizing them. It gives a visual summary of its results [7]. Figure 1 shows the adopted SMR process. We searched the literature looking for primary studies that reported OM applications for the Portuguese. Primary studies that met at least one of the following *exclusion criteria* were removed from the study: (i) written in a language other than English or Portuguese; (ii) not available on online scientific libraries; (iii) keynote speeches, workshop reports, books, theses, and dissertations. Automated[2] and manual[3] searches were combined to achieve high coverage. The automated search query was constructed based on *text mining* and *Portuguese language* terms. Synonyms and translation for both terms were included in the final query. The manual search was performed by reading the conference proceedings. For the 661 potentially relevant studies, the researchers reapplied the inclusion criteria and exclusion criteria after reading the *full* paper. This resulted in a list of 203 studies, from which 25 relate to the use of OM for Portuguese.

---

[2] IEEE Xplore, ACM, Science Direct, Scopus, Portal de Periódicos Capes and SciELO.
[3] Inter. Conf. Comput. Processing of Portuguese (PROPOR), Text Mining and Applications (TEMA), Brazilian Workshop of Social Network Analysis and Mining (BRASNAM), Brazilian Symposium on Information and Human Language Technology (STIL), ACM symposium on Document engineering (DocEng), Linguateca Database (www.linguateca.pt), Message Understanding Conferences (MUC), Text Analysis Conference (TAC), Text REtrieval Conference (TREC), Document Understanding Conference (DUC).

Process Steps

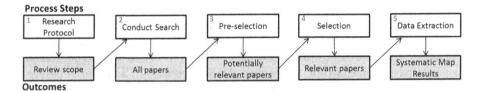

Fig. 1. Systematic mapping process based on [7]

4   Results

In this section, we present the main findings of our review, organized according to the four specific research questions. Due to space constraints, the list of primary studies is available online at http://bit.ly/1LJFD5u.

For the first specific research question it could be observed that, the number of studies has grown over the years (Fig. 2), despite the drop in 2010. Primary studies were classified according to the language variations. Almost 70 % of primary studies evaluated its results using texts written in the Brazilian Portuguese variant. 16 % use European Portuguese. Language variant was not available for 16 % of primary studies.

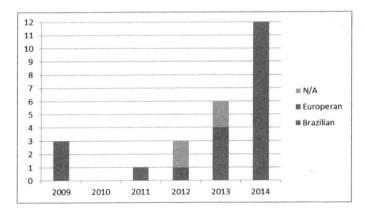

Fig. 2. Temporal distribution of OM primary studies. (Color figure online)

For the second specific research, as expected from Fig. 2, Brazil has a greater number of researchers in the field. Renata Vieira from UNISINOS and Wagner Meira Jr. from UFMG appear as main authors and UFMG appears as the main organization. As stated by [6], research interest on Portuguese processing is shared mainly with Portugal and Brazil. In addition, other countries like USA and Canada proposed Multilanguage approaches (OM06, OM07, OM16, OM25), which included Portuguese. BRASNAM appears as the main venue in the research area.

Answering the third specific research question, we can state that the majority of primary studies has investigated opinions at document level of granularity. A total of 84 % of all 25 primary studies performed at least one type of text pre-processing task

The main tool used for text pre-processing was the Python NLTK. Bayesian Classifiers (BC) were used by 56 % of all primary studies. Support Vector Machine (SVM) represents 36 %, Decision Tree (DT) represents 16 %, and Rule and Pattern-based (RPB) approaches represents 32 %, while the Lexicon-based approaches represents 60 %. More than 50 % have adopted hybrid approaches. WEKA was used by a quarter of primary studies as machine learning tool. Primary studies (OM06, OM07, OM16) have evaluated the sentiment variation over the time, called *Sentiment Drift*.

The *precision, recall, and f-measure* trio was used by 44 % of all primary studies to evaluate their results, followed by *accuracy* (40 %). One primary study (OM06) reported the RAM-Hours measure. Nine primary studies reported the adoption of cross validation for estimating the classifier performance. Most of the primary studies (52 %) manually built their gold standard, and 20 % used both automated and manual approaches. The best results were achieved by the SVM classifiers.

Finally, for the last specific research question, we can state that two primary studies (OM04, OM15) have used publicly available datasets to build and evaluate their OM approaches. Opinion mining for Portuguese aims to discover the opinion/sentiment of consumers (36 %), electors (28 %), sport fans (8 %), stock market (4 %), traffic (4 %), and news headlines polarity (4 %). The domain was not available for 16 % of all primary studies. Twitter was used by 41 % of all studies.

5   Conclusion

We observed an increasing interest in OM, partly due to its potential applications, such as: public relations and political campaign. Portugal and Brazil are the main contributors in processing the Portuguese. Facebook and Twitter are important sources of opinions, however the former is less used as it often contains pictures and the analysis of the text by itself is not effective [19]. As most of OM applications come from social networks, 88 % of text is short, written in an informal way, with grammatical errors, spelling mistakes, as well as ambiguous and ironic. Even when good results are achieved, the used datasets are rarely published. Therefore, less than 50 % of all 25 primary studies have fully answered all research questions. This makes it difficult to implement improvements, as well as comparisons.

The strength of this paper is to promote growth in the research of text mining in the Portuguese Language. Data reported in this paper may help researchers and practitioners to discover what has been achieved and where the gaps are in this field area.

This work is part of an ongoing broader research as shown in the general research question presented in Sect. 1. We are mapping not only the use of OM tasks for the Portuguese language, but for all kind of text mining tasks. To increase coverage, we plan to apply snowball techniques on included primary studies.

References

1. Marine-Roig, E.: Anton Clavé, S.: Tourism analytics with massive user-generated content: a case study of Barcelona. J. Destin Mark. Manag. **4**, 1–11 (2015)
2. Ravi, K., Ravi, V.: A survey on opinion mining and sentiment analysis: tasks, approaches and applications. Knowl. Based Syst. **89**, 14–46 (2015)
3. Balazs, J.A., Velásquez, J.D.: Opinion mining and information fusion: a survey. Inf. Fusion. **27**, 95–110 (2016)
4. Hotho, A., Andreas, N., Paaß, G., Augustin, S.: A brief survey of text mining, pp. 1–37 (2005)
5. Tan, A.: Text mining: the state of the art and the challenges concept-based. In: Proceedings of the PAKDD 1999 Workshop on Knowledge Disocovery from Advanced Databases, pp. 65–70 (1999)
6. Poblete, B., Garcia, R., Mendoza, M., Jaimes, A.: Do all birds tweet the same? Characterizing twitter around the world categories and subject descriptors. In: Society, pp. 1025–1030 (2011)
7. Petersen, K., Feldt, R., Mujtaba, S., Mattsson, M.: Systematic mapping studies in software engineering, pp. 1–10 (2007)
8. Kitchenham, B., Charters, S.: Guidelines for performing systematic literature reviews in software engineering. Technical Report EBSE-2007-01, School of Computer Science and Mathematics, Keele University (2007)
9. Pang, B., Lee, L.: Opinion mining and sentiment analysis. Found. Trends® Inf. Retr. **2**, 1–135 (2008)
10. Varghese, R., Jayasree, M.: A Survey on Sentiment Analysis and Opinion Mining, pp. 312–317 (2013). www.Ijret.Org
11. Vinodhini, G., Chandrasekaran, R.: Sentiment analysis and opinion mining: a Survey. Int. J. Adv. Res. Comput. Sci. Softw. Eng. **2**, 282–292 (2012)
12. Medhat, W., Hassan, A., Korashy, H.: Sentiment analysis algorithms and applications: a survey. Ain Shams Eng. J. **5**, 1093–1113 (2014)
13. Pardo, T., Gasperin, C., Caseli, H., Nunes, M.D.G.V.: Computational linguistics in Brazil: an overview. In: Proceedings of the NAACL HLT 2010 Young Investigators Workshop on Computational Approaches to Languages of the Americas, pp. 1–7 (2010)
14. da Silva Conrado, M., Felippo, A., Pardo, T.S., Rezende, S.: A survey of automatic term extraction for Brazilian Portuguese. J. Braz. Comput. Soc. **20**, 12 (2014)
15. de Freitas, L.A., Vieira, R.: Ontology-based feature level opinion mining for Portuguese reviews. In: WWW 2013 Companion - Proceedings of the 22nd International Conference on World Wide Web, pp. 367–370 (2013)
16. Souza, M., Vieira, R., Chishman, R., Alves, I.M.: Construction of a Portuguese opinion lexicon from multiple resources. In: Proceedings of the 8th Brazilian Symposium in Information and Human Language Technology, pp. 59–66 (2011)
17. Taboada, M., Brooke, J., Tofiloski, M., Voll, K., Stede, M.: Lexicon-based methods for sentiment analysis. Comput. Linguist. **37**, 267–307 (2011)
18. da Fonseca Carvalho, P.C.Q.: Análise e representação de construções adjectivais para processamento automático de texto: adjectivos intransitivos humanos (2007)
19. Evangelista, T.R., Padilha, T.P.P.: Monitoramento de Posts Sobre Empresas de E-Commerce em Redes Sociais Utilizando Análise de Sentimentos (2013)

Finding Compositional Rules for Determining the Semantic Orientation of Phrases

António Paulo Santos[1](✉), Carlos Ramos[1], and Nuno Marques[2]

[1] GECAD, Institute of Engineering,
Polytechnic of Porto, Porto, Portugal
{pgsa, csr}@isep.ipp.pt
[2] DI-FCT, Universidade Nova de Lisboa,
Monte da Caparica, Almada, Portugal
nmm@fct.unl.pt

Abstract. The semantic compositionality principle states that the meaning of an expression can be determined by its parts and the way they are put together. Based on that principle, this paper presents a method for finding the set of compositional rules that best explain the *positive*, *negative*, and *neutral* semantic orientation (SO) of two-word phrases, in terms of the SO of its words. For instance, the phrase *"fake contract"* has a negative SO. A corpus was built for evaluating the proposed method and several experiences are reported. We also use the conditional probability as a reliability measure of the compositional rules. The reliability of the learned rules ranges from 60.44 % for verb-noun phrases to 100 % for adjective-adjective phrases.

Keywords: Semantic orientation · Compositional rules · Sentiment analysis

1 Introduction

The *semantic orientation* (SO) or *polarity* of a word indicates the direction the word deviates from the norm for its semantic group or lexical field [1]. This notion can be generalized to bigger text units (e.g. a word, a sentence, a document). Past work has shown that the SO of two-word phrases has an important role in the determination of the SO of bigger text units [2–4]. Determining the SO of two or more sequences of words (e.g. determining if *"fake contract"* is *positive*, *negative*, or *neutral*) has been researched following different approaches [2, 4–9]. Promising approaches rely on the principle of *semantic compositionality*, implicitly [6, 8] or explicitly [3, 4, 7, 9, 10]. Generally, the SO composition is modeled by a set of rules in several works [6, 7, 10, 11]. These rules are often created manually and defined based on human knowledge of language (human experts). An alternative to manual rules may be to learn the rules from data as we propose in this paper.

This paper is organized as follows. Section 2 describes the proposed approach for finding the compositional rules and describes the resources required to apply it. Section 3 reports the results of applying the previous approach to a set of two-word phrases. We close with the conclusion in Sect. 4.

© Springer International Publishing Switzerland 2016
J. Silva et al. (Eds.): PROPOR 2016, LNAI 9727, pp. 128–133, 2016.
DOI: 10.1007/978-3-319-41552-9_13

2 SO Compositionality

A popular approach for modeling the SO composition of two-word phrases is through a set of rules manually created [3, 7, 10, 11]. For example, a possible hand-coded rule could be:

$$ADJ^-N^0 \rightarrow \text{Phrase}^-$$

Meaning that a two-word phrase composed by a "negative adjective" followed by a "neutral noun", should be classified as negative, such in the "*fake$^-$ contract^0*" phrase. This rule is intuitive. However, this is not always the case. Furthermore, there may be many rules to consider. Our approach instead is to find the best compositional rules from annotated corpus (Sect. 2.1), with help of a lexicon (Sect. 2.2) and applying the algorithm described in Sect. 2.3. Unfortunately, to the best of our knowledge, no corpus currently exists to train and evaluate two-word phrases. So, we built a corpus as described next.

2.1 Corpus of Two-Word Phrases

The corpus of two-word phrases[1] was built by the following steps: First, we collected a set of sentences. We collected all the available Portuguese news headlines on *Wikinews*, in three categories.[2] There were a total of 4,646 news headlines, 982 of them in the *economy and business* category, 2,946 in the *politics and conflicts category* and 718 in the *health* category.

The second step was to apply a part-of-speech tagger[3] to each sentence and extract the two-word phrases using the extraction patterns shown in Table 1. On that table, the first five patterns were used in [2], and the remaining were defined by us. These patterns were chosen because we believe that they are the most relevant to study the semantic orientation of two words phrases in Portuguese language, present in our corpus.

The third step was to manually classify each extracted phrase as *negative* (−), *neutral* (0) or *positive* (+), according the majority polarity of that phrase in the corpus. The total number of annotated phrases is shown in Tables 1 and 2 shows some examples of them. During this third and manual step, we also reviewed the part-of-speech tags assigned to the phrases on the second step. Phrases with wrong part-of-speech were discarded.

2.2 Lexicon

Our approach relies on a lexicon of words, where each entry is associated with its part-of-speech tag and it is associated with a *positive* (+), *negative* (−), or *neutral* (0) a

---

[1] https://github.com/i000313/phd.polarity.compositionalRules.

[2] https://pt.wikinews.org/ - Wikinews in Portuguese.

[3] http://opennlp.sourceforge.net/ - OpenNLP POS Tagger 1.5.2.

Table 1. Statistics of the extracted phrases (N = noun, ADJ = adjective, ADV = adverb, VRB = verb, *prp* = preposition, *art* = article, *num* = numerals)

Pattern	#phrases	Sem. Orientation		
		−	0	+
ADJ N	433	71	300	62
ADV ADJ	38	9	13	16
ADJ ADJ	56	4	50	2
N ADJ	1,753	496	860	397
ADV VRB	240	37	183	20
N prp N	2,112	703	961	448
VRB (prp|art|num)? N	3,147	999	1,151	997
Total:	7,779	2,319	3,518	1,942

Table 2. Example of extracted phrases with their semantic orientation annotated by a human

Pattern	Example of extracted phrases
ADJ N	*[possível aliança]*$^+$ (possible alliance), *[novas regras]*^{0} (new rules)
N prp N	*[acordo de cessar-fogo]*$^+$ (cease-fire agreement)
VRB (prp|art| num)? N	*[sobrevive a tiro]*$^+$ (survives shooting), *[acusado de corrupção]*$^-$ (accused of corruption)

priori semantic orientation. Therefore, we built a lexicon with 21,826 nouns lemmas, 10,500 adjectives lemmas, 1,111 adverbs lemmas, and 8,789 verbs lemmas. The lexicon was built, as described in [12].

2.3 Learning the Compositional Rules

As input, we need a *set of two-word phrases* manually annotated and a *polarity lexicon*. Let *Phrases* = {p_1, p_2, ..., p_m} represent the set of two-word phrases, where each two-word phrase p_i is a sequence of two words. Each two-word phrase p_i should be manually classified as *positive*, *negative*, or *neutral*, and each word should be tagged with its part-of-speech. Let *Lexicon* = {w_1, w_2, ..., w_n} represent the polarity lexicon, where each word w_j is associated with its part-of-speech (i.e. *noun, adjective, adverb*, and *verb*) and SO (i.e. *positive, negative*, and *neutral*).

The first step is to assign the SO found in the lexicon to *word₁* and *word₂* of each p_i phrase. If a word is not found in the lexicon, we should assign an "Unknown" value to that word (represented in this paper by "?"). This last procedure will allow us to have compositional rules capable of dealing with incomplete information. After the assignment of the SO to the words of p_i phrase, a count is incremented. The count represents the number of times we saw each $POS_1^{SO}POS_2^{SO} \rightarrow phrase^{SO}$ combination (or rule), where $phrase^{SO}$ represents the SO assigned by human experts to the phrase p_i and POS_1^{SO}, POS_2^{SO} represents the SO assigned to *word₁* and *word₂* based on the lexicon. For example, in our study, after performing this step for all the two-word phrases, we counted the N^0 ADJ^0 → $Phrase^0$ combination, 267 times. This means that a *neutral* noun followed by a *neutral* adjective (according to the lexicon) was classified as *neutral* by human experts.

The second step is to compute the reliability of each rule. The reliability of each $POS_1^{SO1} POS_2^{SO2}$ → $phrase^{SO}$ rule is the percentage of phrases that contain POS_1^{SO1} POS_2^{SO2} also contain $phrase^{SO}$. In other words, this percentage represents the probability of being assigned the SO to *phrase* given that it was assigned the SO1 and SO2 to the part-of-speech POS_1 and POS_2, respectively. It can be seen as an estimate of the conditional probability $Prob(Phrase^{SO}|POS_1^{SO1} POS_2^{SO2})$. It is computed as follows:

$$reliability = \frac{count(Phrase^{SO} \text{and } POS_1^{SO1} POS_2^{SO2})}{count(POS_1^{SO1} POS_2^{SO2})}$$

For example, in our study, the N^0 ADJ^0 → $phrase^0$ rule had a probability of 77 % (267/345). This value means that 77 % of the phrases composed by a *neutral* noun followed by a *neutral* adjective, where classified as *neutral* by humans experts. Of the remaining 23 % phrases composed of a *neutral* noun followed by a *neutral adjective*, 17 % of them were classified as *positive* such as "[parto^0 normal^0]$^+$" (*natural birth*), and 5 % were classified as *negative* such as "[programa^0 nuclear^0]$^-$" (*nuclear plan*).

The third step is to select the rules with highest reliability, among the rules with the same left-hand side. For example, from the three rules below, the algorithm automatically chooses the first one. This selection is necessary because these rules are conflicting and incompatible. Two or more rules are incompatible if they have the same left-hand side, but a different right-hand side.

$$N^0 ADJ^0 → 0 \, [reliability = 77.39\%]$$
$$N^0 ADJ^0 → - \, [reliability = 5.22\%]$$
$$N^0 ADJ^0 → + \, [reliability = 17.39\%]$$

The rules learned can be used as a probabilistic model to classify unseen examples.

3 Compositional Rules – Reliability Evaluation

Applying the approach described in the previous section, we got a set of compositional rules which best describe the SO of the phrases annotated by human experts. Table 3 illustrates some of those rules. The rules are grouped by part-of-speech pattern. Each rule is associated with a reliability value (the conditional probability). For example, for the pattern "ADJ N", the cell "++" (row: + and column: +) containing the "$^+100$" value represents the rule:

$$ADJ^+ N^+ → + \, [reliability = 100\%]$$

This rule says that 100 % of the phrases with a positive adjective followed by a positive noun were classified as positive by human experts.

In Table 3, "n.f." stands for *not found*, meaning that a certain phrase doesn't exist in the corpus. For instance, in our corpus there are no occurrences of phrases composed by a *positive* ADJ followed by a *negative* N. Table 4 shows the average reliability value of the learned rules, for each extraction pattern. Among these patterns the hardest SO to predict is the SO of two-word phrases formed by a VRB, followed by an optional "prp|art|num", followed by a N, since the conditional probability value is the lowest. The most frequent cases are neutral verbs followed by neutral nouns that together convey a non-neutral SO. E.g. "*[levado para hospital]⁻*" (taken to hospital), "*[sai do hospital]⁺*" (leaves the hospital), "*meter água*" (expression that can be used literally to mean "taking on water" or idiomatically to mean "to screw up").

Table 3. Compositional rules learned for each part-of-speech pattern and their reliability values (in percentage). (?) = word not found in the lexicon. (n.f) = phrase not found in the corpus

ADJ N

ADJ \ N	+	-	0	?
+	$^{+}$100	n.f.	$^{+}$100	$^{+}$71
-	$^{-}$100	$^{-}$100	$^{-}$100	$^{-}$67
0	$^{+}$75	$^{-}$86	^{0}87	^{0}89
?	$^{+}$50	$^{-}$75	^{0}94	^{0}94

ADV ADJ

ADV \ ADJ	+	-	0	?
+	$^{+}$100	n.f.	$^{+}$100	$^{+}$71
-	n.f.	$^{-}$100	$^{-}$100	n.f.
0	$^{+}$85	$^{-}$89	^{0}92	^{0}90
?	$^{+}$67	$^{-}$71	^{0}75	^{0}89

N ADJ

N \ ADJ	+	-	0	?
+	$^{+}$84	$^{-}$80	^{0}61	$^{+}$57
-	$^{-}$57	$^{-}$86	$^{-}$90	$^{-}$89
0	^{0}49	$^{-}$88	^{0}77	^{0}68
?	$^{+}$66	$^{-}$85	^{0}81	^{0}64

ADV VRB

ADV \ VRB	+	-	0	?
+	n.f.	n.f.	$^{+}$100	$^{+}$100
-	n.f.	n.f.	n.f.	$^{-}$100
0	^{0}75	^{0}59	^{0}91	^{0}59
?	^{0}100	$^{-}$75	^{0}100	^{0}100

N prp N

N \ N	+	-	0	?
+	$^{+}$78	$^{+}$62	$^{+}$65	$^{+}$65
-	$^{-}$85	$^{-}$97	$^{-}$94	$^{-}$95
0	$^{+}$48	$^{-}$67	^{0}72	^{0}70
?	^{0}49	$^{-}$67	^{0}76	^{0}80

VRB (prp|art|num)? N

VRB \ N	+	-	0	?
+	$^{+}$77	$^{+}$47	$^{+}$46	^{0}51
-	$^{-}$55	$^{-}$70	$^{-}$65	$^{-}$64
0	$^{+}$63	$^{-}$77	^{0}64	^{0}65
?	$^{+}$68	$^{-}$74	$^{+}$39	^{0}42

Table 4. Average reliability of the learned rules, for each part-of-speech pattern, considering only words present in the lexicon (+/−/0) VS considering all words (+/−/0/?)

Pattern	Average reliability of: +/-/0	Average reliability of: +/-/0/?
ADJ N	93.50%	85.87%
ADV ADJ	95.14%	86.85%
ADJ ADJ	100%	91.43%
N ADJ	74.67%	73.88%

ADV VRB	87.18%	81.25%
N prp N	74.22%	73.13%
VRB (prp|art|num)? N	62.67%	60.44%

4 Conclusions

This paper presents an approach for learning compositional rules from data, which explains the semantic orientation of two-word phrases, considering the a priori semantic orientation of its words and makes three important contributions:

- First, it describes a method to find the set of compositional rules that best describes a set of two-word phrases. Among other things, results have shown that the SO of some part-of-speech patterns are considerably more compositional than others. For instance, the SO of phrases formed by a noun followed by an adjective are much more compositional than phrases formed by a verb followed by a noun.

- Second, it empirically gives evidence that the proposed reliability measure is indeed a good estimator for measure of SO rule quality.
- Finally, it introduces a new corpus consisting of real online data, which will be useful for future studies and research in this area. Namely this corpus can present a first tool for researchers that want to research the semantic dependencies among words and their role for extracting, namely, semantic orientation at sentence level.

Acknowledgments. António Paulo Santos is supported by the FCT grant SFRH/BD/47551/2008 and supported by Department of Informatics of FCT/Universidade de Lisboa.

References

1. Lehrer, A.: Semantic fields and lexical structure. North-holl. Linguist. Ser. **11**, 225 (2008)
2. Turney, P.D.: Thumbs up or thumbs down?: semantic orientation applied to unsupervised classification of reviews. In: Proceedings of the 40th Annual Meeting on Association for Computational Linguistics, pp. 417–424. Association for Computational Linguistics, Morristown (2002)
3. Moilanen, K., Pulman, S.: Sentiment composition. In: Proceedings of the Recent Advances in Natural Language Processing International Conference, pp. 378–382 (2007)
4. Socher, R., Huval, B., Manning, C.D., Ng, A.Y.: Semantic compositionality through recursive matrix-vector spaces. In: Proceedings of the 2012 Joint Conference on Empirical Methods in Natural Language Processing and Computational Natural Language Learning, pp. 1201–1211 (2012)
5. Wilson, T., Wiebe, J., Hoffmann, P.: Recognizing contextual polarity in phrase-level sentiment analysis. In: Proceedings of the Human Language Technology Conference and the Conference on Empirical Methods in Natural Language Processing (HLT/EMNLP), pp. 347–354 (2005)
6. Polanyi, L., Zaenen, A.: Contextual valence shifters. Comput. attitude Affect text Theory Appl. **20**, 1–10 (2006)
7. Choi, Y., Cardie, C.: Learning with compositional semantics as structural inference for subsentential sentiment analysis. In: Proceedings of the Conference on Empirical Methods in Natural Language Processing - EMNLP 2008, p. 793. Association for Computational Linguistics, Morristown (2008)
8. Taboada, M., Brooke, J., Tofiloski, M., Voll, K., Stede, M.: Lexicon-based methods for sentiment analysis. Comput. Linguist. **37**, 267–307 (2011)
9. Yessenalina, A., Cardie, C.: Compositional matrix-space models for sentiment analysis. In: Proceedings of the 2011 Conference on Empirical Methods in Natural Language Processing, pp. 172–182 (2011)
10. Klenner, M., Petrakis, S., Fahrni, A.: Robust compositional polarity classification. In: Proceedings of the International Conference RANLP 2009, pp. 180–184 (2009)
11. Shaikh, M.A.M., Prendinger, H., Mitsuru, I.: SenseNet: a linguistic tool to visualize numerical-valance based sentiment of textual data. In: Proceedings of ICON - 2007 5th International Conference on Natural Language Processing, pp. 147–152, Hyderabad, India (2007)
12. Paulo-Santos, A., Ramos, C., Marques, N.C.: Determining the polarity of words through a common online dictionary. In: Antunes, L., Pinto, H. (eds.) EPIA 2011. LNCS, vol. 7026, pp. 649–663. Springer, Heidelberg (2011)

Sentiment Analysis for Brazilian Portuguese over a Skewed Class Corpora

Henrico Brum[1][(✉)], Filipe Araujo[1], and Fabio Kepler[1,2]

[1] UNIPAMPA - Federal University of Pampa, Alegrete, Brazil
henrico.brum@gmail.com, filipe.santos.araujo@gmail.com
[2] L2F/INESC-ID, Lisbon, Portugal
fabio@kepler.pro.br

Abstract. The goal of this paper is to compare existing sentiment analysis models, namely Doc2Vec and Recursive Neural Tensor Network, when applied to a skewed class corpus. Such setting is not uncommon, but the literature lacks results on it. We used two techniques to create more balance between classes: under-sampling and over-sampling the target corpora. Doc2Vec achieved the best result overall on the skewed classes, but performed poorly over small and sampled configurations. RNTN achieved the best result on the over-sampled corpus. The Naive Bayes baseline was not surpassed in the under-sampled corpus with Pos/Neg classes, which was the smallest corpus configuration.

1 Introduction

The internet contains information on several topics, with several layers of data that can be used by companies for improving services and products. Among data sources are social networks, which are being used by costumers as an unofficial platform for complaining and recommending products. The feedback that companies can retrieve from these websites can be even more useful if they can do it in an automated way, like with tools that classify the opinion (or sentiment) of user generated data, specially commentaries. In this scope, [11] define sentiment analysis (SA) as "the body of work which deals with the computational treatment of opinion, sentiment, and subjectivity in text".

In this work, we analyze different SA techniques over a collection of book reviews manually annotated by [6] with three sentiment polarities: positive, neutral, and negative. The polarities distributions are very unbalanced, thus posing a challenge for machine learning models. Nevertheless, the existing literature usually reports results over large and balanced corpora. Our goal is to evaluate and analyze the behavior of existing SA models over a corpus with skewed classes.

In Sect. 2 we analyze related works, while in Sect. 3 we present the two classifiers we selected for comparing. In Sect. 4 we show more details about the sentiment corpus at hand. Finally, in Sect. 5, we show the results and draw some conclusions.

© Springer International Publishing Switzerland 2016
J. Silva et al. (Eds.): PROPOR 2016, LNAI 9727, pp. 134–138, 2016.
DOI: 10.1007/978-3-319-41552-9_14

2   Related Works

[2] experimented eight approaches for SA using neural networks and supervised learning for *tweets* in English using SentiWordNet [5] and LIWC [13].

A document's domain also plays an important role in SA, since some expressions may refer to positive things in a certain domain and to neutral or even negative ones in others. [7] analyzed sentiments over blogs and news, while [1] focused on child stories and worked with the classes anger, disgust, fear, happiness, sadness, positive surprise and negative surprise.

[12] studied SA and proposed a scale for sentiment polarity: instead of working with positive, neutral or negative sentences, they proposed working with a numeric scale from zero to four because it would better represent the subjectivity of the task. A few years later [11], the same authors investigated machine learning, features engineering, and some unsupervised methods to evaluate opinions.

[14] proposed a recursive neural model for SA: the neural part of the model learns the sentiment of pairs of words, and the recursive part applies it to the whole sentence using a syntactic compositional tree. The model was built for English and achieved 80.7 % of accuracy, becoming the state of the art of SA for English. For Brazilian Portuguese, [3] applied the same model for a Portuguese corpus and obtained 69.08 % of accuracy.

[8] used the Doc2Vec algorithm for SA over two movie reviews corpora. In the first corpus, from Rotten Tomatoes provided by the Stanford University, they achieved error rates of 12.2 % in the 2-way coarse-grained task and 51.3 % in the 5-way fine-grained one. In the second corpus, from IMDb and also provided by the Stanford University, Doc2Vec achieved an error rate of 7.42 %.

3   Selected Classifiers

In [14], the authors describe a powerful model for sentiment analysis, called Recursive Neural Tensor Network (RNTN), that can observe the sentences composition and classify them in five classes – Very Negative, Negative, Neutral, Positive and Very Positive. The RNTN applies the learning characteristics of a neural network to pairs of words recursively, as can be seen in Fig. 1. Each sentence is split in pairs of words according to its syntactic tree, which needs to be known a priori.

Distributed representations of sentences have reached accurate results when used in SA tasks. [8] proposed the Paragraph Vector technique (or Doc2Vec), an unsupervised learning algorithm that "has emerged" as an alternative technique to its predecessor Word2Vec [9]. Paragraph Vector attempts to improve Word2Vec for modeling a variable length sequence of words, like phrases and paragraphs, as we can see in Fig. 2.

4   Resources

The number of available annotated corpora for Portuguese is very scarce and it is caused by the high demand of human efforts and the time needed for annotation.

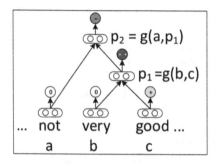

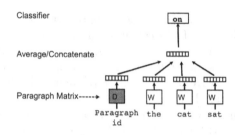

Fig. 1. Recursive Neural Tensor Network proposed by [14].

Fig. 2. Paragraph Vector proposed by [8].

The subjectivity of this specific task also contributes for the lack of resources for some specific languages.

ReLi [6] is a Brazilian Portuguese corpus with sentiment annotation of book reviews. It contains 12,508 sentences extracted from reviews of 14 books from the website *skoob*[1]. It is annotated with three classes, positive, neutral, and negative, but the distribution over them is very unbalanced. There are 9,605 (72%) neutral sentences, 2,854 (22%) positive, and only 593 (4%) negative.

Learning with skewed class distributions is an important issue in supervised learning [10]. An unbalanced number of samples per class can cause skewed results in classification since a classifier can repeatedly guess the most frequent class and still achieve a high accuracy rate.

We tried two approaches for minimizing the skewed distribution: over-sampling, which replicates examples of the less frequent classes, and under-sampling, which eliminates examples of the most frequent classes [10]. We also tried removing the neutral class altogether, and combining this with over- and under-sampling. Table 1 shows the classes frequencies for all configurations.

Table 1. Class frequency for each configuration of the sentiment corpus (OS: over-sampled; US: under-sampled).

Sentences	Pos/Neu/Neg			Pos/Neg		
	Full	OS	US	Full	OS	US
Positive	2,780	2,780	579	2,780	2,780	579
Neutral	9,149	2,780	579	-	-	-
Negative	579	2,780	579	579	2,780	579

---

[1] www.skoob.com.br.

5   Results and Discussions

We used the implementation of the RNTN model available in the Stanford CoreNLP[2] [4], and the implementation of the Paragraph Vector model called Doc2Vec from a third party developer available online[3]. For the baseline we used a Naive Bayes (NB) approach, since it is a simple and efficient technique commonly used in practice. We ran all three models over the six corpora configurations with 10-fold cross validation, averaging the accuracy. Results are shown in Table 2.

Table 2. Accuracy over the six corpus configurations; best result for each configuration is shown in bold (OS: over-sampled; US: under-sampled).

	Pos/Neu/Neg			Pos/Neg		
	Full	OS	US	Full	OS	US
NB	72.99 %	69.08 %	67.18 %	83.09 %	80.80 %	**74.26 %**
RNTN	68.82 %	**74.14%**	55.56 %	67.79 %	**82.85 %**	65.27 %
Doc2Vec	**75.28 %**	65.98 %	**79.29 %**	**83.19 %**	68.72 %	64.04 %

Doc2Vec obtained the best accuracies over both full unbalanced corpora.

On the over-sampled corpus, RNTN performed better for either three or two classes and Doc2Vec showed its lower results on this set compared ot the other techniques. This might indicate that for Doc2Vec the variety of the data is more important than a larger but more uniform data set.

RNTN obtained its lowest accuracies on the under-sampled corpora, showing that it is highly dependent on the amount of data. Overall, the Doc2Vec approach worked better for the unbalanced corpora while not that well when dealing with binary classification. The RNTN performed better when handling a larger and balanced set of data. The Naive-Bayes approach achieved the highest result when classifying a small, unbalanced corpus with two classes.

A deeper analysis of the results through confusion matrices or the usage of a different corpus could shed some light into why the Doc2Vec model achieved such high accuracy with very skewed classes.

Acknowledgments. This work was partially supported by national funds through Fundação para a Ciência e a Tecnologia (FCT) with reference UID/CEC/50021/2013.

---

[2] http://nlp.stanford.edu/software/lex-parser.shtml.
[3] http://radimrehurek.com/gensim/models/doc2vec.html.

References

1. Alm, C.O., Roth, D., Sproat, R.: Emotions from text: machine learning for text-based emotion prediction. In: Proceedings of the Conference on Human Language Technology and Empirical Methods in Natural Language Processing, pp. 579–586 (2005)
2. Araújo, M., Gonçalves, P., Benevenuto, F.: Measuring sentiments in online social networks. In: Proceedings of the 19th Brazilian Symposium on Multimedia and the web, pp. 97–104 (2013)
3. Brum, H., Kepler, F.: Análise de sentimentos para português brasileiro usando redes neurais recursivas. In: IV Workshop de Iniciação Cientifica em Tecnologia da Informação e da Linguagem Humana (TiLic) (2015)
4. Manning, C.D., Surdeanu, M., Bauer, J., Finkel, J., Bethard, S.J., McClosky, D.: The stanford CoreNLP natural language processing toolkit. In: Proceedings of 52nd Annual Meeting of the Association for Computational Linguistics: System Demonstrations, pp. 55–60 (2014)
5. Esuli, A., Sebastiani, F.: Sentiwordnet: a publicly available lexical resource for opinion mining. In: Proceedings of LREC, vol. 6, pp. 417–422 (2006)
6. Freitas, C., Motta, E., Milidiú, R., César, J.: Vampiro que brilha.. rá! Desafios na anotaçao de opiniao em um corpus de resenhas de livros. 11 Encontro de Linguistica de Corpus (2012)
7. Godbole, N., Srinivasaiah, M., Skiena, S.: Large-Scale Sentiment Analysis for News and Blogs. In: ICWSM, vol. 7 (2007)
8. Le, Q., Mikolov, T.: Distributed Representations of Sentences and Documents CoRR (2014)
9. Mikolov, T., Chen, K., Conrrado, G., Dean, J.: Efficient Estimation of Word Representations in Vector Space CoRR (2013)
10. Monard, M.C., Batista, G.: Learning with skewed class distributions. Adv. Logic, Artif. Intell. Robot. LAPTEC 2002 **85**, 173 (2002)
11. Pang, B., Lee, L.: Opinion mining and sentiment analysis. Found. Trends Inf. Retrieval **2**, 1–135 (2008)
12. Pang, B., Lee, L.: Seeing stars: exploiting class relationships for sentiment categorization with respect to rating scales. In: Proceedings of the 43rd Annual Meeting on Association for Computational Linguistics, pp. 115–124 (2005)
13. Pennebaker, J.W., Francis, M.E., Booth, R.J.: Linguistic inquiry and word count: LIWC 2001. Mahway: Lawrence Erlbaum Associates **71**, 2001 (2001)
14. Socher, R., Alex, P., Wu, J.Y., Chuang, J., Manning, C.D., Ng, A.Y., Potts, C.: Recursive deep models for semantic compositionality over a sentiment treebank. In: Proceedings of the Conference on Empirical Methods in Natural Language Processing (EMNLP), pp. 1631–1642 (2013)

Language Processing – Full Papers

Semantic Relation Extraction. Resources, Tools and Strategies

Marcos Garcia[✉]

Grupo LyS, Departamento de Galego-Português, Francês e Linguística Faculdade de Filologia, Universidade da Coruña, Campus da Coruña, Coruña, Spain
marcos.garcia.gonzalez@udc.gal

Abstract. Relation extraction is a subtask of information extraction that aims at obtaining instances of semantic relations present in texts. This information can be arranged in machine-readable formats, useful for several applications that need structured semantic knowledge. The work presented in this paper explores different strategies to automate the extraction of semantic relations from texts in Portuguese, Galician and Spanish. Both machine learning (distant-supervised and supervised) and rule-based techniques are investigated, and the impact of the different levels of linguistic knowledge is analyzed for the various approaches. Regarding domains, the experiments are focused on the extraction of encyclopedic knowledge, by means of the development of biographical relations classifiers (in a closed domain) and the evaluation of an open information extraction tool. To implement the extraction systems, several natural language processing tools have been built for the three research languages: From sentence splitting and tokenization modules to part-of-speech taggers, named entity recognizers and coreference resolution systems. Furthermore, several lexica and corpora have been compiled and enriched with different levels of linguistic annotation, which are useful for both training and testing probabilistic and symbolic models. As a result of the performed work, new resources and tools are available for automated processing of texts in Portuguese, Galician and Spanish.

Keywords: Information extraction · Natural language processing · Named entity recognition · Part-of-speech tagging · Coreference resolution

1 Introduction

In recent years, the amount of data generated by our society increased exponentially, and several studies show that this growth is currently even faster. An important portion of these data is compound by text, published every day in digital media and different languages.

This work has been partially supported by the Spanish Ministry of Economy and Competitiveness through the project FFI2014-51978-C2-1-R, and by a *Juan de la Cierva formación* grant, reference FJCI-2014-22853.

© Springer International Publishing Switzerland 2016
J. Silva et al. (Eds.): PROPOR 2016, LNAI 9727, pp. 141–152, 2016.
DOI: 10.1007/978-3-319-41552-9_15

This huge amount of text contains information that can be very useful for various applications in many areas. However, the size of these data makes impossible their processing through reading.

Aimed at simplifying this process, our work has the main objective of developing linguistically-based resources and tools for the automatic extraction of semantic information as well as their evaluation. The extraction of structured semantic knowledge from free text is useful both for theoretical purposes (it gives information about how semantic relations are represented linguistically) and from a pragmatic point of view (it permits to create structured databases and other useful resources) [19].

Thus, the work carried out involved the development and evaluation of Relation Extraction (RE) tools, capable of automatically extracting semantic knowledge from free text in Portuguese, Galician and Spanish. As an example, from a sentence (in Portuguese) like the following:

John A. Garcia (nascido em 1949 na Galiza) é um dos pioneiros da indústria moderna americana de videojogos e o atual presidente da Novalogic.[1]

a semantic RE system could extract the following structured knowledge:

1. John A. Garcia *BirthDate* 1949
2. John A. Garcia *BirthPlace* Galiza
3. John A. Garcia *PresidentOf* Novalogic

Taking the above into account, we explore various strategies for building RE systems: From symbolic methods relying on a syntactic analysis to different machine learning models that use both weakly-supervised and supervised approaches. Furthermore, in order to implement the RE systems, it was necessary to build and adapt several tools for performing automatic linguistic analysis in the target languages: From modules for sentence boundary identification, morphological analyzers and lemmatizers (that recognize, for instance, *nascido* as a form of the verb *nascer*), to named entity recognizers (that can identify *John A. Garcia* as a single personal name, or *Galiza* as a location) or syntactic parsers.

The results of the performed work bring interesting information about the use of linguistic data in different strategies for RE in Portuguese, Galician, and Spanish. On the one hand, several evaluations showed that linguistic information (namely those produced by lemmatization and semantic classification) is critical for building machine learning systems for relation extraction, even though their performance could be negatively affected if the linguistic analysis produces errors.

On the other hand, rule-based approaches obtained high-quality results in some tasks. For instance, in Named Entity Recognition (NER) —with competitive results when compared to a supervised approach—, or in relation extraction, with precision results between 85 % and 95 % (depending on the relation and language).

---

[1] A possible English translation could be: "John A. Garcia (born in 1949 in Galicia) is one of the pioneers of the modern American computer game industry and the current president of Novalogic.".

Also, we introduce novel strategies for improving tasks such as PoS-tagging (using dedicated parsers), corpus annotation (by means of distant-supervision) and Open Information Extraction (OIE), using automatic Coreference Resolution (CR). Finally, the performed work makes freely available a large set of resources and tools for the three target languages.[2]

Apart from this introduction, this paper is organized as follows: Sect. 2 presents the main hypotheses and objectives of our work as well as the methodological aspects. Then, the structure is outlined in Sect. 3, which also shows some of the main experiments and results. After that, Sect. 4 includes an overview of some of the related work for each of the performed tasks, while the conclusions are shown in Sect. 5.

2   Problems, Hypotheses, and Objectives

Apart from the main problem presented in the introduction (i.e., the difficulty of taking advantage of the vast amount of data that is being produced), two other related issues need to be solved in order to face the mentioned one:

- Lack of resources and tools for RE in Portuguese, Galician and Spanish.
- Need for multilingual Natural Language Processing (NLP) tools and resources for text processing (previous to the extraction).

Aimed at facing these problems, three main hypotheses were formulated, presented here as questions:

- Is it possible to create RE systems capable of extracting accurate biographical knowledge in the three target languages?
- What kind of linguistic information is needed to build symbolic and statistical NLP tools?
- Is it feasible to develop rapidly and to adapt NLP resources and tools that are needed for performing the extractions?

During the work carried out, some other issues were taken into account apart from the mentioned ones. Problems such as the PoS-tagging of several varieties (national and orthographic) of Portuguese, the correction of the critical PoS-tagging errors or the best combination of coreference resolution with open information extraction, among others.

Once formulated the different hypotheses, a primary objective was defined as follows:

- The main goal consists in evaluating different strategies for the extraction of encyclopedic knowledge —mainly biographic— in closed domain, in Portuguese, Galician and Spanish.

To achieve the primary objective, several parallel goals were defined, that can be summarized in a single one:

---

[2] All of them are freely available at http://gramatica.usc.es/~marcos/phd.html.

– The implementation or adaptation of the NLP tools that are required for semantic RE in Portuguese, Galician, and Spanish.

These objectives are gradually achieved during our work, aimed at answering the formulated hypotheses and solving the problems that had been found. In this respect, we introduce the tools and resources that have been developed and their evaluation, as well as the various strategies for semantic RE that have been implemented.

2.1 Methodology

From a methodological point of view, our work is mainly based on the use of linguistic knowledge for NLP, but it combines this information with approaches from other areas (such as machine learning) in order to better achieve the proposed objectives. Thus, during the implementation and adaptation of each tool and resource, the theoretical proposals were taken into account, but the quality of the results was prioritized over the formal consistency. Therefore, the work is essentially pragmatic.

3   Structure, Experiments and Results

Our work can be divided into three well-distinguished parts: (i) natural language processing before the extraction, (ii) strategies for relation extraction and (iii) the combination of coreference resolution with open information extraction. Regarding the chapter structure, it is summarized in Fig. 1.

3.1 NLP Before the Extraction

Before the implementation of RE systems, we adapted and the evaluated different modules for tokenization, sentence identification, lemmatization and morphosyntactic analysis for Portuguese and Galician [20], as well as a strategy for PoS-tagging correction [21]. We also made and adapted several tagged corpora and morphosyntactic dictionaries for different linguistic varieties.

Specially for Portuguese, we addressed the problem of PoS-tagging different varieties of this language. Several lexica and corpora has been combined to train morphosyntactic analyzers for tagging texts in various national varieties (Brazil, Portugal, Angola, etc.) and spelling systems (before and after the *Acordo Ortográfico de 1990*) [30].

Different experiments showed that the resulting tools achieve competitive performance on different tasks, with sentence splitting, tokenization and lemmatization results between >98 % and >99 %, depending on the language.

Concerning PoS-tagging, the evaluation in different varieties of Portuguese showed that a single model (e.g., European Portuguese) trained with consistent resources can achieve better performance than complex models that combine corpora from different varieties.

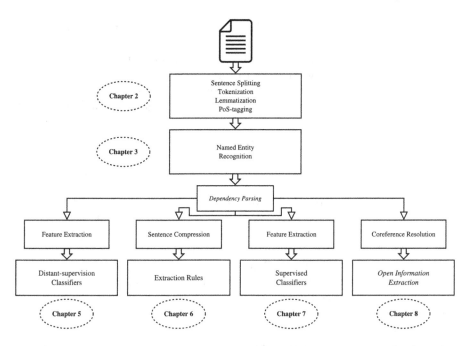

Fig. 1. Diagram of the processes carried out in each chapter. The input (top) is plain text, and there are evaluated four different strategies of information extraction (bottom) in Chaps. 5 to 8. Elements in italic were not implemented specifically in the thesis. Chapter 4 contains a bibliographic review, so it does not appear in the diagram.

Furthermore, we also presented a set of tools (both statistical and based on linguistic rules and resources) for named entity recognition, capable of identifying and semantically classifying named entities in free text automatically [16,31]. The presented tools, which work in the three referred languages, recognize proper nouns such as persons, organizations, and locations, as well as other expressions such as dates, quantities, etc.

Named entity recognition was approached in two steps: the identification of the named entity boundaries (with F1 results of 91% in Portuguese and 94% in Galician) and the semantic classification of each entity (≈75% F1 in both Portuguese and Galician). Several experiments carried out showed that the supervised classifier achieves better results when evaluated in corpora from the same domain as the training data. However, the resource-based method, which takes advantage of large lists of gazetteers, obtains more stable values across different domains.

3.2   Relation Extraction Strategies

Three different strategies were evaluated for performing relation extraction in a closed domain: (i) a distant-supervision approach, (ii) the use of supervised classifiers and (iii) a novel rule-based strategy that takes advantage of text compression techniques.

First we implemented and evaluated strategies for distant-supervision RE [23, 25]. This technique allows the user to obtain labeled data in a semi-automatic way, using semantically related pairs extracted from structured resources. These data are then used for building machine learning models that are capable of automatically extracting new knowledge.

The evaluation was performed with the relation `Profession` in Portuguese and Spanish, obtaining competitive results (78%–83% F1 in Portuguese and Spanish, respectively) when generalizing the patterns by means of the *longest common string algorithm*.

Then, we evaluated the impact of different linguistic knowledge in the training of supervised classifiers for RE [26]. It was performed a detailed analysis of several sets of classifiers that only differ in the linguistic knowledge they use: From basic lexical units to deep syntactic information, combined with semantic and pseudo-syntactic knowledge. Two corpora (one for Portuguese and other for Spanish) were semi-automatically labeled with five biographical relations using a distant-supervision approach. After that, the annotation was manually corrected, and different evaluations showed the benefits of lemmatization and NER for biographical RE.

Finally, we developed a new technique that consists in the application of text compression methods for simplifying the contexts containing semantic relations [22, 24]. This method is combined with a strategy, inspired in distant-supervision, for semi-automatic building of lexico-syntactic rules for RE. The performed experiments showed that the proposed method keeps the high-precision values of the pattern-matching approaches while increasing the recall. Using both encyclopedic and journalistic corpora, the systems extracted tens of thousands of semantically related pairs, with precision values between 85% and 96%, depending on the language, the domain, and the relation.

3.3   Coreference Resolution and Open Information Extraction

After evaluating closed domain RE, we studied the impact of CR in open information extraction in the three target languages. With the mentioned objective, it was developed a deterministic system for automatic CR of person entities in Portuguese, Galician, and Spanish. The tool, evaluated in different corpora also created during our work [29], is capable of identifying and linking various expressions that refer to the same person in a text (*Miguel Gomes, the film director, he, Gomes,* etc.) [27].

Coreference resolution was evaluated together with *DepOE*, a multilingual tool for OIE also developed in our research group [17]. The performed tests show that the combination of these two methods allows the extraction to improve both its precision and its recall, being a promising strategy for further research [28]. In this regard, Table 1 exemplifies how a previous application of a CR tool improves the open information extraction output.

Table 1. Examples of open information extraction without coreference resolution (OIE) and in combination with CR (CR+OIE).

Sentence (Spanish)	"Debutó en la Tercera división"
Sentence (English)	"[-] debuted in the third division"
OIE Extraction	∅
CR+OIE Extract. (Sp)	Ander Herrera *debutó_en* la Tercera división
CR+OIE Extract. (Eng)	Ander Herrera *debuted_in* the third division
Sentence (Portuguese)	"Anderson viajou por Europa"
Sentence (English)	"Anderson traveled in Europe"
OIE Extraction (Port)	Anderson *viajou_por* Europa
OIE Extraction (Eng)	Anderson *traveled_in* Europe
CR+OIE Extract. (Port)	Wes Anderson *viajou_por* Europa
CR+OIE Extract. (Eng)	Wes Anderson *traveled_in* Europe

The results of this combination (Table 2) show the benefits of applying coreference resolution before open information extraction: on average, the number of extractions increased 22.7 %, while the precision was 10.6 % better. Finally, it was calculated an *enrichment* value as follows: we verified, from all the correct extractions, if the personal mention had been correctly solved by the CR tool. These cases were divided by the total number of correct extractions, being these results considered as the *enrichment* value. Although these results are not a direct evaluation of OIE, they suggest that the extraction is ≈79 % better when applied after CR.

Table 2. Results of two runs of the OIE system (OIE and CR+OIE) in the three target languages. *Prec.* means Precision, while the *Wikipedia* and *Journal* values are the number of extractions in these domains. *Enrich.* is the enrichment caused by the previous execution of the CR tool.

Language	OIE Extraction			CR+OIE Extraction			Enrich.
	Wikipedia	*Journal*	*Prec.*	*Wikipedia*	*Journal*	*Prec.*	
Portuguese	82	133	39 %	111	155	56 %	75 %
Galician	168	114	49 %	221	115	54 %	77 %
Spanish	47	82	49 %	80	86	58 %	84 %

4 Related Work

This section briefly presents, following the structure presented in Sect. 3, some of the works that have been considered more relevant to each of the performed tasks.

Concerning the first steps of the NLP pipeline (sentence splitting, tokenization, lemmatization and PoS-tagging), some strategies already presented for Portuguese [5,6] and Galician [32] were used in order to adapt the FreeLing suite

[39] for these languages, following the EAGLES guidelines for morphosyntactic annotation [34].

After that, both probabilistic [10] and knowledge-based [16] models were used for implementing NER tools for Portuguese and Galician. NER was handled taking into account "timex", "numex" and "enamex" expressions as defined in the MUC-7 conference [36], thus differing from other approaches such as the HAREM evaluations [38,43], which propose a more fine-grained entity classification.

About relation extraction, several works (such as [7]) were inspired by the pattern-matching approach used by Hearst [33], while some others proposed strategies for the automatic learning of new extraction patterns [1,14,41].

More recently, new strategies were used to both reduce the effort of annotating training data and increase the number of extracted relations. Thus, techniques such as distant-supervision take advantage of large repositories of structured data for automatically obtaining training examples [37].

As pointed out in the previous section, open information extraction [2] is a new paradigm that extracts triples (with the following structure: *argument_1* verb-based_relation *argument_2*) without the need of previously define the target relations. Some OIE approaches use training data and shallow parsing for building the extractor [15], while others rely on heuristics based on dependency parsing [12].

In Portuguese, three different systems were presented to the ReRelEM task [38], focused on generic relation extraction: REMBRANDT [9], which uses knowledge extracted from the Wikipedia, SEI-Geo [11], that applies patterns similar to those used by Hearst [33], and SeRELeP [8], a rule-based system which identifies relations relying on NER labels.

Finally, regarding coreference resolution, the tool implemented in our work was inspired by the entity-centric approach presented in [35], enriched with some linguistic knowledge such as the proposed in [40].

Also, we followed (with minor differences) the guidelines proposed in [42] for annotating three corpora with coreference information of person entities.

In general, the mentioned papers were carefully studied to build different resources and tools for the research languages, combining strategies proposed by different authors and adapting them to our objectives. Also, it is important to note that out work took advantage of several available resources [3,4,13] and tools [10,18,39] for different languages.

5 Conclusions

The work that we carried out permitted, on one hand, to develop and evaluate novel and promising techniques for different types of semantic relation extraction, both in a closed domain and using OIE. Several tests in various languages proved that it is feasible to obtain automatically structured knowledge from free text.

On the other hand, it has been shown the effectiveness of different strategies for the development and adaptation of resources and tools for NLP in the three target languages.

Taking into account that the presented work takes advantage of both theoretical and applied linguistics, some parts proved that the combination of linguistic knowledge with statistical methods is useful in the different NLP tasks that have been evaluated.

Several tests showed that linguistic information (especially those produced by lemmatization as well as by semantic classification) is essential for building machine learning classifiers for relation extraction.

Also, those approaches based on syntactic dependencies achieved high-quality results in some tasks, such as the combination of coreference resolution with OIE. The promising results of this strategy turn it an interesting approach to further work on open domain information extraction.

As it has been said, it was necessary to implement several NLP tools and resources for the three research languages. This way, some of the developed tools were the first ones for performing various NLP tasks in Galician, and some others were the first ones with open source licenses for Portuguese and Spanish.

5.1   Resources and Tools

Besides the theoretical conclusions, the contributions of our work also include the following resources and tools:

- Sentence splitting modules for Portuguese and Galician.
- Tokenization modules for Portuguese and Galician.
- Morphological analysis modules for Portuguese and Galician.
- Morphosyntactic analysis modules for Portuguese and Galician.
- Adaptation of Bosque 8.0 corpus tags to EAGLES standard.
- Adaptation of LABEL-LEX (SW) lexicon to EAGLES standard.
- PoS-tagged corpora for different varieties of Portuguese and Galician.
- Lexica (and extension of existing lexica) with PoS-tags for different varieties of Portuguese and Galician.
- Named entity identification modules for Portuguese and Galician.
- Named entity classification modules for Portuguese and Galician.
- Modules for recognizing numerical expressions, quantities and hours for Portuguese and Galician.
- Named entity annotation of Bosque 8.0 corpus.
- Testing corpus with named entity annotation for Galician.
- Corpora with annotation of biographical relations for Portuguese and Spanish.
- Coreference resolution tool for person entities for Portuguese, Galician and Spanish.
- Corpora with coreferential annotation of person entities for Portuguese, Galician and Spanish.

All the resources and tools are available under open source licenses (GPLv3) or keeping the original license in the case of adaptations.

References

1. Agichtein, E., Gravano, L.: Snowball: extracting relations from large plain-text collections. In: Proceedings of the 5th ACM International Conference on Digital Libraries, pp. 85–94 (2000)
2. Banko, M., Cafarella, M., Soderland, S., Broadhead, M., Etzioni, O.: Open information extraction from the web. In: Proceedings of the 20th International Joint Conference on Artifical Intelligence (IJCAI 2007), pp. 2670–2676 (2007)
3. Barcala, F.M., Domínguez Noya, E.M., Otero, P.G., López Martínez, M., Moscoso Mato, E.M., Rojo, G., Santalla del Río, M.P., Sotelo Docío, S.: A corpus and lexical resources for multi-word terminology extraction in the field of economy in a in a minority language. In: Human Language Technologies as a Challenge for Computer Science and Linguistics, Proceedings of the 3rd Language & Technology Conference, pp. 359–363 (2007)
4. Bosque 8.0: Uma floresta integralmente revista por linguistas (2008)
5. Branco, A., Silva, J.R.: Contractions: breaking the tokenization-tagging circularity. In: Mamede, N.J., Baptista, J., Trancoso, I., Nunes, M.G.V. (eds.) PROPOR 2003. LNCS (LNAI), vol. 2721, pp. 167–170. Springer, Heidelberg (2003)
6. Branco, A., Silva, J.: Evaluating solutions for the rapid development of state-of-the-art POS taggers for portuguese. In: Proceedings of the 4th International Conference on Language Resources and Evaluation (LREC 2004), pp. 507–510 (2004)
7. Brin, S.: Extracting patterns and relations from the World Wide Web. In: Proceedings of the WebDB Workshop at the 6th International Conference on Extending Database Technology (EDBT 1998), pp. 172–183 (1998)
8. Bruckschen, M., Camargo de Souza, J., Vieira, R., Rigo, S.: Sistema SeRELeP para o reconhecimento de relações entre entidades mencionadas. In: Desafios na avaliação conjunta do reconhecimento de entidades mencionadas: O Segundo HAREM, Chap. 14, pp. 247–260. Linguateca (2008)
9. Cardoso, N.: REMBRANDT - Reconhecimento de Entidades Mencionadas Baseado em Relações ANálise Detalhada do Texto. In: Desafios na avaliação conjunta do reconhecimento de entidades mencionadas: O Segundo HAREM, pp. 195–211. Linguateca (2008)
10. Carreras, X., Márquez, L., Padró, L.: A simple named entity extractor using AdaBoost. In: Proceedings of the 7th Conference on Natural Language Learning at HLT/NAACL 2003, vol. 4, pp. 152–155. ACL (2003)
11. Chaves, M.: Geo-ontologias e padrões para reconhecimento de locais e de suas relações em textos: o SEI-Geo no Segundo HAREM. In: Desafios na avaliação conjunta do reconhecimento de entidades mencionadas: O Segundo HAREM, pp. 231–245. Linguateca (2008)
12. Corro, L.D., Gemulla, R.: ClausIE: clause-based open information extraction. In: Proceedings of the 22nd International Conference on World Wide Web (WWW 2013), pp. 355–366 (2013)
13. Eleutério, S., Ranchhod, E., Mota, C., Carvalho, P.: Dicionários Electrónicos do Português. Características e Aplicações. In: Actas del VIII Simposio Internacional de Comunicación Social, pp. 636–642 (2003)
14. Etzioni, O., Cafarella, M., Downey, D., Kok, S., Popescu, A.M., Shaked, T., Soderland, S., Weld, D., Yates, A.: Web-scale information extraction in Know-ItAll. In: Proceedings of the 13th International Conference on World Wide Web (WWW 2004), pp. 100–110. ACM (2004)

15. Etzioni, O., Fader, A., Christensen, J., Soderland, S., Mausam, M.: Open information extraction: the second generation. In: Proceedings of the 22nd International Joint Conference on Artificial Intelligence (IJCAI 2011), pp. 3–10 (2011)
16. Gamallo, P., Garcia, M.: A resource-based method for named entity extraction and classification. In: Antunes, L., Pinto, H.S. (eds.) EPIA 2011. LNCS (LNAI), vol. 7026, pp. 610–623. Springer, Heidelberg (2011)
17. Gamallo, P., Garcia, M., Fernández-Lanza, S.: Dependency-based open information extraction. In: Proceedings of the Joint Workshop on Unsupervised and Semi-Supervised Learning in NLP at the 13th Conference of the European Chapter of the Association for Computational Linguistics (EACL 2012), pp. 10–18. ACL (2012)
18. Gamallo, P., González López, I.: A grammatical formalism based on patterns of part-of-speech tags. Int. J. Corpus Linguist. **16**(1), 45–71 (2011)
19. Garcia, M.: Extracção de Relações Semânticas. Recursos, Ferramentas e Estratégias. Ph.D. thesis, Universidade de Santiago de Compostela (2014)
20. Garcia, M., Gamallo, P.: Análise Morfossintáctica para Português Europeu e Galego: Problemas, Soluções e Avaliação. Linguamática. Revista para o Processamento Automático das Línguas Ibéricas **2**(2), 59–67 (2010)
21. Garcia, M., Gamallo, P.: Using morphosyntactic post-processing to improve PoS-tagging accuracy. In: Proceedings of the 9th International Conference on Computational Processing of Portuguese Language (PROPOR 2010), Extended Activities Proceedings (2010)
22. Garcia, M., Gamallo, P.: A weakly-supervised rule-based approach for relation extraction. In: Proceedings of the XIV Conference of the Spanish Association for Artificial Intelligence (CAEPIA 2011). Workshop on Knowledge Extraction and Exploitation from Semi-structures Online Sources (KEESOS) (2011)
23. Garcia, M., Gamallo, P.: An exploration of the linguistic knowledge for semantic relation extraction in Spanish. In: Proceedings of the Joint Workshop FAM-LbR/KRAQ 2011. In: Learning by Reading and its Applications in Intelligent Question-Answering at 22nd International Joint Conference on Artificial Intelligence (IJCAI 2011), pp. 7–12 (2011)
24. Garcia, M., Gamallo, P.: Dependency-based text compression for semantic relation extraction. In: Proceedings of the Workshop on Information Extraction and Knowledge Acquisition (IEKA 2011) at 8th International Conference on Recent Advances in Natural Language Processing (RANLP 2011), pp. 21–28 (2011)
25. Garcia, M., Gamallo, P.: Evaluating various features on semantic relation extraction. In: Proceedings of the 8th International Conference on Recent Advances in Natural Language Processing (RANLP 2011), pp. 721–726 (2011)
26. Garcia, M., Gamallo, P.: Exploring the effectiveness of linguistic knowledge for biographical relation extraction. Nat. Lang. Eng. **21**(4), 519–551 (2013)
27. Garcia, M., Gamallo, P.: An entity-centric coreference resolution system for person entities with rich linguistic information. In: Proceedings of COLING 2014, the 25th International Conference on Computational Linguistics: Technical Papers, pp. 741–752 (2014)
28. Garcia, M., Gamallo, P.: Entity-centric coreference resolution of person entities for open information extraction. Procesamiento del Lenguaje Natural **53**, 25–32 (2014)
29. Garcia, M., Gamallo, P.: Multilingual corpora with coreference annotation of person entities. In: Proceedings of the 9th International Conference on Language Resources and Evaluation (LREC 2014), pp. 3229–3233. ELRA (2014)

30. Garcia, M., Gamallo, P., Gayo, I., Pousada Cruz, M.: PoS-tagging the Web in Portuguese. National varieties, text typologies and spelling systems. Procesamiento del Lenguaje Natural **53**, 95–101 (2014)
31. Garcia, M., Gayo, I., González López, I.: Identificação e Classificação de Entidades Mencionadas em Galego. Estudos de Lingüística Galega **4**, 13–25 (2012)
32. Graña, J., Barcala, F.-M., Vilares, J.: Formal methods of tokenization for part-of-speech tagging. In: Gelbukh, A. (ed.) CICLing 2002. LNCS, vol. 2276, pp. 123–144. Springer, Heidelberg (2002)
33. Hearst, M.: Automatic acquisition of hyponyms from large text corpora. In: Proceedings of the 14th Conference on Computational Linguistics, vol. 2, pp. 539–545. ACL (1992)
34. Leach, G., Wilson, A.: Recommendations for the morphosyntactic annotation of corpora. Technical report, Expert Advisory Group on Language Engineering Standard (EAGLES) (1996)
35. Lee, H., Chang, A., Peirsman, Y., Chambers, N., Surdeanu, M., Jurafsky, D.: Deterministic coreference resolution based on entity-centric, precision-ranked rules. Comput. Linguist. **39**(4), 885–916 (2013)
36. Mikheev, A., Grover, C., Moens, M.: XML tools and architecture for Named Entity Recognition. J. Markup Lang. Theory Pract. **1**(3), 89–113 (1998)
37. Mintz, M., Bills, S., Snow, R., Jurafsky, D.: Distant supervision for relation extraction without labeled data. In: Proceedings of the 47th Annual Meeting of the Association for Computational Linguistics (ACL 2009), pp. 1003–1011. ACL (2009)
38. Mota, C., Santos, D. (eds.): Desafios na avaliação conjunta do reconhecimento de entidades mencionadas. O Segundo HAREM. Linguateca (2008)
39. Padró, L., Stanilovsky, E.: FreeLing 3.0: towards wider multilinguality. In: Proceedings of the Language Resources and Evaluation Conference (LREC 2012). ELRA (2012)
40. Palomar, M., Ferrández, A., Moreno, L.: Martínez-Barco, P., Peral, J., Saiz-Noeda, M., Muñoz, R.: An algorithm for anaphora resolution in Spanish texts. Comput. Linguist. **27**(4), 545–567 (2001)
41. Pantel, P., Pennacchiotti, M.: Espresso: leveraging generic patterns for automatically harvesting semantic relations. In: Proceedings of the International Conference on Computational Linguistics and the Annual Meeting of the Association for Computational Linguistics (COLING/ACL 2006), pp. 113–120. ACL (2006)
42. Recasens, M.: Martí, M.: AnCora-CO: coreferentially annotated corpora for Spanish and Catalan. Lang. Res. Eval. **44**(4), 315–345 (2010)
43. Santos, D., Cardoso, N. (eds.): Reconhecimento de entidades mencionadas em português: Documentação e actas do HAREM, a primeira avaliação conjunta na área. Linguateca (2007)

Extracting and Structuring Open Relations from Portuguese Text

Sandra Collovini[1]([✉]), Gabriel Machado[1], and Renata Vieira[2]

[1] Pontifícia Universidade Católica do Rio Grande do Sul, Porto Alegre, Brazil
{sandra.abreu,gabriel.machado.002}@acad.pucrs.br
[2] Universidade Federal de Ciências da Saúde de Porto Alegre,
Porto Alegre, Rio Grande do Sul, Brazil
renata.vieira@pucrs.br

Abstract. The task of Open Relation Extraction from texts faces many challenges, considering the required linguistic knowledge and the sophistication of the language processing techniques employed. This paper presents the extraction and structuring of open relations between named entities from Portuguese texts. We apply the Conditional Random Fields model for the extraction of relation descriptors between named entities belonging to Person, Place and Organisation categories. A 0.64 of F-measure was reached as a result. To make better sense of the output, we structure the extracted relation descriptors using mining configurations.

Keywords: Information extraction · Relation extraction · Named entity · Natural language processing

1 Introduction

Information Extraction systems extract information from texts and represent it in a structured way, for example, as a list, table or graph, amenable to storage, indexing, and query processing by a standard database management system, or processing by a statistical analysis tool, among other applications.

Finding the information automatically from text requires IE applications, such as Named Entity Recognition (NER) and Relation Extraction (RE). NER aims to identify and classify Named Entities (NE) and their categories in texts [26], such as names of Organisation, Place, Person, among others; whereas RE looks for relations that occur between entities [20], for instance, the "*affiliation*" relation between entities of type Person and Organisation. According to [29], the identification of named entities is the first step towards the semantic analysis of a text, being crucial to relation extraction systems. In the literature, we find several works that consider NER to be an integral part of RE systems [4,19,23], given that NER can help the identification of arguments (entities) that are part of a certain relation.

Currently, there are two major types of RE: closed domain and open domain. Closed-domain RE systems consider only a closed set of relations between two

© Springer International Publishing Switzerland 2016
J. Silva et al. (Eds.): PROPOR 2016, LNAI 9727, pp. 153–164, 2016.
DOI: 10.1007/978-3-319-41552-9_16

arguments, while open-domain RE systems do not need a pre-specified definition of the relation and aim at identifying all possible relations from an open-domain corpus [6]. One of the challenges of open-domain RE is that due to the diversity of extracted relations, there is a difficulty in organizing these relations following some criteria. The mining of the resulting data from an RE task may be a promising approach to the discovery/association of important relations contained in texts [1, 2, 30].

Different approaches have been developed for RE tasks depending on the application and the resources available, such as: supervised learning techniques employing annotated corpus; unsupervised approaches based on generic extraction patterns; and semi-supervised methods. In RE systems, the learning methods most commonly applied are: Hidden Markov Models (HMM) [16], Conditional Random Fields (CRF) [13, 22, 24], k-Nearest-Neighbors (KNN) [32], Maximum Entropy Models [21], Support Vector Machines (SVM) [14], Naive Bayes [5]. Specifically, Conditional Random Fields have been efficiently applied to a variety of sequential text processing tasks including part-of-speech tagging, word segmentation, Named Entity Recognition and Relation Extraction [13]. These are undirected graphical models used to calculate the conditional probability of values on designated output nodes given values assigned to other designated input nodes [22].

In this paper, we present the extraction and structuring of open relations between named entities in the Organization domain for Portuguese. The study of the Organization domain was chosen because of its potential applicability to different areas such as Competitive Intelligence, Risk Management, and Marketing. We apply the Conditional Random Fields model for the extraction of any relation descriptor expressing any type of relation between a pair of named entities (Person, Place and Organisation categories). The relation descriptor is defined as the text chunks that describe the explicit relation occurring between these entities in a sentence. We also show that the extracted relation descriptors can be better structured. Specifically, we present a way for organizing the triples resulting from the extraction of open relations between named entities, from the analysis of patterns between them.

This work is organized as follows. In Sect. 2, we present related work on Open Relation Extraction. The Relation Extraction method that we applied is described in Sect. 3. In Sect. 4, we describe the experiments and results obtained. A proposal for structuring the open relations is presented in Sect. 5. Finally, Sect. 6 presents some concluding remarks.

2   Related Work

Traditional Relation Extraction Systems take as input relation names, labeled examples of the relations, and a corpus. In the Open IE task, relation names are not known in advance [5]. The sole input to an Open IE system is a corpus, along with a small set of relation-independent heuristics, which are used to learn a general model of extraction for all relations at once [6].

In general, Open IE systems aim at extracting a large set of related triples (E1, Rel, E2) from a certain corpus without requiring human supervision, where

E1 and E2 are strings meant to denote entities or noun phrases, and Rel is a string meant to denote a relation between E1 and E2.

The first Open IE Sytem was the TextRunner [5], which used a Naive Bayes classifier with POS and NP-chunk features. Banko and Etzioni [6] present the O-CRF system. The authors show that many relations can be categorized using a compact set of lexicon-syntactic patterns. An approach to Open IE which uses Wikipedia as a source of training data is proposed in [31]. They present the WOE system (Wikipedia-based Open Extractor), which generates relation-specific training examples by matching Wikipedia Infobox content with its corresponding patterns. In [15] the ReVerb Open IE system is presented, which is based on syntactic and lexical heuristics that identify verbs expressing relations in English.

Portuguese relation extraction has been boosted mainly by the HAREM evaluation contest. The first HAREM [29] dealt with Named Entity Recognition (NER) mainly whereas the relation extraction task appeared in the Second HAREM in 2008, in the ReRelEM[1] track [9]. A great deal of the literature in this area refers to this evaluation contest. Three relation extraction systems took part in the ReRelEM track [7,8,10], however the relation types were previously defined and labeled relation instances were available. The REMBRANDT system [8] was developed to recognize all categories of named entities and relations between them (*"identity"*, *"inclusion"*, *"placement"* and *"other"*). The SeRELeP system [7] aimed at recognizing three relations: *"identity"*, *"inclusion"* and *"placement"*. SEI-Geo [10] dealt only with the Place category and its relations.

As stated before, there are many Open IE systems for English language. In contrast, there are very few proposals for Portuguese [18,28,30]. A multilingual dependency-based Open IE system (DepOE) has been proposed in [18], it was used to extract triples from the Wikipedia in four languages: Portuguese, Spanish, Galician and English. Santos et al. [28] present the News2Relations system for extracting open relations from titles of news written in Portuguese, dealing with relations of the type (subject, verb, object). In [27,30], the RePort system is presented, it is based on the ReVerb system [15] for English. RePort used syntactic and lexical rules adapted for Portuguese, linguistic knowledge and lexicon of verbal relations extracted from Portuguese corpus. The author also applies data mining to extract other relations, already present, but previously unknown. In [17] a multilingual rule-based Open IE system (ArgOE) is proposed. It is configured for English, Spanish, French, Galician and Portuguese.

In our previous work [12], we extracted relations between named entities in the Organisation domain, using CRF with the BIO schema. Different feature configurations for CRF based on lexical, syntactic and semantic information have been evaluated, based on a subset of HAREM corpus. After that we evaluated IO and BIO representations and concluded that there were advantages in using IO [11]. Now, we consider IO encoding to represent the relation descriptors for pairs of organization and persons or places. The manual annotation went trough

---

[1] Recognition of Relation between Named Entities.

a new round of revision, aiming concordance among the annotators. We extract descriptors that express any type of relation, and we explore these extracted descriptors considering different mining configurations [1,2,30].

3   Relation Extraction Method

In this work, we adopt Conditional Random Fields model to extract relation descriptors between pairs of named entities. Conditional Random Fields are undirected graphical models trained to maximize the conditional probability of a finite set of labels L given a set of input observations o [22]. This probabilistic model is considered to be highly effective to solve the sequence labeling problem.

We applied the linear-chain CRF, which is when output nodes of the graphical model are linked by edges in a linear chain. According to the definition of linear-chain CRF, let $o = (o_1, o_2, ..., o_T)$ be the sequence of observed input data (values on T input nodes); let S be a set of states, in which each state is associated with a label L; and $s = (s_1, s_2, ..., s_T)$ is the sequence of states corresponding to the T output nodes [25].

We consider each word of a sentence as an observation o, which receives a L label as IO notation (I-REL: a word is Inside of a relation descriptor; O: a word is Outside of a relation descriptor) defined in [11]. An illustration is given in Table 1, in which the bold part of the corresponds to the sequence of words indicating the relation descriptor "presidente de" (*president of*), that relates the named entities "Almeida_Henriques" and "Associação_do_Viseu" described in (a). The second column presents the IO encoding. The named entities represent the arguments and receive the label O, because they are not part of the relation descriptor.

Table 1. IO encoding for a fragment of sample sentence described in (a).

Words	Label
Almeida_Henriques	O
,	O
presidente	**I-REL**
de	**I-REL**
o	O
Associação_do_Viseu	O
...	

"*Almeida Henriques,* **presidente da** *Associação do Viseu, é o novo ...*"
(*Almeida Henriques,* **president of** *the Viseu Association, is the new ...*) (a)

Linear-chain CRFs define the conditional probability of state sequence given an input sequence as $p(\mathbf{s}|\mathbf{o})$ [25], described in (1):

$$p(\mathbf{s}|\mathbf{o}) = \frac{1}{Z_o} \exp(\sum_{t=1}^{T} \sum_{k=1}^{K} \lambda_k f_k(s_{t-1}, s_t, \mathbf{o}, t)), \tag{1}$$

where Z_o is the normalization factor over all state sequences; $f_k(s_{t-1}, s_t, \mathbf{o}, t)$ is an arbitrary feature function over its arguments; $\lambda_k \in (-\infty; +\infty)$ is a learned weight for each feature function. In (2) is illustrated that the factor Z_o corresponds to the sum of the scores of all possible state sequences, and the number of state sequences is exponential in the input sequence length T [25].

$$Z_o = \sum_{s} \exp(\sum_{t=1}^{T} \sum_{k=1}^{K} \lambda_k f_k(s_{t-1}, s_t, \mathbf{o}, t)), \mathbf{s} \in S^T \tag{2}$$

Generally, the features functions f_k can ask arbitrary questions about the input sequence, including queries about previous words, next words, and combinations of all these. In this paper, we use specific features for Portuguese described in detail in [12]. An overview of the set of features is presented in Table 2. Feature vectors were generated for pairs of named entities (parameters of the relation) and for the words occurring between them, resulting in a vector with 57 elements for each word.

Table 2. Overview of the set of features.

Features	Explanation
Part-of-Speech	POS tags of the word
Lexical Item	canonic form of the word
Syntactic	syntactic tag of the word head of the segment
Patterns	a noun followed by a preposition a verb
Phrasal Sequence	POS tags of the word sequence between two NEs
Semantic	semantic tag of the word named entity category
Dictionary	list of Person titles/jobs and list of Place words

Table 3. Output of CRF model.

Triples (NE, Relation Descriptor, NE)
(*Almeida_Henriques*<O>, **presidente**<I-REL> de<I-REL>, *Associação_do_Viseu*<O>)

This model is generated from the feature vectors, considering that for every feature a certain weight is attributed, which results in a weight matrix. From this matrix, the CRF model is capable of classifying correctly the words that express an explicit relation in new texts, not tagged yet. An example of the CRF output is presented in Table 3, corresponding to triples extracted from the sentence described in (a).

4   Experiments

In this work we evaluated the performance of the RE task using as measures the number of correct labels ($\#C$), Recall, Precision, and F-measure. For the implementation of the CRF algorithm, we used the NLTK[2] and Mallet[3] libraries, and we applied 10-fold cross validation in the data set described in Sect. 4.1.

We evaluated the performance using two criteria [12]: *exact matching*, when the extracted relation descriptor is exactly the same as the reference; and *partial matching*, when the extracted relation descriptor has at least one word in common with the reference (see Sect. 4.1). An example of the evaluation criteria is presented in Table 4.

Table 4. Examples of exact and partial matching.

Relation instance (reference)	Exact matching	Partial matching
Na *Biblioteca Nacional*, o **presidente da** instituição, *Pedro Corrêa do Lago* (...)	presidente<I-REL> de<I-REL>	presidente<I-REL> de<O>
(In *Biblioteca Nacional*, the **president of** the institution, *Pedro Corrêa do Lago* (...))	president<I-REL> of<I-REL>	president<I-REL> of<O>

We can notice that the relation descriptor "presidente de" (*president of*) is considered complete matching when all sequence of words is annotated, according to the reference[4], with I-REL label, and it is considered partial matching when at least one word matches the reference.

4.1   Data Set

We used a subset of the Golden Collections from the two HAREM conferences[5]. The annotation of the data was performed in two steps: we select the texts from these Golden Collections, and after we add the annotation of relations expressed between named entities (Organisation, Person and Place categories) contained in the selected texts. We only analysed texts that deal with the Organisation domain, such as opinion, journalistic, and politic texts, among others.

We added to these texts the annotation of relation descriptors occurring between pairs of named entities. The annotation was performed by two linguists in the following way: given two named entities occurring in the same sentence, the text sequence (descriptor) that best described an explicit relation between these entities was annotated.

---

[2] http://www.nltk.org/.

[3] http://mallet.cs.umass.edu/.

[4] It is worth noting that preposition-article contraction is split ("da", "do" changes to "de + a", "de + o").

[5] http://www.linguateca.pt/harem/.

A total of 341 relation instances were annotated for the reference data set, organized according to the categories of the pairs of named entities (ORG-PERS, ORG-PLACE). The total number of relation instances and the number of positive and negative instances in each data set are summarized in Table 5.

Table 5. Number of instances of reference data set.

NE categories	Total	Positive	Negative
ORG-PERS	171	95	76
ORG-PLACE	170	97	73
TOTAL	**341**	**192**	**149**

Positive instances are those that have an explicit relation descriptor between two named entities, negative instances are those where the text in between two named entities do not express a relation between them. An example of positive instance is presented in (b), the relation is given in bold according to the reference. An example of negative instance is presented in (c), where the text between Portugal (ORG) and Spain (PLACE) "na" (*in*) does not express a relation between them.

"A *Marfinite* **fica em** *Itaquaquecetuba.*"
(*Marfinite* **is located in** *Itaquaquecetuba.*) (b)

"Sinódio Pais, embaixador de *Portugal* <u>na</u> *Espanha.*"
(Sinódio Pais, ambassador of *Portugal* <u>in</u> *Spain.*) (c)

4.2 Results

In this Section, we present the results of the RE task, considering *exact* and *partial matching*. In Table 6, we show that the developed system classified 113 instances (60 cases of ORG-PLACE, and 53 of ORG-PERS) from a total of 192 positive instances, as *partial matching*. Due to the high number of correct instances, the Recall and F-measure rates are consequently high. Overall, we achieved high rates of Precision in the experiments, seen that CRF is very precise in the process of tagging the relation descriptors, presenting few cases of false positives. As expected, *partial matching* achieved the best results, it occurs due to the difficulty of classifying every element that composes a descriptor. However, in general, it seems that *partial matching* is enough to represent the existing relations. Examples of descriptors classified as *partial matching* are presented in Table 7, where both *partial matching* and reference are presented.

It is difficult to make a comparison with other works, since the resources and data sets are different [3]. In Table 8 we show the results achieved in other open RE systems for Portuguese: RePort [27] and ArgOE [17]. We can see that our results considering any relation descriptors between NEs (open RE) are not distant from these works.

Table 6. Results of the Relation Extraction.

	#C	Recall	Precision	F-measure
Exact matching	81	0.42	0.50	0.46
Partial matching	**113**	**0.58**	**0.71**	**0.64**

Table 7. Examples of output considering *partial matching*.

Relation instance	Partial matching	Reference
As *Forças da Escola de Cavalaria* eram **comandadas pelo** *Salgueiro Maia.*	comandar	comandar por
(*Forças da Escola de Cavalaria* were **commanded by** *Salgueiro Maia.*)	to command	to command by
Na *Biblioteca Nacional*, o **presidente da** instituiçao, *Pedro Corrêa do Lago*	presidente	presidente de
(In *Biblioteca Nacional*, the **president of** the institution, *Pedro Corrêa do Lago*)	president	president of
Goa Tourism Development Corporation Office **organiza excursões à** *Goa*	organizar excursão	organizar excursão a
(*Goa Tourism Development Corporation Office* **organizes excursions to** *Goa*)	to organize excursion	to organize excursion to

Table 8. Results reported by open RE systems.

Works	Recall	Precision	F-measure
CRF (Exact matching)	0.42	0.50	0.46
CRF (Partial matching)	0.58	0.71	0.64
RePort	0.42	0.52	0.46
ArgOE	–	0.53	–

5   Structuring the Open Relations

In this Section, we present a way for organizing the triples resulting of the extraction of open relations between named entities. In the open RE task there is a great diversity of relations, which makes the classification/organization of the relations more difficult. Our approach is based on the mining configurations proposed by [1,2]. Abedjan and Naumann mine a RDF triple structure consisting of a subject, a predicate, and an object. According to their work, a *context* may be any part of the triple, which is used to group one of the two remaining parts as the *target* for mining. A transaction is defined as a set of target elements associated with one context element. Each *context* and *target* combinations is called a *configuration*. In this work, we adapted these concepts, considering *target* as each triple (NE, Relation Descriptor, NE) associated with a *context*, resulting in the configurations described in Table 9.

Table 9. Configurations for different *context* (the *target* being the whole triple).

Configuration	Context
1	NE
2	NE of Place category
3	NE of Person category
4	NE of Organisation category
5	Relation Descriptor

Table 10. Configuration examples.

Configuration	Transaction
(1) *Context: NE*	*Target: Triple*
Portugal	(Legião da Boa Vontade, implantação, **Portugal**)
	(Ministro de Negócios, embaixador de, **Portugal**)
	(PSD, em, **Portugal**)
(2) *Context: NE Place*	*Target: Triple*
Portugal	(Legião da Boa Vontade, implantação, **Portugal**)
	(PSD, em, **Portugal**)
(3) *Context: NE Person*	*Target: Triple*
Almeida Henriques	(**Almeida Henriques**, presidente de, Associação do Viseu)
	(**Almeida Henriques**, de, Conselho Empresarial do Centro)
(4) *Context: NE Organisation*	*Target: Triple*
Creative Commons	(**Creative Commons**, em, Brasil)
	(Ronaldo Lemos, diretor de, **Creative Commons**)
(5) *Context: Relation Descriptor*	*Target: Triple*
presidente de	(Almeida Henriques, **presidente de**, Associação do Viseu)
	(Fernando Gomes, **presidente de**, Câmara do Porto)
	(Antônio Nunes, **presidente de**, Autoridade de Segurança)

In Table 10 we show examples of the five configurations applied in the output of the model. Considering the NE "Portugal", in Configuration (1), we can see three mined transactions, and in Configuration (2) only two were found, that is due to fact that "Portugal" can be classified as PLACE or as ORG, depending on the situation. In configurations (1) to (4), the mined triples express every relation involving the NE in general or in a determined category. Finally, configuration (5) clusters all the named entities involved in a common relation.

6   Concluding Remarks

This paper presented the extraction and structuring of open relations between named entities previously defined (Person, Place and Organisation). We extract relation descriptors that express explicit relations between these entities. In general, previous Portuguese systems do not make use of the machine learning approach, and the relations are specified in advance. We evaluated the CRF classifier considering *exact* and *partial matching*, regarding a reference data set. The best results were achieved for *partial matching*, this occurred due to the difficulty to identify all elements that compose a relation descriptor. We organized the triples resulting from the extraction of open relations between named entities, considering different mining configurations. The relation descriptors organized in these configurations can be useful to classify relation types, to cluster the named entities involved in a common relation, and to populate relational databases, among other uses. In future work, we plan to explore larger data sets.

Acknowledgments. We thank the CNPQ, CAPES and FAPERGS for their financial support.

References

1. Abedjan, Z., Naumann, F.: Context and target configurations for mining RDF data. In: Proceedings of the 1st International Workshop on Search and Mining Entity-Relationship Data, SMER 2011, New York, USA, pp. 23–24 (2011)
2. Abedjan, Z., Naumann, F.: Improving rdf data through association rule mining. Datenbank-Spektrum **13**(2), 111–120 (2013)
3. Abreu, S.C., Bonamigo, T.L., Vieira, R.: A review on relation extraction with an eye on Portuguese. J. Braz. Comput. Soc. **19**(4), 553–571 (2013)
4. Agichtein, E., Gravano, L.: Snowball: extracting relations from large plain-text collections. In: Proceedings of the Fifth ACM Conference on Digital Libraries, pp. 85–94. ACM Press (2000)
5. Banko, M., Cafarella, M.J., Soderl, S., Broadhead, M., Etzioni, O.: Open information extraction from the web. In: Proceedings of the 20th International Joint Conference on Artifical Intelligence, IJCAI 2007, pp. 2670–2676. Morgan Kaufmann Publishers Inc., San Francisco (2007)
6. Banko, M., Etzioni, O.: The tradeoffs between open and traditional relation extraction. In: McKeown, K., Moore, J.D., Teufel, S., Allan, J., Furui, S. (eds.) Proceedings of ACL 2008: HLT, pp. 28–36. Association for Computational Linguistics, Columbus (2008)
7. Brucksen, M., Souza, J.G.C., Vieira, R., Rigo, S.: Sistema serelep para o reconhecimento de relações entre entidades mencionadas. In: Mota, C., Santos, D. (eds.) Segundo HAREM, chap. 14, pp. 247–260. Linguateca (2008)
8. Cardoso, N.: Rembrandt - reconhecimento de entidades mencionadas baseado em relações e análise detalhada do texto. In: Mota, C., Santos, D. (eds.) Segundo HAREM, chap. 11, pp. 195–211. Linguateca (2008)

9. Carvalho, P., Oliveira, H.G., Mota, C., Santos, D., Freitas, C.: Segundo harem: modelo geral, novidades e avaliação. In: Mota, C., Santos, D. (eds.) Desafios na avaliação conjunta do reconhecimento de entidades mencionadas: O Segundo HAREM (2008)

10. Chaves, M.S.: Geo-ontologias e padrões para reconhecimento de locais e de suas relações em textos: o sei-geo no segundo harem. In: Mota, C., Santos, D. (eds.) Segundo HAREM, chap. 13, pp. 231–245. Linguateca (2008

11. Collovini, S., de Bairros Filho, M., Vieira, R.: Analysing the role of representation choices in Portuguese relation extraction. In: Mothe, J., Savoy, J., Kamps, J., Pinel-Sauvagnat, K., Jones, G.J.F., SanJuan, E., Cappellato, L., Ferro, N. (eds.) CLEF 2015. LNCS, vol. 9283, pp. 105–116. Springer, Switzerland (2015)

12. Collovini, S., Pugens, L., Vanin, A.A., Vieira, R.: Extraction of relation descriptors for Portuguese using conditional random fields. In: Bazzan, A.L.C., Pichara, K. (eds.) IBERAMIA 2014. LNCS, vol. 8864, pp. 108–119. Springer, Heidelberg (2014)

13. Culotta, A., McCallum, A., Betz, J.: Integrating probabilistic extraction models and data mining to discover relations and patterns in text. In: Proceedings of the Main Conference on HLT-NAACL, HLT-NAACL 2006, pp. 296–303. Association for Computational Linguistics, Stroudsburg (2006)

14. Culotta, A., Sorensen, J.: Dependency tree kernels for relation extraction. In: Proceedings of the 42nd Meeting of the Association for Computational Linguistics (ACL 2004), Main Volume, Barcelona, Spain, pp. 423–429 (2004)

15. Fader, A., Soderland, S., Etzioni, O.: Identifying relations for open information extraction. In: Proceedings of Empirical Methods in Natural Language Processing, EMNLP, pp. 1535–1545 (2011)

16. Freitag, D., Mccallum, A.: Information extraction with HMM structures learned by stochastic optimization. In: Proceedings of the Seventeenth National Conference on Artificial Intelligence, pp. 584–589. AAAI Press (2000)

17. Gamallo, P., Garcia, M.: Multilingual open information extraction. In: Pereira, F., Machado, P., Costa, E., Cardoso, A. (eds.) EPIA 2015. LNCS, vol. 9273, pp. 711–722. Springer, Heidelberg (2015)

18. Gamallo, P., Garcia, M., Fernández-Lanza, S.: Dependency-based open information extraction. In: Proceedings of the Joint Workshop on Unsupervised and Semi-Supervised Learning in NLP, pp. 10–18. Association for Computational Linguistics, Avignon (2012)

19. Hasegawa, T., Sekine, S., Grishman, R.: Discovering relations among named entities from large corpora. In: Proceedings of the 42nd Annual Meeting on Association for Computational Linguistics (ACL 2004), pp. 415–422. Association for Computational Linguistics, Morristown (2004)

20. Jurafsky, D., Martin, J.H.: Speech and Language Processing: An Introduction to Natural Language Processing, Computational Linguistics and Speech Recognition. Prentice Hall Series in Artificial Intelligence, 2nd edn. Pearson Education Ltd., London (2009)

21. Kambhatla, N.: Combining lexical, syntactic, and semantic features with maximum entropy models for information extraction. In: Proceedings of 42nd Annual Meeting of the Association for Computational Linguistics, pp. 178–181. Association for Computational Linguistics, Barcelona (2004)

22. Lafferty, J.D., McCallum, A., Pereira, F.C.N.: Conditional random fields: probabilistic models for segmenting and labeling sequence data. In: Proceedings of the Eighteenth International Conference on Machine Learning, ICML 2001, pp. 282–289. Morgan Kaufmann Publishers Inc., San Francisco (2001)

23. Li, H., Bollegala, D., Matsuo, Y., Ishizuka, M.: Using graph based method to improve bootstrapping relation extraction. In: Gelbukh, A. (ed.) CICLing 2011, Part II. LNCS, vol. 6609, pp. 127–138. Springer, Heidelberg (2011)
24. Li, Y., Jiang, J., Chieu, H.L., Chai, K.M.A.: Extracting relation descriptors with conditional random fields. In: Proceedings of 5th International Joint Conference on Natural Language Processing, pp. 392–400. Asian Federation of Natural Language Processing, Chiang Mai (2011)
25. Mccallum, A.: Efficiently inducing features of conditional random fields. In: Proceedings of Uncertainty in Artificial Intelligence, pp. 403–410. Morgan Kaufmann, San Francisco (2003)
26. Mota, C., Santos, D., Ranchhod, E.: Avaliação e reconhecimento de entidades mencionadas: princípio do harem. In: Santos, D. (ed.) Avaliação Conjunta: um Novo paradigma no Processamento Computacional da Língua Portuguesa, chap. 14, pp. 161–176. IST Press (2007)
27. Pires, J.C.B.: Extração e mineração de informação independente de domínios da web na língua Portuguesa. Master's thesis, Universidade Federal de Goiás, Goiânia (2015)
28. Santos, A.P., Ramos, C., Marques, N.C.: Extração de Relações em Títulos de Notícias Desportivas. In: INFORUM 2012, Simpósio de Informática, Lisbon, Portugal (2012)
29. Santos, D., Cardoso, N.: Breve introdução ao HAREM, chap. 1, pp. 1–16. Linguateca (2007)
30. Santos, V., Pinheiro, V.: Report - um sistema de extração de informações aberta para língua Portuguesa. In: Proceedings of the X Brazilian Symposium in Information and Human Language Technology (STIL). SBC, Natal (2015)
31. Wu, F., Weld, D.S.: Open information extraction using wikipedia. In: Proceedings of the 48th Annual Meeting of the Association for Computational Linguistics, ACL 2010, pp. 118–127. Association for Computational Linguistics, Stroudsburg (2010)
32. Zhao, S., Grishman, R.: Extracting relations with integrated information using kernel methods. In: Proceedings of the 43rd Annual Meeting of the Association for Computational Linguistics (ACL 2005). The Association for Computer Linguistics (2005)

Towards *Keyphrase* Assignment for Texts in Portuguese Language

Raquel Silveira$^{(\boxtimes)}$, Vasco Furtado, and Vládia Pinheiro

Programa de Pós-Graduação em Informática Aplicada,
Universidade de Fortaleza, Av. Washington Soares, 1321,
Fortaleza, Ceará, Brazil
`raquel.vsilveira@hotmail.com,`
`{vasco,vladiacelia}@unifor.br`

Abstract. *Keyphrase* assignment has often been confounded with *keyphrase* extraction, since the basic hypothesis is that a *keyphrase* of a text must be extracted from this text. Typically, *keyphrase* extraction approaches use a training set restricted to textual terms, reducing the learning capabilities of any inductive algorithm. Our research investigates ways to improve the accuracy of the *keyphrase* assignment systems for texts in Portuguese language by allowing classification algorithms to learn from non-textual terms as well. The basic assumption we have followed is that non-textual terms can be included into the training set by inference from an eventual semantic relationship with textual terms. In order to discover the latent relationship between non-textual and textual terms, we use deductive strategies to be applied in Portuguese common sense bases such as Wikipedia and InferenceNet. We show that algorithms that follow our approach outperform others that do not use the same methods introduced here.

Keywords: *Keyphrase* extraction · *Keyphrase* assignment · Semantic annotation · Information retrieval

1 Introduction

The task of assigning a text with *keyphrases* is important because they enable text categorization [1], advertising [2], or simply for the purpose of summarizing the content to allow a rapid understanding of the subject matter [3]. This task, when done manually, is tedious and time consuming. When there is a need to consolidate a pre-defined vocabulary, this activity is non-trivial and its automation becomes mandatory.

Traditionally, automatic *keyphrase* extraction concerns "the automatic selection of important and topical phrases from the body of a document" [4]. Its goal is to extract a set of phrases that are related to the main topics discussed in a given document [5]. In fact, the task of *keyphrase* assignment (discovery of keyphrases contained or no in the text) has often been confounded with *keyphrase* extraction, whose basic hypothesis is that a *keyphrase* of a text must be extracted from this text.

J. Silva et al. (Eds.): PROPOR 2016, LNAI 9727, pp. 165–176, 2016.
DOI: 10.1007/978-3-319-41552-9_17

Our preliminary analysis from a corpus of news in Portuguese with *keyphrases* assigned by humans has shown that approximately 20 % of them are not in the text. Lately we have fortified the conclusions reached in the preliminary study by exploring a corpus of thesis and dissertations abstracts in Portuguese, which showed us that 55 % of the keyphrases assigned by the authors are not found in the text.

The literature of automatic extraction of *keyphrases* is dominated by inductive learning (typically, classification). This kind of learning discovers patterns based on examples composed of statistical, structural and syntactic features of textual terms such as their frequency, their topological position in the text, and external resource-based features computed based on information gathered from resources other, such as knowledge bases (e.g., Wikipedia), with the goal of exploiting external knowledge [5]. Each example is associated as a positive or negative case of a *keyphrase*.

A natural way to deal with the limitation of the traditional *keyphrase* extraction methods of using only textual terms is to augment the training set with examples of non-textual terms. However, two challenges appear in this way. The first is how to select non-textual terms from an external resource (e.g. a knowledge base or a lexicon) and the second is how to balance the set of examples, since the number of terms considered *keyphrases* and those that are not is often uneven. Silveira et al. [6] proposes a method that allows classification algorithms to learn also from non-textual terms. The basic assumption is that the non-textual terms to be included in the training set must have a semantic relationship with textual terms.

In order to discover the latent relationship between not only non-textual but also textual terms, we propose to apply semantic resources in Portuguese language, such as Wikipedia [7] and InferenceNet [8] for inferring the non-textual keyphrase candidates, aiming to leverage the *keyphrase* assignment task for Portuguese-language applications. Additionally, we use a new feature Inference Semantic Relatedeness (ISR) [6] to capture whether the semantic relatedness of a non-textual term nt_j is higher than the semantic relatedness of a textual term t_i (in respect to other terms of the text). The intuition is that terms that are more semantically related to other terms in the document are more likely to be *keyphrase*. Besides this, we apply techniques for balancing the training set and compare with an unbalanced training set.

Our approach was validated by means of a comparison of the f-measure (i.e. recall and precision) against the f-measure of state-of-the-art systems that do not use the strategy we have followed, in two corpus in Portuguese – a News corpus with news from a major Brazilian Web Newspaper about crime, sports and politics; and a Scientific corpus, with abstracts of thesis and dissertations from University of São Paulo (USP). The experiments allowed us to evaluate the impact of the quality and coverage of the knowledge base, of the new feature ISR and of the balancing methods in the *keyphrase* assignment task. The results demonstrated that the accuracy improves around 16 % in the best scenario - the corpus of crime news and the best coverage situation of the InferenceNet (27 % of non-textual keyphrases).

2 Background Knowledge

2.1 InferenceNet – Common Sense Knowledge Base

The InferenceNet database (www.inferencenet.org) [8] contains the inferential and common sense content of concepts of the Portuguese and English languages and have been used in several NLP tasks [9]. Currently, InferenceNet.BR 2.0 contains 230000 concepts related by 750000 semantic relations divided according to the nature of the relationship: affective – 6 %; agents – 32 %; causal – 2 %; events – 6 %; functional – 29 %; spatial – 10 %; hyperonymy and synonymy- 15 %.

The inferential content of InferenceNet consists of classifying the semantic relations according to the inferentialist view [10], which determines that the use of a concept in inferences or potential inferences in which this concept may participate are: (i) its pre-conditions or premises of use: what gives someone the right to use the concept and what could exclude such a right, serving as premises for utterances and reasoning; and (ii) its post-conditions or conclusions of use: what follows or what are the consequences of using the concept, which let one know what someone is committed to by using a particular concept, serving as conclusions from the utterance *per se* and as premises for future utterances and reasoning.

Formally, InferenceNet is represented in a directed graph $G_c(C, Rc)$. Each inferential relationship $rc_j \in Rc$ is represented by a tuple (*relationName*, c_i, c_k, *type*), where *relationName* is the name of an semantic relation (*CapableOf*, *PropertyOf*, *EffectOf* etc.), c_i and c_k are concepts C of a natural language, and *type* = "Pre" or "Pos" (pre-condition or post-condition for using the concept c_j). For example, the tuple (*EffectOf*, *crime*, *violence*, Pos) expresses a causal relationship between the concepts *crime* and *violence* that is a post-condition of using the concept *crime*.

2.2 Traditional *Keyphrase* Extraction

Classical approaches that recognize keywords in a text are based on statistical, structural, and syntactic features of textual terms and/or features based on knowledge bases. The most common features used in Machine Learning to determine whether a term is likely to be a *keyphrase* are described in [11] – TFxIDF, position of the first occurrence, *keyphraseness*, semantic relatedness, phrase length, spread, wikipedia's *keyphraseness,* inverse Wikipedia linkage, and node degree. This set of features has been traditionally used and have shown their relevance in *keyphrase* extraction.

Wikify! [10] uses a measure called *keyphraseness*, calculated as the number of times the term is a text anchor in links to Wikipedia's internal pages divided by the overall frequency of the term in Wikipedia. This measure quantifies the probability of the term appearing as the tag on Wikipedia and was inspired in the editorial style recommended to Wikipedia's editors. With this strategy to select *keyphrases*, Wikify!'s results showed to be 10 % better when compared with TFxIDF.

A large amount of research work applies machine learning for *keyphrase* extraction from manually assigned *keyphrases* by humans [11]. KEA [14] regarded *keyphrase* extraction as a classification task in which the candidate terms are represented by three features: TFxIDF, position of the first occurrence and *keyphraseness*. KEA uses the Naive Bayes classifier to learn probability that can be used to rank candidates.

Milne and Witten [15] developed a method based on machine learning to reference documents to Wikipedia articles. From disambiguated terms, the procedure extracts statistical evidence, information about the topology of text and information about the meaning of the words of Wikipedia pages to train classifiers. Terms that have links to other Wikipedia pages are presented as positive examples of *keyphrases*, while those that appear only as text are displayed as negative examples. The classifiers try to learn the editorial style adopted in Wikipedia.

MAUI system [11] enhances KEA's successful machine-learning framework with semantic knowledge retrieved from Wikipedia, new features, and a new classification model. In the candidate selection stage, all n-grams up to a maximum length of 3 words that do not begin or end with a stop-word and that appear more than once are selected as candidate tags. Maui adds six new features to KEA's feature set, in which three have not been evaluated before: spread, semantic relatedness, and inverse Wikipedia linkage. Maui combines nine features amongst which there are many obvious relationships, e.g. first occurrence and spread, or node degree and semantic relatedness. The best results are obtained by combining all nine features, again using bagged decision trees, giving a notably improved F-Measure of 47.1 %, whereas the baseline system KEA (three features+Naive Bayes) achieves an F-Measure of 42.1 %.

Graph-based ranking methods follow an unsupervised approach for *keyphrase* extraction. Mihalcea and Tarau [16] proposed a model called *TextRank*, which builds a graph that represents the text and interconnects words or other text entities with meaningful relations. The principle of the model is the fact that when one vertex links to another one, it is basically casting a vote for that other vertex. The higher the number of votes that are cast for a vertex, the higher the importance of the vertex, and this information is also taken into account by the ranking model.

Grineva et al. [17] follow a similar approach by building a graph and analyzing the interconnectivity of the vertices. However, instead of using statistics gathered from a training set, they use semantic information derived from Wikipedia. They applied unsupervised machine learning to uncover communities in the graph. By doing so, it is possible to rank term communities in a way that the highest ranked communities would contain terms semantically related to the main topics of the document (key terms).

All of the works presented here consider the terms that appear in the text. In experiments done on two corpora (news and scientific articles), we identified that around 21 % and 55 % of *keyphrases* tagged by humans, respectively, not appears in the texts. *Keyphrase* extraction methods, that do not consider non-textual *keyphrases,* are not comparable with those assigned by the best performing human taggers.

2.3   Metrics for *Keyphrase* Relevance

Statistical, topological and semantic features capture how elements can be related to determine whether a term is relevant to be a *keyphrase*. We use some features available in the literature to *keyphrase* extraction, described in Table 1.

Table 1. Features for *keyphrase* relevance.

Feature	Description
Position of the first occurrence	Is computed as the relative distance of the first occurrence of the term from the beginning of the document [14]
Phrase length	Is measured in words. Generally, the longer the term, the more specific it is [11]
Spread	Is the distance between its first and last occurrences in a document. Both values are computed relative to the length of the document [11]
TFxIDF	Attributes to the terms a weight proportional to the frequency of occurrence in the document and inversely proportional to the occurrence of the term in corpus [18]
Keyphraseness	Quantifies how often a term appears as a keyphrase in the training corpus [14]
Semantic relatedness	Quantifies the semantic relatedness of a term to others. As with MAUI's Semantic relatedness [11], the semantic similarity of a given term is the average number of hyperlinks in common with other candidate terms from the same document
Wikipedia's keyphraseness	Is the likelihood of a term being a link in the Wikipedia corpus. It divides the number of Wikipedia pages in which the term appears in the anchor text of a link by the total number of Wikipedia pages containing it [11]
Inverse Wikipedia Linkage (IWL)	Counts the number of other Wikipedia articles that link to the most likely Wikipedia article for a given term, and normalizes this value as in inverse document frequency: $IWL = -log_2 \frac{linksTo(A_p)}{N}$. Where $linksTo(A_P)$ is the number of incoming links to the article A representing the candidate to *keyphrase* P, and N the total number of links in the Wikipedia snapshot [11]
Node degree	Quantifies the semantic relatedness of a term to other candidates terms. The node degree of a term is the number of hyperlinks that connect it to other Wikipedia pages that have been identified for another candidate of the document [11]

3   An Approach for Keyphrase Assignment

The task of identifying *keyphrases* can be modeled as the problem of selecting—from a list of candidate terms $C = \{c_1, c_2, ..., c_n\}$—the subset $K = \{k_1, k_2, ..., k_m\}$ terms that summarize the contents of the document $d \in D$, where D is a collection of documents. In this paper, we propose a deductive method to generate part of the set of candidates C. Figure 1 illustrates the steps that our method follows.

First, the set $T = \{t_1, t_2, ..., t_i\}$ of textual candidates is generated. T is formed by tokens that are used in the document d. In the Natural Language Preprocessing step, for any d in D, the text of d is processed for exclusion of punctuation marks. The remaining words are submitted to *tokenizers* and morphological filters for radical reduction and determination of the morphological class of the terms. Then all possible

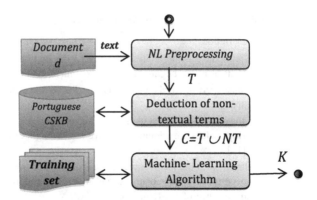

Fig. 1. Deduction of Non-textual Candidates from a Portuguese Common-Sense Knowledge Base (CSKB) and Learning Keyphrases of a document d.

n-grams up to a maximum length of 3 words that do not begin or end with a stop word are selected as textual candidates, forming the set $T = \{t_1, t_2, ..., t_i\}$ of textual terms.

Additionally, a deductive approach generates a set $NT = \{nt_1, nt_2, ..., nt_j\}$ of non-textual candidate terms ($NT \nsubseteq T$), i.e., candidate terms that are not within the text. We assume that it is possible to find, for any nt_j, a semantic relationship with one or more terms t_i from D. This semantic relationship is represented in a Common Sense Knowledge Base (CSKB). The complete set of candidates $C = T \cdot NT$ contains terms that appear in any document of the training set as well as non-textual terms ($nt_j \nsubseteq d$) having a semantic relationship with d.

For example, Fig. 2 shows a text about sports with the corresponding *keyphrases*. The set of textual candidates T is formed by the terms that are in the text such as, those referring to the name of teams and players (**Barcos**, **Palmeiras** and **Brazilian Championship**). In this example, a *keyphrase* that is not explicitly represented in the text refers to the type of sport. A direct relationship between the name of the teams and/or players might help to retrieve from the knowledge base a non-textual term *(in this case Football)*, indicating that this term would be a potential non-textual keyphrase candidate, forming the set non-textual candidate terms *NT*.

Barcos is one of the most important players of **Palmeiras** and his departure from the team would be something very harmful. Therefore, the top scorer of the team in the **Brazilian Championship** with 11 goals and the season, with 25, admitted he has taken care not to take a new yellow card and thus be suspended. "I'm taking care for several games not to take more cards. I will keep calm and avoid any trouble and complaint," said the striker, who won the last card on 22 September, in the victory by 3 to 1 over Figueirense.

Keyphrases: Football, Brazilian Championship, Palmeiras, Barcos.

Fig. 2. An example of sport news and its *keyphrases*.

In the inductive stage, several features are computed for each candidate c_n, which are then input to a machine-learning algorithm to obtain the probability that the candidate is indeed a *keyphrase* in K. To take advantage of the semantic information, it is necessary to connect the textual terms with the concepts belonging to some CSKB. The use of semantic knowledge requires recognizing the sense of the candidate term c_i in a document d. However, it happens that some terms of d may present more than one possible meaning in the base. For this reason, the task of Word Sense Disambiguation must be carried out prior to the extraction of *keyphrases*. We have used the method proposed in [19] that basically verifies the similarity of two concepts from the viewpoint of conceptual proximity in the Wikipedia hierarchy and the proximity between such concepts in terms of the inferences that they can make.

3.1 Deducing and Selecting Non-textual Candidate Terms from InferenceNet

As defined above, the set of candidate *keyphrases* C contains terms (in the form of tokens) that appear in a document d, and non-textual terms that have a semantic relationship with terms that are in d. It is important to note that the set NT can initially provide a large number of tokens, since there are numerous words that can be considered semantically related to a textual token of T.

These textual terms in T are then sent to the step of deduction of non-textual terms. As explained, InferenceNet expresses the semantic relations between two concepts by means of a triplet $c_i, \overset{n}{\overset{\frown}{\Phi}}, c_k$, where c_i and c_k are terms in portuguese and $\overset{n}{\overset{\frown}{\Phi}}$ represents the path of size n that links the two terms by means of relationships between them. The relationships can be of several types, such as *capableOf*, *propertyOf*, *effectOf*, *isA*, *partOf*, etc. In this step, for each textual term from T, one searches the triplets of $n = 1$ with form $x, \overset{1}{\overset{\frown}{\Phi}}, t_i$ or $t_i, \overset{1}{\overset{\frown}{\Phi}}, x$. If $x \notin T$ then x is a non-textual candidate term and is included in the set NT. Formally, this can be written in a logical way by means of inference rules: $\dfrac{x, \overset{1}{\overset{\frown}{\Phi}}, t_i, (t_j \in T), (x \notin T)}{x \in NT}$; $\dfrac{t_i, \overset{1}{\overset{\frown}{\Phi}}, x, (t_j \in T), (x \notin T)}{x \in NT}$.

The intuition behind this heuristic is that non-textual terms, which are semantically related to textual terms by means of a direct relationship ($n = 1$), are potential candidates for *keyphrases*. Of course, it is up to the inductive approach to learn what characterizes a good example of a *keyphrase*. However restricting the training set for only textual terms reduces the learning capabilities of any inductive algorithm.

For each training document, candidate terms in C of a document d are identified and their feature values are calculated. Each candidate is then marked as a positive or negative example, depending on whether users have assigned it as a tag to the corresponding document.

Not all non-textual terms inferred from textual terms deserve to be considered a good candidate to *keyphrase*. We consider that the Semantic Relatedness of a term is a good indication of this. In order to characterize the quality of the non-textual terms, we use a boolean semantic feature called *Inference Semantic Relatedness (ISR)* [6]. This feature is intended to capture whether the semantic relatedness of term nt_j (in respect to other terms of the text) is higher than the semantic relatedness of term t_i. The intuition is that terms that are more semantically related to other terms in the document are more likely to represent the text.

For example, let us consider a text about crime without a mention to the type of crime reported but containing the textual terms *handgun* and *death*. These terms *handgun* and *death* can infer, from a knowledge base, the non-textual term *murder* through a direct relation. If *murder* has a higher semantic relatedness with the text than *handgun* and *death*, then it is a better *keyphrase* of the text.

Finally, it is important to point out that a non-textual term can be inferred from n textual terms. We considered that the textual term t_i that best represents the deduction relationship is one that is more significant to the document. Therefore, given n textual terms that infer one non-textual term, the term with higher TFxIDF is chosen.

4   Experimental Evaluation

In this section, we show the results generated by our approach and by KEA [14] and MAUI [11], using Random Forest as a classifier. It is important to point out that the choice of classifier is not determinant to validate our approach, since our contribution is to boost the accuracy of automatic assignment of *keyphrases* for texts in Portuguese by qualifying the training set with non-textual terms from a Portuguese Common Sense Knowledge Bases – Wikipedia and InferenceNet.

We used two corpora in Portuguese to comparatively evaluate our approach with KEA and MAUI, two state-of-the-art machine learning-based *keyphrase* extraction systems. The first one is a news corpus containing three domains: *crime, sports* and *politics,* with a total of 493 news articles collected from a major Brazilian Web Newspaper. A total of 2,778 *keyphrases* were assigned to the documents by journalists, of which 20.3 % of the *keyphrases* do not appear in the texts. The second is a corpus with abstracts of thesis and dissertation from University of São Paulo (USP), with a total of 2000 documents collected from USP website. A total of 9,453 *keyphrases* were assigned to the documents by authors, of which 55,4 % of the *keyphrases* do not appear in the texts. Table 2 presents the basic statistics of the datasets and the percentage of non-textual tags in each corpus.

Table 2. Statistics of News and Scientific Corpora.

Corpus	Docs	Textual *keyphrases*	Non-textual *keyphrases*
USP Thesis-Dissertations	2000	9,453	5,235
% of non-textual keyphrases			**55,4 %**
News - Crime	206	946	261
News - Sports	186	787	101
News - Politics	101	481	202
% of non-textual keyphrases			**20.3 %**

We set up the experiments with the following resources, tools and parameters:

- Encyclopedic knowledge base – Wikipedia's Brazilian dataset (snapshot in September, 14, 2011 with 1,701,888 articles);

- Common-sense Knowledge Base – *InferenceNet* 2.0 [8];
- Lemmatizer – The Stanford CoreNLP Toolkit [20];
- Weka Framework [21] – The Random Forest algorithm.

The accuracy of our approach depends on the coverage of the Portuguese CSKB (with relationships between non-textual *keyphrases* with textual terms). That is to say, the more the relationship between them is represented in the Portuguese CKSB, the more the training set improves. For evaluating the impact of the coverage of *InferenceNet* in our approach, we tested it with the original relationships that contained 8,3 % of non-textual *keyphrases* to USP Thesis-Dissertations Corpus and to the News Corpus with 27 % to Crime domain, 26 % to Sports and 15 % to Politics. We also varied the number of features that characterize an example. We used three sets of features:

- F1 - has the same features as KEA, with only three features: TFxIDF, First Occurrence, and *Keyphraseness*;
- F2 - has the same features as MAUI, containing five more features (Semantic Similarity, Phrase Length, Spread, Wikipedia's *Keyphraseness*, IWL);
- F3 - has the same features as MAUI, plus the new feature *Inference Semantic Relatedness (ISR)*.

Finally, since *non-keyphrase* terms outnumber the non-textual *keyphrase* terms, we applied three different strategies of balancing the training sets: random over-sampling, Synthetic Minority Over-sampling Technique (SMOTE) [22] and SMOTEBoost [23].

Tables 3, 4, 5, 6 show the comparison of the performance (with respect to f-measure) of the evaluation scenarios for *USP* and News corpora, considering the original knowledge base *InferenceNet*, and the following cases: (i) Datasets Unbalanced; (ii) Datasets Balanced by SMOTE (F1 and F2) and Random Over-sampling (F3) methods; (iii) Datasets Balanced by SMOTEBoost. In the tables, T represents the f-measure calculated only to the textual candidates, NT calculated only to the non-textual candidate terms, and C calculated to the complete set of candidates ($T \cup NT$).

Table 3. Results in terms of F-measure for the USP Thesis-Dissertations corpus.

	K	Unbalanced		Balanced		SMOTEBoost	
		Only text	With NT 8,3 %	Only text	With NT 8,3 %	Only text	With NT 8,3 %
KEA (F1)	T	0.575	0.581	0.535	0.559	0.534	**0.562**
	NT	0.000	0.000	0.000	0.000	0.000	**0.000**
	C	0.334	0.337	0.339	0.341	0.335	**0.342**
MAUI (F2)	T	0.583	0.589	0.605	0.595	0.597	**0.580**
	NT	0.000	0.003	0.000	0.009	0.000	**0.013**
	C	0.339	0.344	0.370	0.378	0.370	**0.382**
MAUI + ISR (F3)	T	0.583	0.585	0.605	**0.600**	0.597	0.576
	NT	0.000	0.002	0.000	**0.014**	0.000	0.011
	C	0.339	0.340	0.370	**0.386**	0.370	0.372

Table 4. Results in terms of F-measure for the News corpus – Crime domain.

	K	Unbalanced		Balanced		SMOTEBoost	
		Only text	With NT 27 %	Only text	With NT 27 %	Only text	With NT 27 %
KEA (F1)	T	0.549	0.566	0.586	0.590	0.593	**0.596**
	NT	0.000	0.176	0.000	0.185	0.000	**0.180**
	C	0.482	0.511	0.518	0.536	0.528	**0.546**
MAUI (F2)	T	0.604	0.640	0.655	0.654	0.663	**0.664**
	NT	0.000	0.227	0.000	0.259	0.000	**0.252**
	C	0.530	0.580	0.582	0.602	0.589	**0.612**
MAUI + ISR (F3)	T	0.604	0.640	0.655	**0.668**	0.644	0.643
	NT	0.000	0.227	0.000	**0.278**	0.000	0.248
	C	0.530	0.580	0.578	**0.612**	0.570	0.589

Table 5. Results in terms of F-measure for the News corpus – Sports domain.

	K	Unbalanced		Balanced		SMOTEBoost	
		Only text	With NT 26 %	Only text	With NT 26 %	Only text	With NT 26 %
KEA (F1)	T	0.656	0.664	0.674	0.676	0.664	**0.674**
	NT	0.000	0.362	0.000	0.362	0.000	**0.365**
	C	0.616	0.640	0.617	0.652	0.628	**0.653**
MAUI (F2)	T	0.740	0.739	0.759	0.747	0.762	**0.759**
	NT	0.000	0.362	0.000	0.362	0.000	**0.369**
	C	0.694	0.708	0.716	0.717	0.720	**0.726**
MAUI + ISR (F3)	T	0.740	0.731	0.756	**0.761**	0.746	0.749
	NT	0.000	0.362	0.000	**0.362**	0.000	0.200
	C	0.694	0.701	0.711	**0.730**	0.703	0.712

Table 6. Results in terms of F-measure for the News corpus – Politics domain.

	K	Unbalanced		Balanced		SMOTEBoost	
		Only text	With NT 15 %	Only text	With NT 15 %	Only text	With NT 15 %
KEA (F1)	T	0.569	0.580	0.600	0.597	0.599	**0.604**
	NT	0.000	0.187	0.000	0.187	0.000	**0.191**
	C	0.468	0.505	0.502	0.525	0.511	**0.536**
MAUI (F2)	T	0.684	0.684	0.694	0.696	0.697	**0.696**
	NT	0.000	0.203	0.000	0.199	0.000	**0.200**
	C	0.563	0.589	0.583	0.609	0.585	**0.612**
MAUI + ISR (F3)	T	0.684	0.676	0.694	**0.701**	0.697	0.699
	NT	0.000	0.202	0.000	**0.202**	0.000	0.177
	C	0.563	0.582	0.583	**0.615**	0.585	0.600

In all scenarios and corpus (of different types and domains) we achieved an improvement of around 10 % (on average) compared to MAUI, which does not consider non-textual *keyphrases*. For example, for the News corpus – Crime domain (Table 4), MAUI presents 53 % in f-measure (only textual terms), while our approach achieves 61.2 % (considering non-textual terms and a balanced training set by SMOTE).

5   Conclusion

In this paper we advocate that the task of *keyphrase* extraction for different domains via classification may be improved by qualifying the training set by deducing non-textual candidate terms and by inferring features for them. An empirical evaluation has shown that this strategy has increased the accuracy of *keyphrase* assignment systems for texts in Portuguese Language - an improvement of around 16 % in the best scenario - the corpus of crime news and the best coverage situation of the knowledge base InferenceNet (27 % of non-textual *keyphrases*).

When treating the issues of unbalanced size between sets of *keyphrases* and *non-keyphrases* after the application of our deductive strategy to enrich the training set, we discovered, also by empirical analysis, that the accuracy can also increase largely.

While we have achieved an improvement in the results, it is evident that accuracy rates depend on the strategies of inferring non-textual candidate terms. As future works, we pretend investigate the deduction of non-textual candidates terms related to textual terms in others levels of the semantic network (n > 1). Furthermore, our intention, following the framework exposed here, is to explore the combination of others common sense knowledge bases, such as *ConceptNet* and *WordNet*.

References

1. Hulth, A., Megyesi, B.: A study on automatically extracted keywords in text categorization. In: Proceedings of the 21st International Conference on Computational Linguistics and the 44th Annual Meeting of the ACL 2006, pp. 537–544 (2006)
2. Yih, W., Goodman, J., Carvalho, V.R.: Finding advertising keywords on web pages. In: Proceedings of the 15th International Conference on World Wide Web, WWW 2006, pp. 213–222. ACM, New York (2006). http://dx.doi.org/10.1145/1135777.1135813
3. Zhang, Y., Zincir-Heywood, N., Milios, E.: World Wide Web site summarization. Web Intell. Agent Syst. **2**, 39–53 (2004)
4. Turney, P.: Learning algorithms for keyphrase extraction. Inf. Retrieval **2**, 303–336 (2000)
5. Hasan, K.S., Ng, V.: Automatic keyphrase extraction: a survey of the state of the art. In: Proceedings of the 52nd Annual Meeting of the Association for Computational Linguistics (Volume 1: Long Papers), pp. 1262–1273 (2014)
6. Silveira, R., Furtado, V., Pinheiro, V.: Using non-textual terms for boosting document keyphrase assignment. In: Proceedings of the 2015 IEEE/WIC/ACM International Conference on Web Intelligence, WI 2015 (2015)

7. Li, Y., Bandar, Z.A., Mclean, D.: An approach for measuring semantic similarity between words using multiple information sources. IEEE Trans. Knowl. Data Eng. **15**, 871–882 (2003)

8. Pinheiro, V., Pequeno, T., Furtado, V., Franco, W.: InferenceNet.Br: expression of inferentialist semantic content of the Portuguese language. In: Pardo, T.A.S., Branco, A., Klautau, A., Vieira, R., Lima, V.L.S. (eds.) PROPOR 2010. LNCS, vol. 6001, pp. 90–99. Springer, Heidelberg (2010)

9. Pinheiro, V., Furtado, V., Pequeno, T., Nogueira, D.: Natural language processing based on semantic inferentialism for extracting crime information from text. In: IEEE International Conference on Intelligence and Security Informatics (ISI), pp. 19–24. IEEE (2010)

10. Brandom, R.: Articulating Reasons: An Introduction to Inferentialism. Harvard University Press, Cambridge (2001)

11. Medelyan, O., Frank, E., Witten, I.H.: Human-competitive tagging using automatic keyphrase extraction. In: Proceedings of the 2009 Conference on Empirical Methods in Natural Language Processing, EMNLP 2009, vol. 3, pp. 1318–1327. Association for Computational Linguistics, Stroudsburg (2009)

12. Mihalcea, R., Andra, C.: Wikify!: Linking documents to encyclopedic knowledge. In: CIKM 2007, Lisbon, Portugal, pp. 233–242 (2007)

13. Turney, P.D.: Coherent keyphrase extraction via web mining. In: Proceedings of the 18th International Joint Conference on Artificial Intelligence, IJCAI 2003, pp. 434–439. Morgan Kaufmann Publishers Inc., San Francisco (2003)

14. Frank, E., Paynter, G.W., Witten, I.H., Gutwin, C., Nevill-Manning, C.G.: Domain-specific keyphrase extraction. In: Proceedings of the 16th International Joint Conference on Artificial Intelligence, Stockholm, Sweden. Morgan Kaufmann Publishers, San Francisco (1999)

15. Milne, D., Witten, I.H.: Learning to link with Wikipedia, pp. 509–518. ACM (2008)

16. Mihalcea, R., Tarau, P.: TextRank: Bringing Order into Texts, pp. 404–411. Association for Computational Linguistics, Barcelona (2004)

17. Grineva, M., Grivev, M., Lizorkin, D.: Extracting key terms from noisy and multi-theme documents. In: Proceedings of 18th International Conference on World Wide Web, New York, USA, pp. 661–670 (2009)

18. Salton, G., McGill, M.J.: Introduction to Modern Information Retrieval. McGraw-Hill, New York (1983)

19. Pinheiro, V., Furtado, V., Freire, L.M., Ferreira, C.: knowledge-intensive word disambiguation via common-sense and wikipedia. In: Barros, L.N., Finger, M., Pozo, A.T., Gimenénez-Lugo, G.A., Castilho, M. (eds.) SBIA 2012. LNCS, vol. 7589, pp. 182–191. Springer, Heidelberg (2012)

20. Manning, C.D., Surdeanum, M., Bauer, J., et al.: The stanford CoreNLP natural language processing toolkit. In: Proceedings of 52nd Annual Meeting of the Association for Computational Linguistics: System Demonstrations (2014)

21. Hall, M., Frank, E., Holmes, G., Pfahringer, B., Reutermann, P., Witten, I.H.: The WEKA data mining software: an update. SIGKDD Explor. **11**(1), 10–18 (2009)

22. Chawla, N.V., Bowyer, K.W., Hall, L.O., Kegelmeyer, W.P.: SMOTE: synthetic minority over-sampling technique. JAIR **16**, 321–357 (2002)

23. Chawla, N.V., Lazarevic, A., Hall, L.O., Bowyer, K.W.: SMOTEBoost: improving prediction of the minority class in boosting. In: Lavrač, N., Gamberger, D., Todorovski, L., Blockeel, H. (eds.) PKDD 2003. LNCS (LNAI), vol. 2838, pp. 107–119. Springer, Heidelberg (2003)

Entity Linking with Distributional Semantics

Pablo Gamallo[1](✉) and Marcos Garcia[2]

[1] Centro Singular de Investigación en Tecnoloxías da Información (CiTIUS),
Universidade de Santiago de Compostela, Galiza, Spain
pablo.gamallo@usc.es
[2] Grupo LyS, Dep. de Galego-Português, Francês e Linguística,
Universidade da Coruña, Galiza, Spain
marcos.garcia.gonzalez@udc.gal

Abstract. Entity Linking (EL) consists in linking name mentions in a given text with their referring entities in external knowledge bases such as DBpedia/Wikipedia. In this paper, we propose an EL approach whose main contribution is to make use of a knowledge base built by means of distributional similarity. More precisely, Wikipedia is transformed into a manageable database structured with similarity relations between entities. Our EL method is focused on a specific task, namely semantic annotation of documents by extracting those relevant terms that are linked to nodes in DBpedia/Wikipedia. The method is currently working for four languages. The Portuguese and English versions have been evaluated and compared against other EL systems, showing competitive range, close to the best systems.

Keywords: Entity linking · Semantic annotation · Term extraction

1 Introduction

Entity Linking (EL) puts in relation mentions of entities within a text with their corresponding entities or concepts in an external knowledge resource. Typically, entity mentions are proper names and domain specific terms which can be linked to Wikipedia pages. Most EL methods include three basic subtasks: (i) extraction of the terms likely to be entity mentions in the input text, by using Natural Language Processing (NLP) techniques such as tokenization and multiword extraction; (ii) selection of the entity candidates: each mention is associated to a set of entities in the external resource; and (iii) selection of the best entity candidate for each mention by making use of disambiguation strategies.

In most cases, two types of approaches are suggested for the selection or disambiguation subtask:

1. *Non-collective approaches*, which resolve one entity mention at each time on the basis of local and contextual features. These approaches generally rely on supervised machine learning models [11,14,23,24].

This research has been partially funded by the Spanish Ministry of Economy and Competitiveness through project FFI2014-51978-C2-1-R.

J. Silva et al. (Eds.): PROPOR 2016, LNAI 9727, pp. 177–188, 2016.
DOI: 10.1007/978-3-319-41552-9_18

2. *Collective approaches*, which semantically associate a set of relevant mentions by making use of the conceptual density between entities through graph-based approaches [1,2,6,7,12,14,15,17,18,20,21,25,26].

Many applications can benefit from the EL systems, namely educational applications. Text annotated with EL allows students to have fast access to additional encyclopedic knowledge relevant to the study material, by linking proper names and terms to the corresponding pages in Wikipedia or other external encyclopedic sources. In the research community oriented to educational applications, EL is better known as the task of *semantic annotation* [28]. Given a source text, the semantic annotation task is generally restricted to those mentions in the text referring to the same conceptual category. In fact, the main goal of semantic annotation is to semantically categorize a text by identifying the main concepts or subconcepts the text content is about. As a result, only those mentions that are semantically related are annotated in the text with links (e.g., DBpedia URIs) to their corresponding entities/concepts in an external knowledge database. In [22], the authors describe the *DBpedia Spotlight* system, which can be configured to detect topic pertinence. In order to constrain annotations to topically related entities, a higher threshold for the topic pertinence can be set. This way, texts can be annotated by *DBpedia Spotlight* using semantically related entities.

In this article, we will describe an EL system for the task of semantic annotation. For this specific task, the collective approach, which identifies those mentions associated to conceptually related entities, seems to be the most appropriate strategy.

The main drawback of collective approaches is the fact that the conceptual graph generated is too large and very difficult to explore in an efficient and scalable way. The graph can grow dramatically as the set of entities associated to the different mentions in the text is expanded by making use of different types of semantic relations, including the hierarchical ones (hyperonymy).

To minimize this problem, a collective method is proposed and implemented. Our EL method relies on a distributional similarity strategy to select a restricted set of conceptual relations/arcs between entities. In particular, it only selects relations between the most similar entities. Distributional similarity was computed using Wikipedia articles, as in [8]. The conceptual relations between entities that are not similar in distributional terms are removed from the graph. So, the conceptual graph used to search for the entity candidates is dramatically simplified and, then, can be explored in a more efficient way.

The remainder of the article is organized as follows. In the next section (Sect. 2), we describe the method: It starts by sketching a brief overview of the proposed strategy. Subsect. 2.2 describes how we build an entity database computing distributional similarity. Next, Subsect. 2.3 is focused on the NLP approaches to term extraction. In Subsect. 2.4, the entity linking strategy is described. Then, we evaluate and compare our method in Sect. 3 and, finally, some conclusions are addressed in Sect. 4.

2   The Method

2.1   Overview

Our EL method consists of three modules:

Distributional Similarity: This module builds the main encyclopedic resource used by the Entity Linking module. Each Wikipedia entity is put in relation with its most similar entities in terms of distributional similarity. This is the main contribution of our work, since, to our knowledge no EL method relies on such a sort of resource. This is described in Sect. 2.2.

Term Extraction: This module makes use of NLP strategies to extract the most relevant terms from the text. It is described in Sect. 2.3.

Entity Linking: This is the core of the system. It makes use of Wikipedia-based resources (such as that built by distributional similarity) and of the terms previously extracted from the text. It consists of two tasks. First, it identifies those relevant terms that are linked to Wikipedia entities and, then, it selects, for each term, the best entity candidate by making use of a disambiguation strategy. This module is described in Sect. 2.4.

2.2   Distributional Similarity

We use a distributional similarity strategy to select only semantic relations between very similar entities. This strategy allows us to dramatically simplify the number of relations/arcs to be explored in a collective approach.

Let us see an example. In the English DBpedia, the entity *Aníbal Cavaco Silva* (President of Portugal between 2006–2016) is directly related to 17 categories by means of the hyperonymy relationship: for instance, *Living People*, *Prime Ministers of Portugal*, *People from Loulé Municipality*, etc. If we explore these 17 categories going down to obtain their direct child (or hyponyms), the results are 619, 406 new entities, which are in fact co-hyponyms of *Cavaco Silva*. Most of these co-hyponyms have a very vague conceptual relation (e.g. being a living person) with the target entity. In order to remove vague conceptual relations, we only select those entities that can be somehow considered as similar to *Cavaco Silva*. Similarity between two entities is computed by taking into account both the *internal links* appearing in the Wikipedia articles of the two entities, and the set of categories directly classifying them. More precisely, two entities are considered to be similar if they share at least one direct category and a significant amount of internal links.

In our experiments, the target entity *Cavaco Silva* is associated with its most similar entities (first column in Table 1), and for each similar entity we also select the most frequent internal links with which they co-occur (second column of the table). The entities in the second column represent the conceptual context with regard to which two entities are similar. As a result, we obtain a very restricted and very similar set of entities related to *Cavaco Silva*, which includes other Presidents and Primer Ministers of Portugal. Notice that the target entity is also similar to former Finance Ministers (*Ferreira Leite* and *Vítor Gaspar*), since *Cavaco Silva* also had that political function before becoming Prime Minister. In addition, he shares with these two individual the fact of being Economist and having been working at the same universities.

In our experiments, both *similar* and *contextual* entities are all considered in the same way: all are directly related to the target entity. As the list of co-hyponyms for each entity is reduced from some hundred thousands candidates to a few entities (similar and contextual ones), the resulting database is easy to explore by most searching strategies.

Table 1. Entities related to *Aníbal Cavaco Silva* using distributional similarity

Similar entities	Contextual entities
Mário Soares	President of Portugal, Ordem Nacional do Cruzeiro do Sul Ordem do Libertador
Jorge Sampaio	António Guterres, Timor-Leste Portuguese Presidential Election Ordem de Amícar Cabral
Diego Freitas do Amaral	Prime Minister of Portugal, New University of Lisbon Catholic University of Portugal
Manuela Ferreira Leite	National Assembly of the Republic, Economist, Bank of Portugal Fundação Calouste Gulbenkian
Vítor Gaspar	Economist, Francico Louçã, Bank of Portugal, Professor

Let e_1 and e_2 be two entities with the corresponding articles in Wikipedia. They are comparable if they share at least one Wikipedia category. Distributional similarity is only computed on entity pairs sharing at least one category. So, if entities e_1 and e_2 share at least one category, they are actually comparable and similarity is computed. Distributional similarity is computed using the following version of the *Dice* coefficient [3]:

$$Dice(e_1, e_2) = \frac{2 * \sum_i min(f(e_1, link_i), \ f(e_2, link_i))}{f(e_1) + f(e_2)} \tag{1}$$

where $f(e_1, link_i)$ represents the number of times the entity e_1 co-occurs with the internal link $link_i$. Internal links stand for the distributional contexts of the compared entities. As a result, each entity is assigned a set of similar entities ranked by Dice similarity and a set of internal links ranked by frequency. The resulting entity database is the main knowledge base considered by our semantic annotation strategy. This resource is called *Similarity Knowledge Base*. In [9], *Dice* turned out to be one of the most reliable similarity measures for distributional semantics.

2.3 NLP Techniques for Term Extraction

We distinguish two different types of terms: *basic terms* and *multiword expressions*. Basic terms are lexical units codified as common nouns, adjectives, verbs, or proper names which are considered as relevant for a given text. Except proper names, which can be composite expressions (e.g., New York, University of South California), basic terms are just single words. Multiwords are relevant expressions codified as compounds that instantiate specific patterns of PoS tags. For instance, `discussion forums`, `natural language`, `cells of plants` or `professor at New University of Lisbon` can be multiwords within a text.

For the specific task of semantic annotation, we assume that not all terms within a text which are linked to an entity (or concept) in DBpedia are semantically relevant. There are frequent mentions, e.g. `concept`, `term`, `red`, etc., which are linked to concepts in DBpedia, but which may not be relevant in some texts. So, terms must be ranked according to their relevance in a text and should be considered as entity candidates only the most relevant ones.

Our approach to extract basic terms and multiwords requires PoS tagging, which is performed with the multilingual NLP suite *Citius Tool* [10].[1] For extracting basic terms, we use a different strategy that the one used for multiword extraction. The strategy we follow to extract basic terms is slightly different from that used for multiwords. In the case of basic terms, their extraction relies on the notion of *termhood*, that is, the degree that a linguistic unit is related to domain-specific concepts [19]. In the case of multiwords, the extraction is based on the notion of *unithood*, which concerns with whether or not sequences of words should be combined to form more stable lexical units. More formally, unithood refers to "the degree of strength or stability of syntagmatic combinations and collocations" [19]. The concept of unithood is only relevant to complex units (multiwords).

Extraction of Basic Terms. The first step consists in identifying and selecting common nouns, adjectives, verbs, and proper names from a given text. Proper names are selected by using named entity recognition. The result is a list of term candidates.

The second step consists in providing the term candidates with a statistical weight, representing the conceptual relevance of the term within the input text. The weight of a term is computed by considering the frequency observed in the input text (observed data) with regard to its frequency in a large collection of texts taken as a corpus of reference (expected data). More precisely, the weight of a term is the *chi-square* value, which measures the divergence between the observed data and the values that would be expected. Expected values are provided by the reference corpus. Finally, all weighted terms are ranked according to their score and the N most relevant are selected for semantic annotation. This way, terms very frequent in the reference corpus (common concepts such as for instance `person`, `thing`, `object`, etc.) tend to be assigned low chi-square values. By contrast, very frequent terms in the input text but rare in the reference corpus have high values and, then, are considered as relevant for the given text.

---

[1] Freely available at http://gramatica.usc.es/pln/tools/CitiusTools.html.

Multiword Extraction. The proposed strategy relies on the notion of unit-hood and has common aspects with similar work requiring linguistic patterns [27,29]. Our extraction of multiwords also consists of two steps: candidates selection and statistical ranking. In the first step, candidates are extracted using a set of patterns of PoS tags. This is the set we use for our four languages:

$noun - adj$	$adj - noun$
$noun - noun$	$noun - prep - noun$
$noun - prep - adj - noun$	$noun - prep - noun - adj$
$adj - noun - prep - noun$	$noun - adj - prep - noun$
$adj - noun - prep - noun - adj$	$noun - adj - prep - noun - adj$
$adj - noun - prep - adj - noun$	$noun - adj - prep - adj - noun$

In the second step, the candidates are ranked according to the notion of unithood: A lexical measure, *chi-square*, provides a test of association between the constituents of a multiword, in order to verify whether the constituents are or are not put together by random. More precisely, the observed values of a multiword stands for its frequency in the input text, while the expected values are derived from the single occurrences of its constituents in the same text.

2.4   The Entity Linking Strategy

Resources and Terms. Our strategy makes use of three resources, which represent three different linguistic relations:

Similarity Knowledge Base (SIM). This stands for similarity relationships between Wikipedia entities. Wikipedia entities correspond to the titles of articles in Wikipedia (dump file of December 2014). This resource was built based on distributional similarity (see Sect. 2.2 above).[2]

Categories of Wikipedia entities (HYPER). This database contains hierarchical (hyperonymy) relations between Wikipedia entities and their direct parent categories. This resource is provided by DBpedia[3].

Redirects of Wikipedia entities (REDIR). This database contains synonymous relations between Wikipedia entities and their different names. This resource is also provided by DBpedia.

The union of Wikipedia entities and categories gives rise to the set of (conceptual) entities of our ontology. Indeed, some categories are not Wikipedia entities.

Besides these three resources, our EL strategy also relies on term extraction (see Sect. 2.3). The output of this task, which is a ranked list of relevant terms (both single words and multiwords) is the input of the following EL tasks: searching for candidates and disambiguation. According to [13], the most efficient EL systems divide the process of entity linking in these two tasks. During the search phase the system proposes a set of candidates for an entity mention to be linked to, which are then ranked by the disambiguator.

---

[2] This resource is available from the authors upon request.
[3] http://downloads.dbpedia.org/3.8/.

Searching for Entity Candidates. We verify whether the relevant terms extracted from the input text are actually mentions of entities. For this purpose, they are expanded in two different ways: (1) Each term is expanded with its lemma, for instance the term `databases` is expanded with the singular form `database`. (2) Terms are expanded with their synonymous stored in the resource REDIR. All the inflected forms and synonyms of a term occurring in the input text are joined in a single terminological unit. Then, we search for semantic links between expanded terms (terminological units) and entities. The search for links between terms and entities is performed using our external resources: SIM, REDIR, and HYPER. The main problem arising when terms are intended to be linked to entities is term ambiguity.

One term (hereafter we use interchangeably "term" and "terminological unit") can be associated to several entities, which represent their different senses. A natural way of accessing the different entities/senses of an ambiguous term is to use Wikipedia disambiguation pages. However, these pages include many odd senses which should not be linked to the ambiguous term. For instance, the French town *Barcelonnette* is considered as one of the senses of the term `Barcelona`, which is clearly odd. Instead of using the entities listed in the disambiguation pages, we select the entities/senses of an ambiguous term by taking into account some regular expressions related to the syntax of the Wikipedia titles. In Wikipedia, different entities with the same name are individualized by making use of brackets, commas or hyphens. For instance, the ambiguous term `Paris` is associated to entities like *Paris, Paris,_Ohio, Paris,_Arkansas, Paris_(mythology), Paris_(song)*, etc. All of them can be considered different senses of the original term. Even if our use of regular expressions in Wikipedia titles does not always include all possible senses of an ambiguous term, most extracted senses are apparently correct ones. So, our technique is more precise than that based on disambiguation pages but has lower coverage.

The output of this task is a list of entity candidates associated with all relevant terms extracted from the input text.

Weighting Candidates and Entity Disambiguation. In this task, we select the best entity candidate of each term by making use of a disambiguation strategy. This strategy relies on selecting the entity with the highest weight for each term.

Given a term, the process starts by assigning the same weight to all its entity candidates. Then, it explores the semantic relationships (similarity and hyperonymy) of each entity candidate and searches for related entities that are also semantically related to the candidates of the other terms in the input text. The procedure of exploring and searching common related entities is performed on the two knowledge resources: SIM (similarity) and HYPER (hyperonymy).

The weighting process is just a summary of semantically related entities that are shared by both the target entity and the rest of entity candidates of all input terms. Given a terminological unit t_1 and an entity candidate e_1, the final weight of this entity with regard to t_1 is computed as follows:

$$weight(t_1, e_1) = \sum_{i=1}^{k} sim(e_1, e_i) + hyper(e_1, e_i) \qquad (2)$$

where $sim(e_1, e_i)$ stands for the number of similar entities which are shared by e_1 and each member (e_i) of the set of entity candidates; $hyper(e_1, e_j)$ represents the number of categories which are shared by e_1 and each member of the set of entity candidates. The former function is computed on SIM while the latter works on HYPER. The set of entity candidates is constituted by those entities associated to all terminological units extracted from the text, where k is the size of the set. Finally, for each term, the entity with the highest *weight* value is selected.

Let us take an example. Suppose we have selected the term `Cavaco Silva`. To compute *weight* (`Cavaco Silva`, *Aníbal Cavaco Silva*) given the pool of entities {*Aníbal Cavaco Silva, Jorge Sampaio, Lisbon*}, we compute first the *sim* function, which consists in counting the number of entities in the pool which are similar to the target entity *Aníbal Cavaco Silva* according to the SIM database. Given the Table 1 above, only one of these entities is linked by similarity to the target entity. So, the result of the *sim* function is just 1. A similar procedure is performed to comput the *hyper* value, but using the HYPER resource.

System Implementation. The method was implemented in Perl giving rise to the system called *CitiusLinker*. So far, it works for four languages: English, Portuguese, Spanish, and Galician.[4] In order to facilitate its integration into external web processes, we also implemented a RESTful web service with Dancer.[5] The web service interface can be used to annotate the text with the selected terms and their linked entities. Besides, it also gives as output a set of semantically related DBpedia entities to those found in the text (semantic enrichment), as well as a set of DBpedia categories that can be used to classify the text (semantic categorization). The web service returns HTML, XML, YAML or JSON output documents. It can be configured to select one of the four languages, the output format, and the number of relevant basic terms.

3   Evaluation

In order to provide an evaluation of our system in the task of semantic annotation, we performed two experiments with English and Portuguese texts, using manually annotated test corpora.

For English, we used the DBpedia Spotlight Evaluation Dataset [22]. The test corpus consists of 10 randomly selected excerpts from New York Times news, and each excerpt/document was manually annotated with DBpedia concepts. For Portuguese, we created a similar dataset from 10 different *Jornal de Notícias* news, which were manually annotated by two linguists using the Portuguese

---

[4] A demo is available at http://fegalaz.usc.es/~gamallo/demos/semantic-demo/.
[5] http://fegalaz.usc.es/nlpapi.

DBpedia. To build the gold standard dataset, we selected the concepts identified by both annotators. As a result, we obtained 130 concepts for the 10 documents.

Both annotated datasets are freely available.[6]

Notice that the evaluated task is different from that defined in the different TAC-KBP Entity Linking Tracks [16]. In those tracks, the objective is not to identify the relevant concepts of a given document, but identifying the correct node/concept in DBpedia given a name mention in a document. Besides, the test datasets are just focused on named entities of type PER (person), ORG (organization), or GPE (geopolitical entity). In [5], the author describes the construction of two datasets for entity linking in the Portuguese and Spanish languages, by making use of the cross-lingual XLEL-21 dataset. This dataset is equivalent to the one used in TAC-KBP, and contains just person names.

In the English evaluation, we compare our results with those of several publicly available annotation services. The results of all systems were obtained by using the same gold standard: DBpedia Spotlight Evaluation Dataset. Except *CitiusLinker* and *Alchemy*, whose F_1 scores were obtained from our own experiments, the scores of the remainder systems were taken from [22].

Table 2. F_1 scores reached by different EL systems using the DBpedia Spotlight Evaluation Dataset (for English)

Systems	F_1-score
The Wiki Machine[a]	59.5 %
DBpedia Spotlight (best configuration)	56.0 %
CitiusLinker (best configuration)	55.9 %
Zemanta[b]	39.1 %
Alchemy[c]	21.1 %
Open Calais[d]	14.7 %
Ontos[e]	10.6 %

[a] http://thewikimachine.fbk.eu
[b] http://www.zemanta.com
[c] http://www.alchemyapi.com
[d] http://www.opencalais.com
[e] http://www.ontos.com

Table 2 shows that the performance of our strategy, *CitiusLinker*, is in a competitive range for English, close to the two best systems: *Wiki Machine* and *DBpedia Spotlight*.

Concerning the Portuguese evaluation, results are depicted in Table 3. Unfortunately, we only could compare our system to DBpedia Spotlight and that provided by *Alchemy*. To the best of our knowledge, no further EL systems for Portuguese are available yet. The scores reached by *CitiusLinker* and *DBpedia*

---

[6] http://gramatica.usc.es/~gamallo/datasets/el_dataset.tar.gz.

Table 3. F_1 scores reached by three systems using the Portuguese dataset

Systems	Precision	Recall	F_1-score
CitiusLinker (best configuration)	45.3 %	56.2 %	50.9 %
DBpedia Spotlight (best configuration)	45.6 %	51.2 %	48.4 %
Alchemy	12.8 %	5.38 %	7.56 %

Spotlight are slightly lower than those got in the English evaluation. Both systems achieve similar F_1-score values after having set their parameters to find the best configuration. By contrast, *Alchemy* system dramatically drops performance. In this case, no parameter configuration has been done since the experiments were performed from the API server provided by the company. The Portuguese *DBpedia Spotlight* version belongs to a multilingual system which is described in [4].

The F_1-score of our system has been obtained with the best configuration: 60 most relevant basic terms (only nouns) and all multiwords. When using adjectives and verbs, the F_1-score decreases. Notice also that no multiword was filtered out. Unlike basic terms, which can refer to very generic concepts in some cases, multiwords linked to DBpedia entities are likely to be domain-specific terminological expressions referring to specific concepts. By default, *CitiusLinker* selects all multiwords found in the text.

4   Conclusions

In this article, we proposed a method for a specific entity linking subtask, namely semantic annotation with DBpedia concepts. The main contribution of our method is the use of an external entity base built by means of distributional similarity. This entity base is structured with similarity relationships between entities which are not directly related by means of the DBpedia resources. In the disambiguation process, our method only explores the similarity relations found in this entity base, as well as the direct hyperonymy relationships provided by DBpedia. This way, the weighting process used to disambiguate becomes simpler and more efficient than those based on exploring several levels of organization through DBpedia or any other ontology. Another important contribution of our method is the use of different NLP techniques for term extraction. We defined a specific strategy for the extraction of basic terms, which is different from multiword extraction. Our approach achieved competitive performance over the traditional methods in English, while kept similar performance in Portuguese. In future work, we will evaluate the results obtained for languages other than English and Portuguese. A deep qualitative error analysis is also required in order to find the main drawbacks of our approach. It will also be adapted to be applied on TAC-KBP tasks in order to be compared to other EL systems.

References

1. Cassidy, T., Ji, H., Ratinov, L.-A., Zubiaga, A., Huang, H.: Analysis and enhancement of wikification for microblogs with context expansion. In: Proceedings of the 24th International Conference on Computational Linguistics (COLING 2012): Technical Papers, pp. 441–456 (2012)
2. Cucerzan, S.: TAC entity linking by performing full-document entity extraction and disambiguation. In: Proceedings of the Text Analysis Conference (TAC 2011) (2011)
3. Curran, J.R., Moens, M.: Improvements in automatic thesaurus extraction. In: Proceedings of the ACL 2002 Workshop on Unsupervised Lexical Acquisition, Philadelphia, vol. 9, pp. 59–66 (2002)
4. Daiber, J., Jakob, M., Hokamp, C., Mendes, P.N.: Improving efficiency and accuracy in multilingual entity extraction. In: Proceedings of the 9th International Conference on Semantic Systems (I-Semantics), pp. 121–124. Association for Computing Machinery (2013)
5. dos Santos, J.T.L.: Linking entities to Wikipedia documents. PhD thesis, Instituto Superior Técnico, Lisboa (2013)
6. Fernández, N., Fisteus, J.A., Sánchez, L., Martín, E.: WebTLab: a cooccurrence-based approach to KBP 2010 entity-linking task. In: Proceedings of the Text Analysis Conference (TAC 2010) (2010)
7. Ferragina, P., Scaiella, U.: TAGME: on-the-fly annotation of short text fragments (by Wikipedia entities). In: Proceedings of the 19th ACM International Conference on Information and Knowledge Management (CIKM 2010), Toronto, pp. 1625–1628 (2010)
8. Gamallo, P.: Evaluating two different methods for the task of extracting bilingual lexicons from comparable corpora. In: LREC 2008 Workshop on Comparable Corpora, Marrakesh, pp. 19–26 (2008)
9. Gamallo, P., González, I.: A grammatical formalism based on patterns of part-of-speech tags. Int. J. Corpus Linguist. **16**(1), 45–71 (2011)
10. Garcia, M., Gamallo, P.: Yet another suite of multilingual NLP tools. In: Sierra-Rodríguez, J.-L., et al. (eds.) SLATE 2015. CCIS, vol. 563, pp. 65–75. Springer, Heidelberg (2015). doi:10.1007/978-3-319-27653-3_7
11. Guo, S., Chang, M.-W., Kiciman, E.: To link or not to link? a study on end-to-end tweet entity linking. In: Proceedings of the 2013 Conference of the North American Chapter of the Association for Computational Linguistics: Human Language Technologies (NAACL-HLT 2013), pp. 1020–1030 (2013)
12. Guo, Y., Che, W., Liu, T., Li, S.: A graph-based method for entity linking. In: Proceedings of the 5th International Joint Conference on Natural Language Processing (IJCNLP 2011), pp. 1010–1018 (2011)
13. Hachey, B., Radford, W., Nothman, J., Honnibal, M., Curran, J.R.: Evaluating entity linking with Wikipedia. Artif. Intell. **194**, 130–150 (2013)
14. Han, X., Sun, L.: A generative entity-mention model for linking entities with knowledge base. In: Proceedings of the 49th Annual Meeting of the Association for Computational Linguistics: Human Language Technologies (HLT 2011), Portland, Oregon, vol. 1, pp. 945–954 (2011)
15. Han, X., Zhao, J.: Named entity disambiguation by leveraging Wikipedia semantic knowledge. In: Proceedings of the 18th ACM Conference on Information and Knowledge Management (CIKM 2009), Hong Kong, China, pp. 215–224 (2009)

16. Ji, J.N.H., Hachey, B.: Overview of TAC-KBP2014 entity discovery and linking tasks. In: Proceedings of the Text Analysis Conference (TAC 2014), pp. 539–545 (2014)
17. Hoffart, J., et al.: Robust disambiguation of named entities in text. In: Proceedings of the Conference on Empirical Methods in Natural Language Processing (EMNLP 2011), pp. 782–792 (2011)
18. Huang, H., Cao, Y., Huang, X., Ji, H., Lin, C.-Y.: Collective tweet wikification based on semi-supervised graph regularization. In: Proceedings of the 52nd Annual Meeting of the Association for Computational Linguistics (ACL 2014), Volume 1: Long Papers, pp. 380–390. Association for Computational Linguistics (2014)
19. Kageura, K., Umino, B.: Methods of automatic term recognition: a review. Terminology **3**(1), 259–289 (1996)
20. Kozareva, Z., Voevodski, K., Teng, S.-H.: Class label enhancement via related instances. In: Proceedings of the Conference on Empirical Methods in Natural Language Processing (EMNLP 2011), pp. 118–128. Association for Computational Linguistics (2011)
21. Kulkarni, S., Singh, A., Ramakrishnan, G., Chakrabarti, S.: Collective annotation of wikipedia entities in web text. In: Proceedings of the 15th ACM SIGKDD International Conference on Knowledge Discovery and Data Mining (KDD 2009), Paris, pp. 457–466. Association for Computing Machinery (2009)
22. Mendes, P.N., Jakob, M., García-Silva, A., Bizer, C.: DBpedia spotlight: shedding light on the web of documents. In: Proceedings of the 7th International Conference on Semantic Systems, Graz, pp. 1–8. Association for Computing Machinery (2011)
23. Mihalcea, R., Csomai, A.: Wikify!: linking documents to encyclopedic knowledge. In: Proceedings of the 16th ACM Conference on Information and Knowledge Management (CIKM 2007), Lisbon, pp. 233–242 (2007)
24. Milne, D., Witten, I.H.: Learning to link with Wikipedia. In: Proceedings of the 16th ACM Conference on Information and Knowledge Management (CIKM 2008), Napa Valley, pp. 509–518 (2008)
25. Pennacchiotti, M., Pantel, P.: Entity extraction via ensemble semantics. In: Proceedings of the Conference on Empirical Methods in Natural Language Processing (EMNLP 2011), pp. 238–247 (2009)
26. Radford, W., Hachey, B., Nothman, J., Honnibal, M., Curran, J.R.: Document-level entity linking: CMCRC at TAC 2010. In: Proceedings of the Text Analysis Conference (TAC 2010) (2010)
27. Sánchez, D., Moren, A.: A methodology for knowledge acquisition from the web. J. Knowl.-Based Intell. Eng. Syst. **10**(6), 453–475 (2006)
28. Vidal, J.C., Lama, M., Otero-García, E., Bugarín, A.: Graph-based semantic annotation for enriching educational content with linked data. Knowl.-Based Syst. **55**, 29–42 (2014)
29. Vivaldi, J., Rodríguez, H.: Improving term extraction by combining different techniques. Terminology **7**(1), 31–47 (2001)

Syntax Deep Explorer

José Correia[1,2], Jorge Baptista[2,3]([✉]), and Nuno Mamede[1,2]

[1] Instituto Superior Técnico, Universidade de Lisboa, Lisbon, Portugal
[2] L2F – Spoken Language Lab, INESC-ID Lisboa, Rua Alves Redol 9,
1000-029 Lisbon, Portugal
{jcorreia,Nuno.Mamede}@inesc-id.pt
[3] Universidade do Algarve, Campus de Gambelas, 8005-139 Faro, Portugal
jbaptis@ualg.pt

Abstract. The analysis of the co-occurrence patterns between words
allows for a better understanding of the use (and meaning) of words
and its most straightforward applications are lexicography and linguist
description in general. Some tools already produce co-occurrence information about words taken from Portuguese *corpora*, but few can use
lemmata or syntactic dependency information. *Syntax Deep Explorer* is
a new tool that uses several association measures to quantify several
co-occurrence types, defined on the syntactic dependencies (*e.g.* subject,
complement, modifier) between a target word *lemma* and its co-locates.
The resulting co-occurrence statistics is represented in *lex-grams*, that
is, a synopsis of the syntactically-based co-occurrence patterns of a word
distribution within a given *corpus*. These *lex-grams* are obtained from
a large-sized Portuguese *corpus* processed by STRING [19] and are presented in a user-friendly way through a graphical interface. The *Syntax
Deep Explorer* will allow the development of finer lexical resources and
the improvement of STRING processing in general, as well as providing
public access to co-occurrence information derived from parsed *corpora*.

Keywords: Natural Language Processing (NLP) · Co-occurrence ·
Collocation · Association measures · Graphic interface · Lex-gram ·
Portuguese

1 Introduction

The analysis of the co-occurrence patterns between words in texts shows the
differences in use (and meaning), which are often associated with the different
grammatical relations in which a word can participate [7,33]. The quantifica-
tion of these patterns is a powerful tool in modern lexicography as well as in
the construction of basic linguistic resources, like *thesauri*. The stakeholders
on the study of those co-occurrence patterns are linguists, language students,
translators, lexicographers and Corpus Linguistics' researchers, who study the
behaviour of linguistic expressions in *corpora*. For all of these, co-occurrence data
are essential for a better understanding of language use. Furthermore, comparing different association measures, each capturing different linguistic aspects of

© Springer International Publishing Switzerland 2016
J. Silva et al. (Eds.): PROPOR 2016, LNAI 9727, pp. 189–201, 2016.
DOI: 10.1007/978-3-319-41552-9_19

the co-location phenomena, enables the user to achieve a broader understanding of the distribution of a given word, hence its different meanings and uses. For a better analysis of co-occurrence patterns in a *corpus*, it is also important to provide the user with an interface that helps him/her to explore the patterns thus extracted, namely, by accessing concordances or moving on from an initial query to another interesting collocate.

To date, only one system, *DeepDict* [3], is known to produce co-occurrence information based on syntactic dependencies between (simple) word *lemmata*, but it only features a single association measure, Pointwise Mutual Information (PMI) [7]. Other systems available for Portuguese *corpora*, like *Sketch Engine* (see below), do not benefit from syntactic information nor lemmatization and also use a single association measure, LogDice [18].

This paper presents *Deep Syntax Exporer*[1]. Its main goal is to provide the general public a tool to explore syntactically-based collocations from Portuguese *corpora*, using a broad set of association measures, in view of an enhanced understanding of words' meaning and use. The information on these collocation patterns is made available through a web interface by way of *lex-grams*, a synopsis of the syntactically-based co-occurrence patterns of a word's distribution within a given *corpus*. This mode of presentation organizes the information for a simpler analysis by users. For now, only the four main lexical part-of-speech categories (adjectives, nouns, verbs and adverbs) are targeted by the *Deep Syntax Exporer*.

This paper is organised as follows: Next, in Sect. 2, key concepts and related work are presented; first, the Portuguese processing STRING [19] underlying the tool is briefly sketched; then, the main existing systems, that provide co-occurrence information for Portuguese *corpora*, are succinctly described; finally, the association measures implemented in *Deep Syntax Explorer* are briefly sketched. Section 3 presents the system architecture in detail, beginning with the database, then the co-occurrence information extraction module, and, finally, the web application and the way the *lex-grams* display this information to the end user. Section 4 presents the evaluation of the tool's performance. Section 5 concludes the paper and presents future directions of development.

2   Related Work

This section starts by briefly presenting the STRING system underlying the *Syntax Deep Explorer*, from whose output the co-occurrence information is extracted. Then, it describes three co-occurrence information extraction systems already existing for processing Portuguese *corpora*. Finally, it provides a quick overview of association measures implemented in *Syntax Deep Explorer*.

The STRING NLP System
STRING [19] is a hybrid, statistical and rule-based, natural language processing (NLP) chain for the Portuguese language, developed by Spoken Language

---

[1] string.l2f.inesc-id.pt/demo/deepExplorer (last visit 29/02/2016).

Systems Lab (L^2F) from INESC-ID Lisboa[2]. STRING has a modular structure and performs all basic text NLP tasks. The first module is the *LexMan* [35], which is responsible for text segmentation (sentence splitting and tokenization) for assigning to each token its part-of-speech (POS) and any other relevant morphosyntactic features. The module *RuDriCo* [10] is a rule-driven converter, used to revolve ambiguities and to deal with contractions and certain compounds words. The *MARv* [11,28] is a statistical part-of-speech disambiguator that chooses the most likely POS tag for each word, using the Viterbi algorithm. Finally, the *XIP* (Xerox Incremental Parser) [1] uses a Portuguese rule-base grammar [20] to structure the text into chunks, and to establish syntactic dependencies between their heads. This parser also performs other NLP tasks such as named entity recognition (NER) [15,25], anaphora resolution (AR) [22,26] and temporal expressions normalization [13,14,16,23]. The processing produces a XML output file.

The *Syntax Deep Explorer* uses the following (higher level) syntactic dependencies, produced by the XIP parser: (1) the SUBJ dependency, which links a verb and its subject; (2) the CDIR dependency, linking a verb and its direct complement; (3) the CINDIR dependency, which links the verb with a dative essencial complement; (4) the COMPL dependency, which links a predicate (verb, noun or adjective) to its essential PP complements; and (5) the MOD dependency that links a modifier with the element it modifies (*e.g.* adjectives as modifiers of nouns, or adverbs as modifiers of verbs); this is an umbrella dependency, as it also connects any PP complement to its governor, iff it has not been already captured by COMPL or another dependency); and (6) Named Entities, which also captured by the XIP parser by an unary NE dependency that delimits and assigns them to several general categories (PERSON, ORGANIZATION, PLACE, etc.); these categories are then used for co-occurrence statistics, instead of the individual entities themselves.

Current Systems Providing Co-occurrence Information

Nowadays, some tools already make it possible to obtain information on the co-occurrence patterns of a given word from Portuguese *corpora*, which will now be described briefly.

The *AC/DC* platform available through *Linguateca*[3] produces raw, quantitative data on words' co-occurrences from a large variety of Portuguese *corpora*, allowing the user to query complex combinatorial patterns by way of regular expressions, though it does not make use of any association measure.

The *DeepDict* [3] is a tool developed by GrammarSoft[4] and presented in the *Gramtrans* platform[5]. For Portuguese *corpora*, this tool uses the NLP analyzer PALAVRAS [2] and it collects dep-grams from the syntactical dependencies produced in the parsed texts. A dep-gram is a pair of *lemmas* of two words that feature a dependency relation between them. The *dep-grams* are quantified by

---

PMI measure. In the database, the tool stores the `dep-gram`, the value of PMI, the absolute frequency and the ID of the sentence where these `dep-gram` occurred in the *corpus*. DeepDict has a simple form as an interface. The result is presented as a *lexicogram* of the searched *lemma*, which shows the different dependency relations that the word establishes with other lexical elements, sorted in PMI value descending order. The lexicogram displays the co-locates in the natural reading order of the related words. Words having different POS have different lexicograms. At this stage, DeepDict does not provide access to concordances of a given *dep-gram*.

The *Sketch Engine* [17] is a tool developed by Lexical Computing[6] This system uses Manatee [29] to manage *corpora*. The tool stores for each word its *lemma* and POS tag. Co-occurrences are identified by the tool and are quantified by the LogDice measure. The system has a graphical interface that is generated by the tool Bonito [29] and allows access to stored data. The system main features are: (i) *concordance*, that is, access to examples taken from the *corpora* but without any analysis; (ii) *word sketches*, which show the words more closely related to the target word; (iii) *sketch-diff*, which depicts the differences between two different word sketches; and (iv) *thesaurus*, which shows similar words for a given target word. The system allows the users to create and manage their own *corpora*.

The *Wortschatz* [27] system has been developed at Leipzig University, and it creates lexical similarity networks from *corpora*. The system identifies two types of co-occurrences: (i) words occurring together in a sentence and (ii) words occurring next to each other. The system uses tree-based algorithms that allows a quick co-occurrence analysis of the entire *corpus* [4]. However, it does not use the words *lemmata* for these calculations. Each co-occurrence is quantified by a Significance Measure (see below) and this data is stored in a MySQL database. The system interface provides, for a target word, its frequency class; examples of its use (sentences); significant co-occurrences and a co-occurrence chart.

In sum, all these systems are able to list the co-occurrences of a word, though Wortschatz does not use *lemmas*. The Sketch Engine is the system that offers more features, particularly the *corpus* management tools and the sketch-diff. On the other hand, DeepDict shows co-occurrence statistics based on syntactic dependencies between *lemmata* and does so in a more readable, user-friendly way.

Association Measures

To understand how two words relate to each other, it is essential to quantify the co-occurrence patterns found between them. Six different association measures, commonly used in *corpus*-based linguistic analysis, were adopted and implemented in *Syntax Deep Explorer*. The first association measure is *Pointwise Mutual Information* (PMI) [7], which links the number of co-occurrences between two words with the number of occurrences of each word. The *Dice Coefficient* [9,34] is another association measure that calculates the degree of cohesion between two words. The *LogDice Coefficient* [30] is a variant of the Dice Coefficient and it is said to fix the problem of very small calculated values.

---

[6] http://www.sketchengine.co.uk/ (last visit 29/02/2016).

The *Pearson's Chi-Squared* measure [21] is a statistical method of hypothesis testing. This association measure compares the observed frequencies with the expected frequencies for a given pattern in order to verify if there is independency between events (null hypothesis). If the difference between observed and expected frequencies is larger than a statistical significance threshold, then the null hypotheses is rejected and the frequency of the pattern is deemed to be statistically relevant, given the *corpus*. This measure, however, should not be used in small *corpora* nor with patterns with low frequency. The *Log-likelihood Ratio* [12] is another approach to hypothesis testing, and it is considered to be more suitable for sparse data than the previous method. Finally, the so-called *Significance Measure*, used by the *Wortschatz* system [4], is comparable to the statistical *G-Test* method for Poisson distribution. This measure associates the co-occurrences of two words with the number of sentences in which they occurred.

3 *Deep Syntax Explorer* Architecture

This section presents the *Syntax Deep Explorer* architecture, sketched in Fig. 1. The tool consists of three modules that will be briefly presented below: (i) the database, (ii) the co-occurrence information extraction module, and (iii) the web interface application.

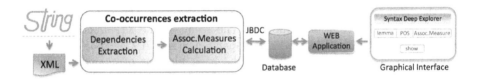

Fig. 1. *Syntax Deep Explorer* architecture.

Database

To store the co-occurrence statistics extracted from *corpora*, a consistent data model with low information redundancy is required. An Entity-Relationship (ER) Model [6] was developed, allowing for a more natural perception of the problem. Since typical database systems use a relational model [8], it is necessary to convert the ER model. The relational model is defined by a set of relations, represented in tables, where the columns are attributes of the relationship and each row is a tuple (entry), which is unique. To identify each tuple, primary keys are used to distinguish them. If a relation refers to another one, it has reference attributes called *foreign key*, where the values of these attributes have the *primary key* value of a tuple in the referenced relation [32].

The conversion to the relational model must follow the defined conversion rules and the previous data integrity constraints. SQLite[7] was used to implement

---

[7] http://www.sqlite.org/about (last visit 29/02/2016).

the obtained relational model. SQLite works locally, and it allows a direct and faster access to data, without a remote server. Access to the data is made through SQL language, making it possible to manipulate the tables in the database and to obtain the desired results.

Co-occurrence Information Extraction Module
Figure 2 introduces the UML (Unified Modelling Language) packages diagram representing the tool that is responsible for the extraction of dependency co-occurrence information, the *DeepExtractor*. The diagram shows the main components used.

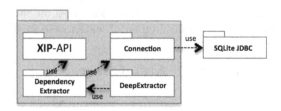

Fig. 2. Packages diagram for dependency co-occurrence extraction.

The XIP-API [5,24] enables the transformation of the XML files content from the STRING parsing module, the XIP parser, into Java structures and it organizes the information imported. The result is the `XIP-Document` that consists in a set of `XIP-Nodes` and `XIP-Dependency`(ies). A `XIP-Node` represents a node from the XIP parser's output, the basic element of *chunking tree*. It contains other `XIP-Nodes` (child nodes) and their `Features`, with the properties of each node. A `XIP-Dependency` contains the information about a dependency detected by the XIP parser and the `XIP-Nodes` to which it applies. To store the data in the database, the JBDC (Java Database Connectivity) API is used. The SQLiteJDBC[8], the JDBC for SQLite, is used.

The package `Connection` facilitates the access to the *Syntax Deep Explorer* database. In this package, a connection is established to the database through the SQLiteJDBC. Here, a group of classes has been developed, where the methods are used to manipulate the database information. Each class represents a database table and has methods to verify if an entry already exists and to insert new entries to that table. Other classes have additional methods to compute the values of the association measures.

The package `DependencyExtractor` is the core of the extraction tool. With the help of the XIP-API, the package analyses the texts produced by STRING to collect the statistics on co-occurrence syntactic dependencies. The `DependencyExtractor` method `Extract` accepts a `XIP-Document`, and for each sentence (indicated by a `XIP-Node` named `TOP`) in this document, it obtains within this

---

[8] https://bitbucket.org/xerial/sqlite-jdbc (last visit 29/02/2016).

sentence: (i) the set of named entities; (ii) set of XIP-Dependency(ies), each one linking two XIP-nodes; and (iii) the sentence string.

Afterwards, for each XIP-Dependency, the analysis process depends on the type of dependency it belongs to. A set of classes was developed, each one representing a XIP-dependency (*e.g.* SUBJ, CDIR, CINDIR, COMPL, and MOD); and it runs the correspondent code. In these classes, the method getDepInformation accepts a XIP-Dependency and a set of named entities in the sentence. In this method, to prevent the analysis of dependencies and properties that are not relevant to the goals of *Syntax Deep Explorer*, the dependency-property pattern of XIP-Dependency is checked against a list of predefined patterns. If that pattern is present, the analysis proceeds normally. The list of dependency-property patterns present in the system indicates for each word POS which are the relevant relations stored in the database. Then, for each XIP-Node the word *lemma* and POS are obtained. In the case a XIP-Node is present in the set of named entities, the word *lemma* is replaced by the correspondent named entity category (*e.g.* *João Silva* by PERSON, *NATO* for ORGANIZATION, *Lisboa* for PLACE).

During the process, it is necessary to store the information that is being found. Once a processed *corpus* is divided into several files, for each file, the information extracted is gathered in Java structures. The DeepStorage class works as a local database and helps to organize the information until this is stored in the database. After the extraction from the *corpus* file is completed, all gathered information is then stored in the database. When the *corpus* extraction is completed, it is necessary to calculate the values of the association measures for all syntactic dependency co-occurrences existing between two words. The DeepMeasures calculates the six association measures described above.

Web Application
Figure 3 shows the architecture of the web application.

Fig. 3. Web application architecture.

This application has two main components, the server-side and the client-side (browser). The first runs the code that implements the processing of information in the database and the second runs the code that implements the graphical interface. The communication between them is performed through asynchronous AJAX (Asynchronous Javascript and XML) requests (via HTTP) and the transmitted data is a JSON (JavaScript Object Notation) object. On the server-side, the code was developed in PHP. A request from the client consists in asking the co-occurrences of a word *lemma* and its POS, or a request for the sentences that exemplify a given co-occurrence. From the client request, several SQL queries are made in the database to get the information about that word or that co-occurrence. This information is organised in a JSON object and sent to the client. On the client-side, the code was developed from the AngularJS[9] framework, which allows the use of HTML and CSS, as declarative and presentation languages respectively.

To begin the process, the user is presented a form where s/he enters the target word *lemma* and chooses its POS; the user also indicates the association measure to be used. Some additional options can also be selected, such as the minimum frequency of each co-occurrence and the maximum number of words to be shown in each dependency-property pattern. By default, these values are set to '2' and '10', respectively. The collected information is organised in a JSON object and sent to the server. The user is informed when the target word does not exist in the *corpus* or no results were found for it. Otherwise, the system produces what we designate as a *lex-gram*, that is, a synopsis of a word distribution in a given *corpus*. This information is presented in two groups: words appearing to the left and the right of the target word. Figure 4 shows the *lex-gram* for the noun *país* (country), using the *LogDice* association measure on the CETEMPúblico *corpus* [31].

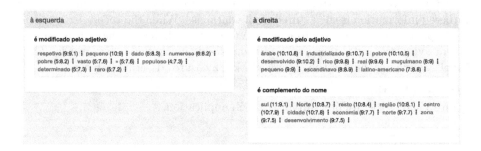

Fig. 4. *Lex-gram* of the noun **país** ordered by the LogDice measure.

For each dependency pattern in which the target word appears, the words that co-occur with it are presented in descending order of the values obtained with selected association measure value. Each word that co-occurs is displayed in the format *lemma (l:m)*, where *l* is the base 2 logarithm of the co-occurrence frequency, and *m* the value of the selected measure. The user may change the association measure without having to re-entering the information about the current word.

When the target word is a verb, the corresponding *lex-gram* shows the words appearing as its subject, direct complement, other essential complements, preposicional complements (eventually adjuncts) and the adverbs that modify the verb. In the case of an adjective, the *lex-gram* shows the adverbs that modify

---

[9] https://angularjs.org (last visit 29/02/2016).

it and the nouns that it modifies. For an adverb, the *lex-gram* presents other adverbs that modify it and the adjectives, nouns, verbs, and other adverbs that the target adverb modifies. For this category, it also shows how many times the adverb modifies an entire sentence. Finally, from the *lex-gram*, and by clicking on any co-occurrent word, it is also possible to obtain the details of this specific collocation and view a set of sentences, taken from the *corpus*, that illustrate that pattern.

4   Evaluation

For evaluating the *Deep Syntax Explorer* performance, the CETEMPúblico [31] *corpus*, processed by STRING, was used. The processed *corpus* occupies 237 GB, divided into 20 parts with 12 GB each. Each part has 210 XML files with 60 MB each. The extraction tool was executed in a x86_64 Unix machine, with 16 CPUs Intel(R) Xeon(R)E5530 2.40 GHz and 48 390 MB of memory.

For a single XML file from the *corpus* (with 60.8 MB), the extraction tool is executed in 32.5 s, where 9.76 s are consumed by the XIP-API to import the file, 10.65 s by the co-occurrence information extraction and 11.69 s are used to store this information and the sentences in the database. After the execution, that information occupies 4.33 MB. For the following parts of the *corpus*, only the time taken to store the information increases, as the database response time also increases.

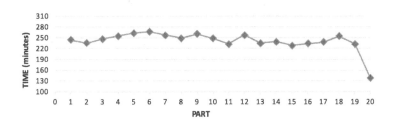

Fig. 5. Time consumed by each part of the CETEMPúblico *corpus*.

Figure 5 depicts the performance of the tool, presenting the time spent on each part of the *corpus*. The entire *corpus* is processed in 4,878 min (3 days, 9 h and 18 min), which corresponds to, on average, 243 min (4 h and 3 min) by each part of the *corpus* and 1 min and 9.4 seconds per file. Figure 5 also shows that the execution time is constant along the *corpus*, except in part 20, which is smaller. The association measures are calculated in 146 min (2 h and 26 min). The evolution of the database during the process is not constant. Initially, the growth is more pronounced, but it tends to decrease as the process evolves, because, in the beginning, the database is empty, and after processing some files fewer words have to be added. At the end of the process, the database has 6.8 GB. With the association measures data (0.5 GB), this value grows to

Table 1. Time the application takes to show the *lex-gram* to a word.

	Noun		Verb		Adverb		Adjective	
Lemma	caneta	país	otimizar	ser	nitidamente	não	lindo	grande
Frequency	998	196,140	972	2,533,385	964	1,362,286	997	263,518
Time (s)	0.102	0.111	0.103	0.760	0.037	0.174	0.067	0.256

7.3 GB. During the process of data extraction from the *corpus*, 308,573 different word *lemmas* were collected, which produced 11,244,852 different co-occurrence patterns occurring 51,321,751 times in the *corpus*.

The evaluation of the web application consisted in measuring the time the application took to show the *lex-gram* for a word. It consists mainly in server-side execution time, running the different SQL queries for each word POS. Table 1 shows the results from this evaluation. Two *lemmas* were used for each POS, one with the highest frequency and another with a median frequency (around 1,000 instances). The time for the word with greater frequency is slightly lower than for the word with smaller frequency. In order to obtain a faster system response, indexes were implemented in the database tables to reduce the time for each SQL query.

Table 2 shows the time the application takes to show the examples of sentences for a co-occurrence. The higher the frequency of the co-occurrence, the lower is the waiting time. This is due to the size of the examples' table, since, for a co-occurrence with low frequency, much of the table may need to be iterated.

Table 2. Time the application takes to show the examples for a co-occurrence pattern.

Co-occurrence	Frequency	Time (s)
(centro,país)	789	0.024
(país,praticamente)	46	0.025
(país,populoso)	16	0.035

5   Conclusion

This paper presented *Syntax Deep Explorer*, a tool created for extracting co-occurrence patterns from *corpora* processed by STRING. The tool takes advantage of the rich lexical resources of STRING, as well as of its sophisticated syntactic and semantic analysis, and finds the syntactically-based co-occurrences patterns of a given word *lemma* storing that information in a database. Then, the tool calculates different association measures, producing a *lex-gram* with the co-occurrence statistical information, a snapshot representing the main distributional patterns of a given word. The *lex-gram* also makes it possible to move from any given pattern to another co-locate of interest found within. Furthermore, STRING rich lexical resources feature a large number of multiword expressions, which are processed in the same way as any simple (single-word) unit. Thus, it is also possible to analyse multiwords' distribution and their co-occurrence patterns. Results from evaluation show that the runtime of the extraction tool remains constant throughout the *corpus*, while the size of the database does not grow linearly, indicating that information is not being repeated. The web application response time also allows fast queries to the database.

In the future, it is necessary to increase the number of *corpora* present in the database and to allow the comparison of different *lex-gram* for the same *lemma* across *corpora*. The automatic creation of *thesauri* from the stored distributional information is also envisaged.

Acknowledgment. This work was supported by national funds through FCT–Fundação para a Ciência e a Tecnologia, ref. UID/CEC/50021/2013. Thanks to Neuza Costa (UAlg) for revising the final version of this paper.

References

1. Art-Mokhtar, S., Chanod, J.P., Roux, C.: Robustness beyond shallowness: incremental deep parsing. Nat. Lang. Eng. **8**, 121–144 (2002)
2. Bick, E.: The Parsing System PALAVRAS. Automatic Grammatical Analysis of Portuguese in a Constraint Grammar Framework. Aarhus University Press, Aarhus (2000)
3. Bick, E.: DeepDict - a graphical corpus-based dictionary of word relations. In: Proceedings of NODALIDA 2009. NEALT Proceedings Series, vol. 4, pp. 268–271. Tartu University Library, Tartu (2009)
4. Biemann, C., Bordag, S., Heyer, G., Quasthoff, U., Wolff, C.: Language-independent methods for compiling monolingual lexical data. In: Gelbukh, A. (ed.) CICLing 2004. LNCS, vol. 2945, pp. 217–228. Springer, Heidelberg (2004)
5. Carapinha, F.: Extração Automática de Conteúdos Documentais. Master's thesis, Instituto Superior Técnico, Universidade Técnica de Lisboa, June 2013
6. Chen, P.: The entity-relationship model—toward a unified view of data. ACM Trans. Database Syst. **1**(1), 9–36 (1976)
7. Church, K., Hanks, P.: Word association norms, mutual information, and lexicography. Comput. Linguist. **16**(1), 22–29 (1990)
8. Codd, E.: A relational model of data for large shared data banks. Commun. ACM **26**(6), 64–69 (1983)
9. Dice, L.: Measures of the amount of ecologic association between species. Ecology **26**(3), 297–302 (1945)
10. Diniz, C., Mamede, N., Pereira, J.: RuDriCo2 - a faster disambiguator and segmentation modifier. In: INFORUM II, pp. 573–584, September 2010
11. Diniz, C., Mamede, N., Pereira, J.D.: RuDriCo2 - a faster disambiguator and segmentation modifier. In: Simpósio de Informática - INForum, pp. 573–584. Universidade do Minho, Portugal (2010)
12. Dunning, T.: Accurate methods for the statistics of surprise and coincidence. Comput. Linguist. **19**(1), 61–74 (1993)

13. Hagège, C., Baptista, J., Mamede, N.: Identificação, Classificação e Normalização de Expressões Temporais em Português: a Experiência do Segundo HAREM e o Futuro. In: Mota, C., Santos, D. (eds.) Desafios na Avaliação Conjunta do Reconhecimento de Entidades Mencionadas: o Segundo HAREM, chap. 2, pp. 33–54. Linguateca (2008). http://www.inesc-id.pt/ficheiros/publicacoes/5758.pdf/

14. Hagège, C., Baptista, J., Mamede, N.: Portuguese temporal expressions recognition: from TE characterization to an effective TER module implementation. In: 7th Brazilian Symposium in Information and Human Language Technology, STIL 2009, pp. 1–5. Sociedade Brasileira de Computação, São Carlos (2009)

15. Hagège, C., Baptista, J., Mamede, N.J.: Reconhecimento de entidadesmencionadas com o xip: Uma colaboração entre o inesc-l2f e a xerox. In: Mota, C., Santos, D. (eds.) Desafios na avaliação conjunta doreconhecimento de entidades mencionadas: Actas do Encontro do Segundo HAREM (Aveiro, 11 de Setembro de 2008). Linguateca (2009)

16. Hagège, C., Baptista, J., Mamede, N.J.: Caracterização e processamento de expressões temporais em português. Linguamática **2**(1), 63–76 (2010)

17. Kilgarriff, A., et al.: The sketch engine: ten years on. Lexicography **1**(1), 7–36 (2014)

18. Kilgarriff, A., Rychly, P., Tugwell, D., Smrz, P.: The sketch engine. In: Proceedings of Euralex. vol. Demo Session, pp. 105–116. Lorient, France, July 2004

19. Mamede, N., Baptista, J., Diniz, C., Cabarrão, V.: STRING: an hybrid statistical and rule-based natural language processing chain for Portuguese. In: PROPOR 2012, vol. Demo Session, April 2012

20. Mamede, N.J., Baptista, J.: Nomenclature of chunks and dependencies in Portuguese XIP Grammar 4.5. Technical report, L2F-Spoken Language Laboratory, INESC-ID Lisboa, Lisboa, January 2016

21. Manning, C., Schütze, H.: Foundations of Statistical Natural Language Processing. MIT Press, Cambridge (1999)

22. Marques, J.S.: Anaphora Resolution. Master's thesis, Instituto Superior Técnico, Universidade Técnica de Lisboa, Lisboa (2013)

23. Maurício, A.: Identificação, Classificação e Normalização de Expressões Temporais. Master's thesis, Instituto Superior Técnico, Universidade Técnica de Lisboa, Lisboa, November 2011

24. Nobre, N.: Resolução de Expressões Anafóricas. Master's thesis, Instituto Superior Técnico, Universidade Técnica de Lisboa, June 2011

25. Oliveira, D.: Extraction and Classification of Named Entities. Master's thesis, Instituto Superior Técnico, Universidade Técnica de Lisboa (2010)

26. Pereira, S.: Linguistics Parameters for Zero Anaphora Resolution. Master's thesis, Universidade do Algarve and University of Wolverhampton (2010)

27. Quasthoff, U., Richter, M., Biemann, C.: Corpus portal for search in monolingual corpora. In: Proceedings of the 5th LREC, pp. 1799–1802 (2006)

28. Ribeiro, R.: Anotação Morfossintática Desambiguada do Português. Master's thesis, Instituto Superior Técnico, Universidade Técnica de Lisboa, March 2003

29. Rychly, P.: Manatee/Bonito - a modular corpus manager. In: Sojka, P., Horák, A. (eds.) RASLAN 2008, pp. 65–70. Masaryk University, Brno (2007)

30. Rychly, P.: A lexicographer-friendly association score. In: RASLAN 2008, pp. 6–9. Masarykova Univerzita, Brno (2008)

31. Santos, D., Rocha, P.: Evaluating CETEMPúblico, a free resource for Portuguese. In: Proceedings of the 39th Annual Meeting of ACL, ACL 2001, pp. 450–457. Association for Computational Linguistics, Stroudsburg (2001)

32. Silberschatz, A., Korth, H., Sudarshan, S.: Database System Concepts. Connect, learn, succeed. McGraw-Hill Education (2010)
33. Sinclair, J.: Corpus, Concordance, Collocation. Oxford University Press, Oxford (1991)
34. Smadja, F., McKeown, K., Hatzivassiloglou, V.: Translating collocations for bilingual lexicons: a statistical approach. Comput. Linguist. **22**(1), 1–38 (1996)
35. Vicente, A.M.F.: LexMan: um Segmentador e Analisador Morfológico com Transdutores. Master's thesis, Instituto Superior Técnico, Universidade de Lisboa, Lisboa, June 2013

Automatic Semantic Role Labeling on Non-revised Syntactic Trees of Journalistic Texts

Nathan Siegle Hartmann[✉], Magali Sanches Duran, and Sandra Maria Aluísio

Interinstitutional Center for Computational Linguistics (NILC),
Institute of Mathematical and Computer Sciences,
University of São Paulo, São Paulo, Brazil
{nathansh,sandra}@icmc.usp.br, magali.duran@uol.com.br

Abstract. Semantic Role Labeling (SRL) is a Natural Language Processing task that enables the detection of events described in sentences and the participants of these events. For Brazilian Portuguese (BP), there are two studies recently concluded that perform SRL in journalistic texts. [1] obtained F1-measure scores of 79.6, using the PropBank.Br corpus, which has syntactic trees manually revised; [8], without using a treebank for training, obtained F1-measure scores of 68.0 for the same corpus. However, the use of manually revised syntactic trees for this task does not represent a real scenario of application. The goal of this paper is to evaluate the performance of SRL on revised and non-revised syntactic trees using a larger and balanced corpus of BP journalistic texts. First, we have shown that [1]'s system also performs better than [8]'s system on the larger corpus. Second, the SRL system trained on non-revised syntactic trees performs better over non-revised trees than a system trained on gold-standard data.

Keywords: Semantic Role Labeling · Non-revised syntactic trees · Brazilian Portuguese

1   Introduction

Semantic Role Labeling (SRL) is a Natural Language Processing (NLP) task responsible for detecting events described in sentences and the participants of these events [11]. The events are held by predicators, such as verbs and eventive names (some nouns, adjectives and adverbs) and the participants are called arguments. This work focuses on verbs.

To automatically annotate a text with semantic roles, most current SRL systems employ Machine Learning (ML) techniques. When using ML, the SRL task is generally performed on syntactic trees due to the extensive set of features that have been identified in the syntactic structure of a sentence, such as those presented by [11].

However, as the syntactic trees of a sentence are generated by automatic parsers and these tools are subject to errors, problematic trees are often manually revised by linguists. For Brazilian Portuguese SRL, the best performance

© Springer International Publishing Switzerland 2016
J. Silva et al. (Eds.): PROPOR 2016, LNAI 9727, pp. 202–212, 2016.
DOI: 10.1007/978-3-319-41552-9_20

was obtained by [1], with F1-measure scores of 79.6 when annotating revised syntactic trees, without reported outcomes for non-revised trees. However, the use of corrected trees does not represent a real scenario of application. In this sense, Fonseca's work [8], following an approach that does not use syntactic features, obtained F1-measure scores of 68.0 in Portuguese sentences. It is known that SRL systems using syntactic features perform better than those that do not use them. However, there are no SRL results for Portuguese on non-revised syntactic trees.

This work evaluates the use of revised and non-revised syntactic trees for manual and automatic SRL tasks in Brazilian Portuguese. We demonstrate that human annotation errors are directly related to annotation errors made by machines. We attribute these errors to problematic syntactic trees generated by the parser. We also show that, for a good performance in automatic SRL on non-revised syntactic trees, it is necessary to train the SRL system with the same type of data and/or invest in improving the parser used to preprocess the corpus.

In Sect. 2, we show the corpora collected and compiled in this work. In Sect. 3, we present the methodology for manual annotation of the corpus whose non-revised syntactic trees are annotated with semantic roles. In Sect. 4, we present the state-of-art SRL system for Portuguese and a system whose methodology does not rely on syntactic errors. In Sect. 5, we show conducted experiments and the obtained results. At last, Sect. 6 presents this work's conclusions.

2   Selection of the Corpora

To evaluate the SRL task on syntactic trees with errors, we annotated semantic roles in a new corpus compiled for this work, whose syntactic trees had not been revised, and also used a corpus annotated with semantic roles, whose syntactic trees had been manually revised, so that we could compare the results. The syntactic trees of both corpora were generated by the PALAVRAS parser [2]. The corpora used in this work are PropBank.Br version 1.1[1] [6], referred to as PB-Br.v1 and a selection of the PLN-Br, corpus of texts from Folha de São Paulo [3], referred to as PB-Br.v2[2].

In the PB-Br.v1 corpus, we observed that many verbs have only one annotation instance and this data sparsity is undesirable for machine learning purposes. When it comes to learning annotation of semantic roles, we have to consider three aspects: (i) which verbs are represented in the corpus; (ii) which meanings of verbs are represented in the corpus; and (iii) which syntactic alternations are represented in the corpus. Alternation is changing the order of constituents (syntactic, semantic, or both at once). For example, the passive voice is a syntactic alternation marking a semantic alternation (the patient takes the place of the agent in the syntactic subject position). Normally, the number of meanings for a

---

[1] Available at http://nilc.icmc.usp.br/portlex/images/arquivos/propbank-br/Prop-BankBr_v1.1.xml.zip.

[2] Available at http://nilc.icmc.usp.br/semanticnlp/propbankbr/pbbr-v2.html.

single verb is associated with the amount of syntactic alternations that it admits, but there may be verbs that admit a large number of alternations for the same meaning. Our goal in compiling the PB-Br.v2 was to get as much representation as possible in the three items above with the lowest number of annotation instances, since manual annotation is costly. The sentences were divided into classes according to linguistic criteria and considering the ML, as we chose to start the annotation process with the most frequent verbs. Using corpus statistics, we found that verbs with a frequency higher than 1000 represent 90 % of verb occurrences in the corpus. We assume that they are good material for training a classifier of semantic roles, because if the classifier learns to rank well 90 % of the corpus it should have good accuracy. In addition, the highly frequent verbs, excluding auxiliary, copula verbs and verbs that require clausal complements, are probably the most polysemous in the language. More information about the selection of sentences for the PB-Br.v2 can be found in [7][3].

The PB-Br.v1 corpus contains 5,931 annotated instances of 3,348 sentences and the PB-Br.v2 contains 7,661 annotated instances of 7,442 sentences. We generate instances of a sentence for each verb contained in it. Furthermore, we only selected instances whose syntactic trees generated by the parser are related, i.e. all elements of the syntactic tree are connected. Section 3 presents the results of the manual annotation of semantic roles on the selection made for the PB-Br.v2.

3   Manual Annotation of the PB-Br.v2

The manual annotation of semantic roles was performed on 7,661 selected instances of the PB-Br.v2, following the annotation of the PropBank project [10], but based on annotation guidelines[4] customized to the Portuguese language enriched with wrong syntactic trees annotation process. The Tiger XML output of the PALAVRAS syntactic parser was used, in the same format of the PB-Br.v1 corpus. Furthermore, Tiger XML syntactic trees can be processed with the SALTO tool [4] that was used in annotating the corpus in question.

A group of seven annotators and one adjudicator participated in the annotation process. The annotators were trained and received a copy of the annotation guidelines. In addition, a repository of verbs and their meanings, called Verbo-Brasil, was available during the annotation[5]. The task consisted of annotating the meaning of the verb – to identify the set of expected semantic roles – and assigning semantic role labels chosen from a set of six numbered arguments (ArgN) and 12 modifiers (ArgM). The description of each role is detailed in the annotation manual. The annotation scheme followed the double-blind standard, in which two annotators annotate the same portion of instances and, afterwards, the adjudicator solves disagreements. We distribute the annotation of instances by blocks, gathering instances of the same verb in the same task. Separately, the

---

[3] Available at http://nilc.icmc.usp.br/semanticnlp/propbankbr/relat.html.

[4] Available at http://nilc.icmc.usp.br/semanticnlp/propbankbr/manual.html.

[5] Available at http://143.107.183.175:12680/verbobrasil/.

sets of copula verbs and verbs that require clausal complements were distributed. The same annotation instance was never assigned to more than two annotators.

After concluding the annotation, we calculated the Kappa statistics [5]. As the SRL task traditionally consists of two steps: identification of arguments and classification of semantic roles of each argument, we calculated the individual Kappa for each step: 0,96 and 0,90 for copula verbs; 0,96 and 0,95 for clausal complement verbs; and 0,79 and 0,75 for verbs with no syntactic pattern. We also calculated the Kappa statistics of the identification of verb meaning: 1,0 for copula verbs, 1.0 for clausal complement verbs and 0,92 for verbs with no syntactic pattern.

The set of verbs that require clausal complements obtained important Kappa results, considered almost perfect by the Kappa scale of Landis and Koch [9]. As the annotation of this set is very predictable and has little syntactic diversity, the identification and classification obtained almost maximum values. The annotation of copula verbs also obtained almost perfect results. In this scenario, however, there is a greater syntactic diversity, which led to a drop in the agreement of argument classification compared to the set of verbs that require clausal complements. This means that the annotators agreed in identifying that a particular syntactic node is the verb argument, but disagreed in selecting its semantic role. Finally, the set of plain verbs that present several alternations obtained a substantial Kappa that, in the used Kappa scale, is lower than that of the other annotated verb sets. For this set, we found that there was disagreement in selecting the meaning of verbs, which triggered disagreements about the detection of arguments and the selection of semantic roles. Additionally, this is the set of verbs whose annotation is far from being predictable as they present a lot of syntactic alternation, i.e., syntactic constituents may appear in different orders, which affects the distribution of roles. After analyzing human annotation, we separated the portion in which there was disagreement between the annotators and sent it to the adjudicator, who solved the disagreements and generated a final version of the annotation. Table 1 shows the number of instances in which there was full agreement between annotators and those that had to be revised. We noted there was disagreement in over 50 % of the instances. This means that, despite the high Kappa values, most instances had to be revised. However, this does not imply a high rate of disagreement – as suggested by the Kappa values. For example, an instance is sent for adjudication even if annotators have agreed on 9 out of 10 annotated arguments and disagreed on just one.

Table 1. Proportion of the PB-Br.v2 annotation in which there was full agreement.

	Full agreement	Some agreement	Total
Copula verbs	46 (64.7 %)	25 (35.2 %)	71
Verbs of clausal complements	—	—	602
Other plain verbs	3,130 (40,7 %)	4,542 (59,2 %)	7,672
Total	**3,176**	**4,567**	**8,345**

It is important to note that many syntactic trees contain errors generated by the parser. As the annotation guidelines do not cover how to deal with all possible errors generated by the parser, the free interpretation of how to annotate these inconsistent trees may have contributed to the number of disagreements obtained.

4   SRL Systems

This paper analyzes the performance of two newly developed systems to annotate semantic roles of texts written in Brazilian Portuguese. Fonseca [8] developed a system for annotation of semantic roles for Brazilian Portuguese, avoiding dependence on external NLP tools, such as syntactic parser. Its results on revised syntactic trees of the PB-Br.v1 corpus were a F1-measure of 68.0. The work of Alva-Manchego [1], which uses the PB-Br.v1 corpus for training, employed a supervised approach for automatic annotation of semantic roles on syntactic trees of Brazilian Portuguese sentences. The system performance on revised syntactic trees of the PB-Br.v1 corpus was a F1-measure of 79.6. The main difference between the systems is the use of syntactic features (the former does not use them). In addition, none of the systems evaluated their performance on non-revised syntactic trees. The contrast on the SRL performance on revised and non-revised syntactic trees has never been done for Portuguese. Yet, few studies relate the causes of errors in human annotation with the errors generated by the machine in the SRL task, and this is one of the contributions of our work.

5   Experiments

In this section, we present the experiments developed to contrast the performance of SRL systems trained on syntactic trees revised by humans (treebanks or gold-standard data) and non-revised syntactic trees. We used Alva-Manchego's system [1] in most experiments because it is the state-of-art system in SRL for Brazilian Portuguese. We also used Fonseca's system [8] in a final experiment to contrast its performance with the one of Alva-Manchego's system. Fonseca's system does not use syntactic trees and, therefore, it does not suffer from the syntactic errors generated by the parser. In the scenario of the PB-Br.v2, it is interesting to note whether the errors generated by the parser are significant to cause the Alva-Manchego's system to have a SRL performance lower than Fonseca's.

The experiments reported below evaluate Alva-Manchego's system on different sets of the PB-Br.v2: (i) the Agreement set, composed of instances where the annotators fully agreed, therefore, not sent for adjudication; (ii) the Adjudication set, composed of instances that were sent for adjudication; and (iii) the Full set, comprising Agreement and Adjudication.

First, we conducted a 10-fold cross-validation for the set where there was agreement between PB-Br.v2 annotators. The following results are categorized by "corr." indicating correct labeling, "excess" indicating a mislabeled annotation, "missed" indicating an argument that was not selected as a candidate to annotation, "prec." as precision, "rec." as recall and "F1" as F-measure. The results in Table 2 show a reasonable indication of the SRL quality, since the Agreement set has approximately 50 % of the amount of PB-Br.v1 instances used by Alva-Manchego. To contrast it with the Agreement set, we also checked the quality of the SRL on the Adjudication set. The results in Table 3 show that the Adjudication set, in spite of being approximately 80 % greater than the PB-Br.v1 and approximately 40 % greater than the Agreement set, does not allow an easy (or simple) automatic annotation of semantic roles. We can interpret that if humans find it hard to annotate, the machine will have the same problem – which justifies the difference of F1 scores of 6.98 on the performance of the Agreement (F1 = 72.71) and Adjudication (F1 = 65.73) sets. We can also speculate that the parsing errors, which caused different annotation interpretations to the human eye, hindered the automatic learning of the task because it increases data sparsity.

Table 2. 10-fold cross-validation results for the PB-Br.v2 Agreement set.

Overall	corr.	excess	missed	prec.	rec.	F1
	3,640	1,401	1,332	72.23	73.22	72.71
A0	807	230	146	77.98	84.64	81.09
A1	1,853	491	600	79.13	75.56	77.28
A2	367	294	223	55.48	62.14	58.42
A3	10	17	43	45.67	26.82	27.83
A4	2	1	15	20.00	8.33	11.67
AM-ADV	9	12	13	28.00	26.67	24.56
AM-CAU	2	4	16	13.33	8.33	9.00
AM-DIR	0	0	0	–	–	–
AM-DIS	6	8	16	21.67	16.67	18.16
AM-EXT	16	4	11	75.00	60.83	64.33
AM-LOC	103	113	50	48.69	67.53	55.54
AM-MNR	80	75	57	52.57	60.55	54.88
AM-NEG	134	14	19	90.30	87.80	88.83
AM-PRD	0	0	1	0.00	0.00	0.00
AM-PRP	39	28	17	58.89	71.23	62.89
AM-REC	0	0	0	0.00	–	–
AM-TMP	212	110	105	65.75	67.07	66.03

Table 3. 10-fold cross-validation results for the PB-Br.v2 Agreement set.

Overall	corr.	excess	missed	prec.	rec.	F1
	7,164	4,001	3,469	64.16	67.37	65.73
A0	1,873	491	377	79.26	83.24	81.18
A1	2,703	1,208	1,142	69.12	70.29	69.68
A2	489	708	517	40.82	48.76	44.32
A3	14	36	82	28.39	14.14	18.00
A4	8	5	23	36.00	19.05	23.00
AM-ADV	157	134	119	54.37	57.50	55.49
AM-CAU	47	67	75	42.80	38.17	39.43
AM-DIR	0	5	13	0.00	0.00	0.00
AM-DIS	150	76	172	66.76	46.85	54.69
AM-EXT	56	35	43	62.70	58.19	58.55
AM-LOC	413	374	161	52.45	72.60	60.79
AM-MNR	200	281	219	41.81	47.78	44.48
AM-NEG	226	22	39	90.92	86.16	88.10
AM-PRD	46	125	124	27.44	27.53	26.91
AM-PRP	97	86	69	53.52	59.89	56.06
AM-REC	0	0	1	0.00	0.00	0.00
AM-TMP	684	338	271	67.24	71.64	69.26

In the 10-fold cross-validation experiment on the Full set of the PB-Br.v2, we obtained F1 of 69.12 (Table 4), which is between the 65.73 (Adjudication set) and 72, 71 (Agreement set) values. We can speculate that despite the Adjudication set has more instances and results lower than the Agreement set, the latter supplies the system with syntactic trees without errors or trees that have frequent errors for which a standard treatment is predicted in the guidelines, and therefore can be identified by ML. It is also interesting to mention that the F1 difference between the system trained on the PB-Br.v1 gold-standard data and the system trained on the non-revised syntactic trees of the PB-Br.v2 is 10.48 F1 scores. The difference of around 10.0 F1 scores between a system trained on revised trees and on non-revised trees has already been investigated by [10, 12] for the English language.

Table 4. 10-fold cross-validation results for the Full set of the PB-Br.v2, comprising both Agreement and Adjudication sets, when using Alva-Manchego's system.

Overall	corr.	excess	missed	prec.	rec.	F1
	11,047	5,335	4,537	67.45	70.89	69.12
A0	2,703	671	500	80.10	84.37	82.15
A1	4,630	1,713	1,664	73.01	73.57	73.28
A2	898	974	702	48.07	56.09	51.67
A3	25	64	124	28.62	16.72	20.85
A4	18	10	30	68.00	36.07	45.21
AM-ADV	159	150	139	50.00	52.37	50.64
AM-CAU	54	69	86	43.71	39.87	40.99
AM-DIR	0	4	13	0.00	0.00	0.00
AM-DIS	183	116	161	60.95	52.57	56.20
AM-EXT	72	41	54	63.05	57.06	59.51
AM-LOC	529	460	198	53.73	72.96	61.75
AM-MNR	302	352	254	46.42	54.34	49.84
AM-NEG	360	42	58	88.91	85.90	87.23
AM-PRD	37	123	134	24.33	20.79	21.96
AM-PRP	147	99	75	59.75	66.54	62.48
AM-REC	0	0	1	0.00	0.00	0.00
AM-TMP	930	447	342	67.53	73.14	70.12

Table 5. 10-fold cross-validation results for the Full set of the PB-Br.v2 selection, comprising the Agreement and Adjudication sets, using Fonseca's system.

Overall	corr.	excess	missed	prec.	rec.	F1
	8,038	6,475	7,464	55.44	51.87	53.58
A0	2,318	1,058	911	68.72	71.82	70.18
A1	3,497	2,979	2,755	54.13	55.93	55.00
A2	440	938	11,16	31.97	28.26	29.94
A3	0	6	150	0.00	0.00	0.00
A4	0	0	48	0.00	0.00	0.00
AM-ADV	90	63	208	62.09	30.84	39.95
AM-CAU	5	23	134	10.19	4.36	6.04
AM-DIR	0	0	13	0.00	0.00	0.00
AM-DIS	121	94	222	57.06	35.73	42.93
AM-EXT	53	34	72	61.13	43.78	49.81
AM-LOC	282	469	444	37.57	38.93	37.98
AM-MNR	143	180	412	45.09	26.13	32.86
AM-NEG	355	40	59	89.81	86.34	87.89
AM-PRD	6	46	163	16.00	3.33	5.02
AM-PRP	64	100	154	38.15	29.89	33.18
AM-REC	0	0	1	0,00	0.00	0.00
AM-TMP	664	445	602	59,92	52.51	55.83

The contrast between the F1 values for the PB-Br.v1 and the PB-Br.v2 is noteworthy. Although the selection made on the PB-Br.v2 respects a distribution that should favor the SRL and this corpus is approximately 24 % larger than the PB-Br.v1, the results show a system with 10.48 F1 scores lower than the result obtained from the system trained by Alva-Manchego on the PB-Br.v1 corpus (F1 69.12 against 79.6). The evidence that the use of automatic parser without human revision adversely affects the performance of SRL systems become more evident when we contrast the results in Table 6 with those of Table 7. The question, however, is whether this drop on performance actually represents a system with worse overall performance, or if the annotation in the automatic parser scenario is difficult to the point that the F1 value of the trained system in PB-Br.v2 (Table 4) is considered good. To this end, we conducted two experiments: one annotating the PB-Br.v1 corpus with the system trained on the Full set of the PB-Br.v2 and another in the opposite direction - annotating the Full set of the PB-Br.v2 with the system trained on the PB-Br.v1 corpus.

We can see in Table 6 that the SRL in revised syntactic trees, using Alva-Manchego's system trained on non-revised syntactic trees, is feasible, even though its quality is not comparable to that of a system trained on revised trees (F1 72.62 versus F1 79.6). We also noted that the value obtained of 72.62 is greater than the 10-fold cross-validation value of 69.12 of the system trained on the PB-Br.v2. Thus, we realized that in addition to a system trained on non-revised syntactic trees being able to annotate syntactic trees containing errors, it could annotate perfect trees with a performance superior than what it annotates in its own scenario. This means that training a SRL system on non-correct syntactic trees makes the system capture syntactic patterns of correct trees and learn to deal with the noise contained in not syntactically correct trees.

Doing the opposite direction of annotation, we can see in Table 7 that the quality of the SRL of the PB-Br.v2 problematic trees using the system trained on PB-Br.v1 is 14.36 F1 scores lower than the result of the 10-fold cross-validation for the Full set of PB-Br.v2 (F1 of 69.12 against 54.76). This result shows that a system trained on perfect syntactic trees faces greater difficulty when annotating faulty syntactic trees. As the PB-Br.v1 gold standard corpus does not contain noisy examples, a minimum deviation from the correct pattern of syntactic tree is enough to cause the system to make annotation errors.

Table 6. Annotation results for PB-Br.v1 using the SRL system trained on the Full set of the PB-Br.v2.

Overall	corr.	excess	missed	prec.	rec.	F1
	9,622	2,958	4,298	76.49	69.12	72.62
A0	2,281	463	778	83.13	74.57	78.61
A1	4,187	863	1142	82.91	78.57	80.68
A2	772	526	613	59.48	55.74	57.55
A3	20	60	127	25.00	13.61	17.62
A4	35	26	79	57.38	30.70	40.00
A5	0	0	1	0.00	0.00	0.00
AM-ADV	162	95	212	63.04	43.32	51.35
AM-CAU	75	55	77	57.69	49.34	53.19
AM-DIR	0	0	13	0.00	0.00	0.00
AM-DIS	180	87	129	67.42	58.25	62.50
AM-EXT	32	20	46	61.54	41.03	49.23
AM-LOC	523	257	190	67.05	73.35	70.06
AM-MNR	168	194	239	46.41	41.28	43.69
AM-NEG	315	47	28	87.02	91.84	89.36
AM-PRD	14	62	169	18.42	7.65	10.81
AM-PRP	97	52	56	65.10	63.40	64.24
AM-REC	0	0	8	0.00	0.00	0.00
AM-TMP	761	151	388	83.44	66.23	73.85

Table 7. Annotation results for the Full set of PB-Br.v2 using the SRL system trained on PB-Br.v1.

Overall	corr.	excess	missed	prec.	rec.	F1
	7,972	5,559	7,612	58.92	51.16	54.76
A0	2,113	1,040	1,090	67.02	65.97	66.49
A1	3,426	1,543	2,868	68.95	54.43	60.84
A2	478	526	1,122	47.61	29.88	36.71
A3	11	24	138	31.43	7.38	11.96
A4	8	13	40	38.10	16.67	23.19
A5	0	0	0	–	–	–
AM-ADV	100	179	198	35.84	33.56	34.66
AM-CAU	36	106	104	25.35	25.71	25.53
AM-DIR	0	2	13	0.00	0.00	0.00
AM-DIS	106	74	238	58.89	30.81	40.46
AM-EXT	63	45	63	58.33	50.00	53.85
AM-LOC	412	520	315	44.21	56.67	49.67
AM-MNR	242	472	314	33.89	43.53	38.11
AM-NEG	204	49	214	80.63	48.80	60.80
AM-PRD	8	245	163	4.66	4.68	3.77
AM-PRP	101	126	121	44.49	45.50	44.99
AM-REC	0	25	1	0.00	0.00	0.00
AM-TMP	664	570	608	53.81	52.20	52.99

To strengthen the analysis of results, we annotated the PB-Br.v2 Agreement and Adjudication sets with the system trained on the PB-Br.v1. Table 8 shows the result of 58.96 F1 in the annotation of the Agreement set of the PB-Br.v2 by the system trained in PB-Br.v1 and Table 9 presents F1 of 52.66 for annotation of the Adjudication set. The difference in the performance of annotation of the sets was 6.3 F1 scores, while the difference of the 10-fold cross-validation performed on these sets was 6.98 F1 scores. As the difference in the performance of annotation of the sets is close, we understand that, indeed, there is a syntactic complexity threshold that distinguishes the Agreement and the Adjudication sets of the PB-Br.v2. Tables 10 and 11 show confusion matrices for the results presented on Tables 6 and 7. These matrices show that our SRL system trained on noisy data performs better when annotating revised trees (Table 9) than the system trained on revised trees when annotating noisy data (Table 10).

For example, there are 32 % of AM-TMP syntactic nodes not identified by the SRL trained on PB-Br.v1 and tested on noisy data (PB-Br.v2) whereas there are only 6 % of AM-TMP not identified by the SRL trained on PB-Br.v2 and tested on the treebank (PB.Br-v1).

Finally, we conducted a 10-fold cross-validation experiment with Fonseca's system and the PB-Br.v2 full. The F1 value of 53,58 for the system that does not use syntactic features presented in Table 5 shows that even with the use of syntactic trees with errors, the SRL approach for syntactic trees is better.

Table 8. Annotation results for the PB-Br.v2 Agreement set.

Overall	corr.	excess	missed	prec.	rec.	F1
	2,803	1,718	2,184	62.00	56.21	58.96
A0	631	372	322	62.91	66.21	64.52
A1	1,455	491	998	74.77	59.32	66.15
A2	210	144	380	59.32	35.59	44.49
A3	4	10	49	28.57	7.55	11.94
A4	4	4	13	50.00	23.53	32.00
AM-ADV	11	58	11	15.94	50.00	24.18
AM-CAU	5	24	13	17.24	27.78	21.28
AM-DIR	0	1	0	0.00	0.00	0.00
AM-DIS	11	27	11	28.95	50.00	36.67
AM-EXT	14	14	13	50.00	51.85	50.91
AM-LOC	92	144	61	38.98	60.13	47.30
AM-MNR	73	130	64	35.96	53.28	42.94
AM-NEG	81	14	72	85.26	52.94	65.32
AM-PRD	0	58	1	0.00	0.00	0.00
AM-PRP	27	44	29	38.03	48.21	42.52
AM-REC	0	6	0	0.00	0.00	0.00
AM-TMP	185	177	132	51.10	58.36	54.49

Table 9. Annotation results for the PB-Br.v2 Adjudication set.

Overall	corr.	excess	missed	prec.	rec.	F1
	5,166	3,844	5,444	57.34	48.69	52.66
A0	1,482	668	768	68.93	65.87	67.36
A1	1,972	1,051	1,873	65.23	51.29	57.43
A2	264	386	742	40.62	26.24	31.88
A3	7	14	89	33.33	7.29	11.97
A4	4	9	27	30.77	12.90	18.18
AM-ADV	89	121	187	42.38	32.25	36.63
AM-CAU	31	82	91	27.43	25.41	26.38
AM-DIR	0	1	13	0.00	0.00	0.00
AM-DIS	95	47	227	66.90	29.50	40.95
AM-EXT	49	31	50	61.25	49.49	54.75
AM-LOC	320	376	254	45.98	55.75	50.39
AM-MNR	169	342	250	33.07	40.33	36.34
AM-NEG	123	35	142	77.85	46.42	58.16
AM-PRD	8	187	162	4.10	4.71	4.38
AM-PRP	74	82	92	47.44	44.58	45.96
AM-REC	0	19	1	0.00	0.00	0.00
AM-TMP	479	393	476	54.93	50.16	52.44

Table 10. Confusion Matrix of SRL on PB-Br.v1 using system trained on PB-Br.v2.

		-1	0	1	2	3	4	5	6	7	8	9	10	11	12	13	14	15	16	17
-1	-NONE-	0	615	605	168	26	21	0	110	31	1	81	4	81	70	24	81	28	2	186
0	A0	130	2,281	265	33	2	1	1	0	9	0	1	4	1	5	0	6	0	1	3
1	A1	292	140	4,187	247	26	21	0	8	5	2	5	11	20	33	0	9	5	5	33
2	A2	76	10	196	772	48	25	0	5	9	5	4	5	36	50	0	17	14	0	26
3	A3	5	1	19	22	20	1	0	0	0	1	0	2	1	4	0	1	0	0	3
4	A4	6	0	2	3	2	35	0	0	1	2	0	0	0	4	0	0	3	0	3
5	A5	0	0	0	0	0	0	0	0	0	0	0	0	0	0	0	0	0	0	0
6	AM-ADV	31	4	1	1	0	0	0	162	1	0	18	3	1	14	3	4	1	0	13
7	AM-CAU	16	0	1	8	3	1	0	7	75	0	1	0	3	10	0	2	1	0	2
8	AM-DIR	0	0	0	0	0	0	0	0	0	0	0	0	0	0	0	0	0	0	0
9	AM-DIS	34	0	1	3	0	0	0	19	3	0	180	0	3	2	0	4	2	0	16
10	AM-EXT	7	0	1	2	0	0	0	4	0	0	2	32	0	2	0	0	0	0	2
11	AM-LOC	43	0	24	69	3	6	0	7	3	1	3	0	523	18	0	26	0	0	54
12	AM-MNR	48	2	9	22	7	2	0	39	5	1	5	7	12	168	1	11	1	0	22
13	AM-NEG	30	0	0	0	0	0	0	0	1	0	1	0	0	0	315	0	0	0	15
14	AM-PRD	20	1	1	4	0	0	0	2	4	0	3	1	7	9	0	14	0	0	10
15	AM-PRP	6	0	7	22	7	0	0	2	3	0	2	0	1	0	0	2	97	0	0
16	AM-REC	0	0	0	0	0	0	0	0	0	0	0	0	0	0	0	0	0	0	0
17	AM-TMP	55	1	9	9	3	1	0	9	2	0	3	9	24	18	0	6	1	0	761

Table 11. Confusion Matrix of SRL on PB-Br.v2 using system trained on PB-Br.v1.

		-1	0	1	2	3	4	5	6	7	8	9	10	11	12	13	14	15	16	17
-1	-NONE-	0	961	2,242	564	61	16	143	76	4	206	2	20	231	193	195	101	90	1	485
0	A0	600	2,113	365	46	3	1	2	6	1	2	0	1	1	4	1	6	0	0	1
1	A1	1,139	93	3,426	236	10	2	1	0	1	0	0	19	8	11	0	5	1	0	17
2	A2	307	11	96	478	28	7	0	1	0	0	0	0	27	16	0	9	11	0	13
3	A3	11	2	3	6	11	1	0	0	0	1	0	0	0	0	0	0	0	0	0
4	A4	5	0	3	2	0	8	0	0	0	0	0	0	0	1	0	0	0	0	2
5	AM-ADV	114	1	6	1	0	0	100	2	0	8	0	3	3	14	2	11	3	0	11
6	AM-CAU	76	4	7	6	1	1	1	36	1	0	0	0	1	4	0	3	0	0	1
7	AM-DIR	1	0	0	1	0	0	0	0	0	0	0	0	0	0	0	0	0	0	0
8	AM-DIS	60	0	1	1	0	0	3	0	0	106	0	0	2	6	0	0	0	0	1
9	AM-EXP	0	0	0	0	0	0	0	0	0	0	0	0	0	0	0	0	0	0	0
10	AM-EXT	33	0	6	2	1	0	1	0	0	0	0	63	0	1	0	0	0	0	0
11	AM-LOC	271	3	30	97	6	2	5	4	0	7	0	2	412	24	1	9	6	0	53
12	AM-MNR	237	1	32	72	14	2	24	8	4	3	0	13	24	242	1	11	5	0	21
13	AM-NEG	40	0	8	1	0	0	0	0	0	0	0	0	0	0	204	0	0	0	0
14	AM-PRD	169	10	17	24	0	3	2	5	0	1	0	0	5	7	0	8	0	0	2
15	AM-PRP	74	0	0	37	6	3	2	1	1	0	0	0	0	1	0	0	101	0	1
16	AM-REC	4	1	17	2	0	0	0	0	0	1	0	0	0	0	0	0	0	0	0
17	AM-TMP	397	3	35	24	8	2	14	1	1	9	0	5	13	32	14	7	5	0	664

6 Conclusions

The obtained results show that a SRL system responsive to syntactic errors, trained on noisy data (non-revised syntactic trees), performs a better SRL on noisy data than when it is trained on revised trees (treebank). We also noted that the SRL system obtained better results when tested on the set in which there was full agreement between annotators and lower results when tested on the set in which there was disagreement between annotators. We did not include in PB-Br.v2 any annotation mark to distinguish well-formed parse trees from those containing parsing errors. For this reason, it was not possible to verify whether the existence of parsing errors are correlated to the drop in inter-annotator agreement rates. During the adjudication process, however, we realized that this correlation probably exists and it would be worthwhile to explore this hypothesis in future work. Based on confusion matrices' results, however, we speculate that the learning from non-revised trees allows the system to better identify the SRL candidates. We also noted that Fonseca's system [8] performs a SRL inferior to Alva-Manchego's [1], even in an unfavorable scenario for the latter system. We believe that, for Brazilian Portuguese, the use of syntactic trees for the SRL task is still the most promising means and, therefore, efforts to the improvement of syntactic parsers are welcome. We also understand that in a real scenario of application (on-the-fly processing), the data is passed directly from the syntactic parser to the SRL system without human intervention to correct the trees. Therefore, we can say that in a real scenario of application, the training of a SRL system on non-revised syntactic trees corpus, such as the PB-Br.v2, provides better annotation of journalistic texts than the system trained on revised syntactic trees corpus (PB-Br.v1).

Acknowledgments. Part of the research developed for this work was sponsored by *Samsung Eletrônica da Amazônia Ltda.* under the terms of Brazilian federal law number 8.248/91. Part of the results presented in this paper were obtained through research activity in the project titled "Semantic Processing of Brazilian Portuguese Texts", sponsored by *Samsung Eletrônica da Amazônia Ltda.* under the terms of Brazilian federal law number 8.248/91.

References

1. Alva-Manchego, F.E., Rosa, J.L.G.: Semantic role labeling for Brazilian Portuguese: a benchmark. In: Pavón, J., Duque-Méndez, N.D., Fuentes-Fernández, R. (eds.) IBERAMIA 2012. LNCS, vol. 7637, pp. 481–490. Springer, Heidelberg (2012)
2. Bick, E.: The Parsing System "Palavras": Automatic Grammatical Analysis of Portuguese in a Constraint Grammar Framework. Aarhus University Press, Aarhus (2000)
3. Bruckschen, M., Muniz, F., Souza, J., Fuchs, J., Infante, K., Muniz, M., Gonçalves, P., Vieira, R., Aluísio, S.: Anotação Lingüística em XML do Corpus PLN-BR. NILC-TR-09-08. Technical report, University of São Paulo, Brazil (2008)
4. Burchardt, A., Erk, K., Frank, A., Kowalski, A., Pado, S.: SALTO - a versatile multi-level annotation tool. In: Proceedings of the 5th International Conference on Language Resources and Evaluation (LREC-2006), pp. 517–520 (2006)
5. Carletta, J.: Assessing agreement on classification tasks: the kappa statistic. Comput. Linguist. **22**(2), 249–254 (1996)
6. Duran, M.S., Aluísio, S.M.: Propbank-Br: a Brazilian treebank annotated with semantic role labels. In: Proceedings of the International Conference on Language Resources and Evaluation (LREC-2012), pp. 1862–1867 (2012)
7. Duran, M.S., Sepúlveda-Torres, L., Viviani, M.C., Hartmann, N.S., Aluísio, S.M.: Seleção de sentenças do córpus PLN-Br para compor o córpus de anotação de papéis semânticos Propbank-Br.v2. NILC-TR-14-07. Technical report, University of São Paulo, Brazil (2014)
8. Fonseca, E.R., Rosa, J.L.G.: A two-step convolutional neural network approach for semantic role labeling. In: Neural Networks (IJCNN), The 2013 International Joint Conference on Neural Networks, pp. 1–7 (2013)
9. Landis, J.R., Koch, G.G.: The measurement of observer agreement for categorical data. Biometrics **33**(1), 159–174 (1977)
10. Palmer, M., Gildea, D., Kingsbury, P.: The proposition bank: an annotated corpus of semantic roles. Comput. Linguist. **31**(1), 71–106 (2005)
11. Palmer, M., Gildea, D., Xue, N.: Semantic Role Labeling, Synthesis Lectures on Human Language Technologies, vol. 3. Morgan & Claypool Publishers (2010)
12. Toutanova, K., Haghighi, A., Manning, C.D.: A global joint model for semantic role labeling. Comput. Linguist. **34**(2), 161–191 (2008)

Improving Coreference Resolution with Semantic Knowledge

Evandro Fonseca[1]([⊠]), Renata Vieira[1], and Aline Vanin[2]

[1] Pontifícia Universidade Católica do Rio Grande do Sul, Porto Alegre, Brazil
evandro.fonseca@acad.pucrs.br, renata.vieira@pucrs.br
[2] Universidade Federal de Ciências da Saúde de Porto Alegre,
Porto Alegre, Rio Grande do Sul, Brazil
aline.vanin@ymail.com

Abstract. This paper evaluates the impact of semantic features in coreference resolution for the Portuguese language. We show that the new proposed features obtained on the basis of currently available Portuguese semantic resources improve results in precision, recall and f-measure.

Keywords: Coreference resolution · Semantic knowledge · Information extraction · Machine learning

1 Introduction

The problem of coreference resolution has received a great deal of attention from the computational linguistics community. This problem usually requires previous language processing on many levels (POS-tagging, parsing, semantic analysis). Less resourceful languages may experience greater difficulty in advancing towards this kind of task. Usually, there is also a shortage of data. While, for English, we have the Ontonotes corpus [21], which contains 34.290 coreference chains, for Portuguese language we are only aware of the Summ-it corpus [5], which contains 560 coreference chains; the Garcia and Gamallo corpus [15] which is annotated only with regards to entities of type person; and the HAREM corpus [12] which contains named entities and their relations (including identity, which may considered as a form of coreference annotation). In spite of the smaller size, Summ-it is a high quality resource which has been used in many research initiatives ([6–8,11]). Regarding semantic resources, only recently we see large semantic databases as an alternative if one wants to go beyond the limits of the usual string matching and morphological heuristics for coreference resolution.

In this paper, we propose a model based on supervised machine learning for Portuguese coreference resolution, where semantic knowledge is considered in the elaboration of features. Our initial set of features is based on traditional previous work in the field, such as Soon et al.'s [25], enriched with features presented in more recent work such as Lee et al.'s [17], also following previous research on coreference for Portuguese [11]. On top of that we evaluate the impact of new features that makes use of semantic knowledge.

J. Silva et al. (Eds.): PROPOR 2016, LNAI 9727, pp. 213–224, 2016.
DOI: 10.1007/978-3-319-41552-9_21

The paper is organized as follows: Sect. 2 discusses the problem of coreference and the relevance of semantic knowledge for this problem; Sect. 3 presents related work; in Sect. 4, we describe the main Portuguese resources that we used in our system; Sect. 5 describes our model; in Sect. 6, we describe the experiments that measure the impact of the inclusion of semantic features; Sect. 7 presents our conclusions and future work.

2 Semantics for Coreference Resolution

Coreference resolution is an important task and also a great challenge for Natural Language Processing. It basically consists of finding different references to a same entity in a text, as in the example: [Schumacher] sofreu um acidente. [O ex-piloto] permanece em coma. ([*Schumacher*] *suffered an accident.* [*The ex-pilot*] *is still in coma*). In this case, the noun phrase [O ex-piloto] ([*The ex-pilot*]) is coreferent to [Schumacher].

There are cases where the coreference relation is simple to grasp, such as in [Barack Obama] and [Obama], in which both NPs share some identical part, in this case, "Obama". In other situations, establishing a coreference relation between two noun phrases is more complex. In cases such as [A abelha] and [O inseto] ([*The bee*] and [*The insect*]) there is a hyponymic relation between the two referents which is usually part of the readers' common sense. Besides, in this example, for Portuguese the two NPs differ in gender, a commonly used feature to deal with coreference. For a system to recognize this kind of relation, it would require an adequate semantic resource.

Although the Portuguese coreference resolution area is at an early stage of development, research in this task should be pursued, as it is quite relevant for many other tasks. In [13] it is shown that coreference resolution may provide meaningful gains for the area of entity relation extraction, since coreference links may be useful for extracting implicit relations. Consider the following sentence: [O presidente dos Estados Unidos], [Barack Obama], afirmou hoje que as alterações climáticas são a maior ameaça ao planeta. ([*The United States president*], [*Barack Obama*], *said today that the climate changes are a great threat for the planet*). When identifying and creating a coreference relation between [Barack Obama] and [o presidente] ([*the president*]), it is possible to infer a relation between the entities [Barack Obama] and [Estados Unidos] ([*United States*]) (in which Barack Obama is the president of the United States). In other words, when we say that Barack Obama is the president, it is possible to classify him as a person, as well as to say that he has a relation with the United States.

In general, previous coreference resolution systems are usually limited to non semantic lexical features, such as: string matching, heuristics for alias recognition, gender, number among others. We propose and evaluate a coreference resolution system for Portuguese with the addition of features based on entity categories and also world knowledge, considering semantic resources which were recently made available. Basically, our model combines known features, adapted from previous work with new semantic features based on semantic resources for

Portuguese. The development of more robust coreference resolution systems may help other NLP tasks for Portuguese, such as relation extraction and sentiment analysis.

3   Related Work

Coreference Resolution represents a great challenge in NLP, given the many levels of language that must be taken into account in this task. Analyzing the current state of the art, we see that it is hard to achieve good results, independent of the language. Most works go around without semantic resources, perhaps because the problem of domain independent semantic resources, that covers a broad range of semantic phenomena and that would fit into a particular processing need, was not solved yet.

One recent and relevant coreference resolution system is Lee et al.'s [17]. Their approach to coreference resolution combines the global information and precise features appointed by machine learning models with the transparency and modularity of deterministic, rule-based systems. Their Entity-Centric Model architecture applies a set of 10 deterministic sieves, where each sieve builds on the previous model's clusters output. In order to increase the recall, they combine several variations of matching.

Based on Lee's work, Garcia et al. [14] proposes Link-People: a model for coreference resolution which is tailored to person entities (what we may consider an initial semantic orientation). They considered three languages: Portuguese, Spanish and Galician. Their model combines the multi-pass architecture and a set of constraints and rules. The authors use some matching rules from [17]; in addition, they use a set of specific rules to dealing with pronouns, anaphora, cataphora for person entities. To detect person entities, the authors uses FreeLing NER [4]. In an error analysis, the authors mention the problem of lack of rich semantic resources, showing that their model could be improved by detecting semantic relations like synonymy, hyponymy and hyperonymy: [the boy] and [the youngster]. For Portuguese coreference, Garcia et al. built their own corpus [15] considering only entities of type person.

Although the semantic problem is mostly unattended, there is previous coreference resolution research that considers semantic knowledge for English. The authors of [22] evaluated the utility of world knowledge using a mention-pair and cluster-ranking model. For world knowledge, the authors used two knowledge bases: Yago and FrameNet. Their strategy consists into identifying relations like "Means", "IS-A" and "Type". Each relation is represented in YAGO as a triple. (AlbertEinstein, Type, physicist), for instance, denotes the fact that Albert Einstein is of type physicist. The relation "Means" provides different ways of expressing an entity, and therefore allows dealing with synonymy and ambiguity, i.e. for the two triples: (Einstein, Means, AlbertEinstein), and (Einstein, Means, AlfredEinstein) denotes the fact that Einstein may refer to the physicist Albert Einstein or the musician Alfred Einstein. From FrameNet, the authors used the semantic role related to verbs. For example: "Peter Anthony decries the program trading as limiting the game to a few, but he is not sure whether he wants to denounce it because [...]". Note that the semantic role may help to establish a coreference link between "program trading" and "it", since with FrameNet it is possible to retrieve a relation between "decry" and "denounce", because these words appear in the same frame and the two noun phrases have the same semantic role. The authors show that each semantic source may offer some small gains but that their cumulative benefits can be substantial. This happens because coreference may be detected either through a semantic relation, like in [the boy] and [the youngster] connection, or through factual world knowledge, such as in [the United States president] and [Obama].

Dealing with English and deterministic rules (like Lee et al.'s), Hou et al. [16] proposes a rule based system to solve anaphora and bridging. Different from our work, which tries to identify coreference (identity relation), bridging resolution consists into recognizing and linking entities through non-identity relations. An example is the meronymyc relation ("Part_of") as in [the house] [the chimney]. To identify this type of relations, the authors used WordNet [19].

For Brazilian Portuguese, Silva [24] proposed a coreference resolution system based in the same Harem [12] semantic categories, using an unsupervised learning algorithm. To detect these categories, Silva used the parser PALAVRAS [1] and the named entities recognizer Rembrandt [3]. Regarding semantic processing, the author uses the synonymy relation based on Tep2.0 [18], a thesaurus containing synonymy and antonymy for Portuguese language. Silva reports that the semantic knowledge did not show improvements in his experiments. However, he considered a small corpus, containing just nine texts, which may be considered a limitation.

Coreixas [6] proposes a coreference resolution for Portuguese also focusing on the main categories of named entities: Person, Organization, Location, Work, Thing and Other. Resources and tools were the HAREM corpus [12], the parser PALAVRAS [1] and the Summ-it corpus [5]. In order to prove which of the semantic categories may help to solve coreferences, Coreixas compares two versions of her system. The author showed that the use of categories of entities resulted in an improvement in determining whether a pair is coreferent or not. Also, the importance of world knowledge for this line of research was mentioned, emphasizing the importance of databases with synonyms, such as Wordnet, to complement and support coreference resolution.

Following Coreixas, Fonseca et al. [11] proposes a machine learning system to solve coreference in Portuguese but considering only proper names. To detect named entity categories, the authors used resources such as Repentino [23] and NERP-CRF [9], plus auxiliary lists containing common nouns, referring to certain categories, such as professions "advogado, agrônomo, juiz..." (*lawyer, agronomist, judge...*) for person and "avenida, rua, praça, cidade..." (*avenue, street, square, city...*) for location.

We see that the previous work combines the use of named entity categories and semantic knowledge resources. However, the only Portuguese semantic resource considered for this task was Tep2.0, which contains 8.528 synonym and antonym relations. There are currently available more comprehensive semantic data bases. Onto-PT [20], contains 168.858 synonymy relations, 91.466 hyperonymy/hyponymy, 9.436 meronymy and 92.598 antonymy relations.

4   Corpus and Semantic Resources

In this section, we cite the main resources used to build our coreference resolution system: the coreference annotated corpus and semantic resources.

4.1   Summ-it Corpus

Summ-it [5] is a corpus consisting of fifty journalistic texts from the Science section of Folha de São Paulo newspaper. It is part of the PLN-BR corpus [2]. The texts were annotated with syntactic, coreference and rhetorical structure information. Summ-it also includes summaries constructed manually and automatically. The corpus has a total of 560 coreference chains with an average of 3 mentions (noun phrases for each chain). The largest chain has 16 mentions. Summ-it has been used in previous coreference resolution research for Portuguese ([6–8,11]) and has had an important role in the training and validation of classification models. The coreference annotation scheme of Summ-it is distributed in tree distinct files: Markables, Words and POS.

A Markables file indicates the mentions. Each markable contains an id, a pointer to the corresponding set of words (span), a link to a coreference set (member), status (new or old) and type of NP (definite, indefinite, pronoun, etc.), for example:

– < *markable id=* "markable_95" span="word_23..word_25"
member="set_2" status="new" np_n="yes" np_form="def-np" / >

In this example, the element "span" indicates word ids (23 to 25), that correspond to a reference to the Words file, which contains the list of tokens. As in the example below, they refer to the noun phrase [o agrônomo Miguel Guerra] ([*the agronomist Miguel Guerra*]).

– < *word id* = "word_23" > o < /word >
< *word id* = "word_24" > agrônomo < /word >
< *word id* = "word_25" > Miguel_Guerra < /word >

Summ-it also contains POS files, which contains the word id, the grammatical class of the word, gender and number (singular/plural), as seen below:

– < *word id* = "word_25" >< *prop canon* = "Miguel_Guerra" *gender* = "M" *number* = "S" >< *secondary_prop tag* = "hum" / >< /prop > < /word >

4.2 Semantic Resources

As semantic resources we used the semantic annotation provided by the Palavras parser [1] and Onto-PT [20]. From Palavras we used the semantic tags for identifying references to person, organization and location. Onto-PT was considered as a general semantic base. Similarly to WordNet [19], it is structured in synsets (groups of synonymous word senses that can be seen as possible lexicalizations of a natural language concept) and semantic relations connecting synsets, including not only hyperonymy (a concept is a kind of another) and part-of (a concept is part of another), but also others, such as causation (a concept causes another) or purpose-of (a concept is used for another). To extract the relations, we utilized Onto-PT API which, for a given pair of words, retrieves all relations between the given elements, as in the examples given in Table 1. Although Onto-PT has several relations, in this work, we focused only in hyponymy, hyperonymy and synonymy relations.

Table 1. Onto-PT: examples of semantic relations returned for a pair of words.

Word pairs	Relations
estudo, pesquisa *study, research*	sinonimoDe/synonymOf
abelha, inseto *bee, insect*	hiponimoDe/hyponymOf
animal, cachorro *animal, dog*	hiperonimoDe/hyperonymOf

5 Enriched Semantic Model

Our model is based on features inspired by Lee et al.'s [17] and Soon et al. [25]. We converted some Lee et al.'s rules into features, such as: Relaxed_String_Match and Word_Inclusion; and adopted some Soon et al.'s features, like Number, Gender and Alias (which have been widely used by many related work). Some features were simply reimplemented, others had to be adapted. In special, to these initial features, we added five new semantic features: two based on entity categories (person, place and organization) and three using Onto-PT. To provide the candidate-pairs, we use the same strategy of Fonseca et al.'s [10], each noun phrase makes pair with their followers. In order to balance the data set, we choose randomly n negative pairs (where n is the quantity of positive pairs). Then we run one hundred times a ten fold cross validation for each model.

5.1 Basic Features

Next we present the basic (non-semantic) features of our model. The first six features were straightforward re-implementations, while features 7 to 12 required some form of adaptation for Portuguese, as described below.

1. Exact String Match: if the current NP and antecedent are equal.
2. Relaxed String Match: if the strings up to the head nouns are equal.
3. Word Inclusion: if there are no different words (nouns, verbs, adjectives, adverbs) in the NP, when compared to the antecedent.
4. Alias: if a NP is acronym of the other.
5. NP Distance: The distance of two NPs is given in terms of the number of NPs between them.
6. Sentence Distance: This feature explores a similar idea but now we count the distance in number of sentences.
7. Embedded Nps: This feature explores the NPs structure, checking if a noun phrase is not identical to a constituent part of the other. To recognize constituents, we observe the presence of prepositions like "de" and "em". In (a), as the noun phrases are embedded, they could no be coreferent.

   (a) [O garoto da casa ao lado] ... [a casa ao lado] ([*The boy of the next house*]) ... ([*the next house*]).

8. Proper Noun Word Match: This feature returns true if three conditions are satisfied[1]: both noun phrases must contain proper nouns; the proper nouns must be equal; and these NPs are not embedded. Example (b) shows a violation of the third condition.

   (b) [Califórnia]... [região sul da Califórnia] ([*the southern of California*]).

9. Gender: Nouns in Portuguese have gender. While in English the NP [the teacher] may link to female or male name, in Portuguese, we would have either [o professor] or [a professora]. But this may be tricky: [a abelha] and [o inseto] ([*the bee*] and [*the insect*]) with different genders, may be coreferent.
10. Number: If the phrases agree in number. Here we have a difference in the articles. For Portuguese, for example, we have [os professores], where both noun and article have plural forms. Like in the Gender feature, we may have coreference links between NPs different in number, as in (c):

    (c) [Um fóssil de tiranossauro Rex] foi encontrado no oceano Atlântico. [Os ossos]... ([*A fossil of tyrannosaur Rex*] *was found in Atlantic Ocean.* [*The bones*])...

5.2 Semantic Features

To these previous features, we added five semantic features: Entity Category Equal, Entity Category Different, Hyponymy, Hyperonymy and Synonymy:

11. Entity Category Equal: This feature sees if the NPs have the same semantic category. To extract these categories we use the parser PALAVRAS [1]. We consider Person, Location and Organization semantic categories.
    (d) A opinião é de [Miguel Guerra], da UFSC (Universidade Federal de Santa Catarina). Para [o agrônomo] ... (*The opinion is from* [*Miguel Guerra*],

---

[1] Different from [17], we did not implement rules "location and numeric mismatches".

of UFSC (Universidade Federal de Santa Catarina). To [the agronomist] ... In text fragment (d), the pairs of NPs [Miguel Guerra] and [o agrônomo] ([*the agronomist*]) are both NPs of the category "Person". As not all NPs have a category associated to it, we consider another feature, "Entity Category Different";

12. Entity Category Different: This feature sees whether the semantic categories of the NPs are different. This feature is used here for the cases where there is no category associated to an NP, in which case both features: "Entity Category Equal" and "Entity Category Different" return false.

13. Hyponymy: This feature basically extracts the lemma from the noun phrases head word and check if they are in hyponymy relation in OntoPT. This feature helps to identify relations as in (e):

    (e) Já se perguntou como [as abelhas] fabricam mel? [Os insetos] saem em busca de... (*Ever wonder how [the bees] make honey? [The insects] seek out* ...)

    To avoid the incorrect links (f), we combine the pre and post modifier techniques in this feature.

    (f) Foi o tempo em que decifrar [o genoma].... [o quebra-cabeça genético]... Isso é [um problema ambiental]... (*There was a time in which to decipher [the genome] ... [the genetic puzzle] ... This is [an environment problem]* ...) In "f" [o quebra-cabeça genético] ([*the genetic puzzle*]) is not coreferent with [um problema ambiental] ([*an environment problem*]). But if we analyze just the relation between their head words "puzzle" and "problem", the semantic relation is found. In other words, the feature Hyponymy returns "true", if the following constraints are satisfied:

    – the lemma of head words must be in an hyponymy relation;
    – the Word_Inclusion feature must return "false".

14. Hyperonymy: Similar to Hyponymy, this feature extracts the lemma from the noun phrases head word and search for their hypernymy relation in OntoPT. Like Hyponymy, this feature also combines pre and post modifier restrictions.

15. Synonymy: Like before, the synonymy relation is verified in OntoPT and pre and post modifier restrictions are considered. This feature helps to link mentions as in (g).

    (g) O trabalho de pesquisadores da USP está revelando uma série de novas espécies de [um tipo especial de fungo]. [Pequenos cogumelos]... (*The work of researchers from USP is revealing a number of new species of [a special type of fungus]. [Small mushrooms]...*)

6 Experiments

Next we describe our experiments in which six models are evaluated. The Baseline considers only the non semantic features. The second model, EntityCat, adds the features related to entity categories to the Baseline. The next model adds the Synonymy feature to EntityCat. Then we add Hyponymy to EntityCat. The fifth model adds Hyperonymy to EntityCat and, finally, the last model includes all the five semantic features.

We extracted the candidate pairs, initially retrieving 3022 positive pairs and 94889 negatives. In order to generate a balanced model, we utilized the random undersampling technique, which consists into choosing randomly a sample from negative pairs, the same number of positive samples (3022). This method was used in [10], presenting satisfactory results. To build the classifiers, we used J48 implementation with ten-fold cross validation. To compare the models we built one hundred versions of each (due to the randomly undersample) and present the resulting average in Table 2.

The data was submitted to a Tukey's test, using 5 % of probability. We can see that the all variations including semantic features presented a significant improvement in recall for the positive class, without loss of precision. Note that the Full Semantic model or the EntityCat+Hyponymy are always better than some of the other models in all measures and classes. Perhaps synonymy and hyperonymy are not so influential, due to the fact that the quantity of positive examples involving these features was small in our data set, so that the learning algorithm do not consider it relevant.

Table 2. Semantic models evaluation

Model	Avg Prec Pos	Avg Prec Neg	Avg Recall Pos	Avg Recall Neg	Avg F-Pos	Avg F-Neg
Baseline	79.16 % abc	64.99 % b	53.81 % b	85.66 % a	63.96 % b	73.87 % c
EntityCat	78.57 % c	66.37 % a	57.26 % a	84.24 % b	66.16 % a	74.21 % b
EntityCat +Synonymy	78.61 % c	66.36 % a	57.20 % a	84.30 % b	66.13 % a	74.22 % b
EntityCat +Hyponymy	79.73 % ab	66.59 % a	57.13 % a	85.41 % a	66.52 % a	74.82 % a
EntityCat +Hyperonymy	79.04 % bc	66.36 % a	56.99 % a	84.73 % ab	66.13 % a	74.39 % b
Full Semantic	79.92 % a	66.56 % a	56.95 % a	85.62 % a	66.45 % a	74.88 % a
CV	2.68 %	1.17 %	4.65 %	3.12 %	1.85 %	1.06 %
σ	2.13	0.77	2.63	2.65	1.22	0.79

When the numbers across the same collumn are followed by the same letters, their difference is not significant (accoring to the Tukey's test (p < 0.05)). Contrasted along the vertical line, 'a' means a better result than 'b', and 'b' than 'c' (a>b>c). 'CV' represents the coefficient of variation and 'σ', the standard deviation.

In general, Table 2 shows that the semantic knowledge improves the coreference resolution in several aspects, allowing the generation of coreference chains which are not only based on morpho-syntatctic and string matching features. Another point to consider is that the occurrence of coreference links between semantic related NPs (without string similarity) is less frequent, so it is not a surprise that the improvements are not expressed in larger numbers.

In other words, with the semantic model, it was possible to identify a few new, but more interesting and difficult cases of coreference relations, such as:

[o fungo], [pequenos cogumelos] ([*the fungus*], [*small mushrooms*]); [o álcool], [o etanol] ([*the alcohol*], [*the ethanol*]).

7   Conclusion

In this paper, we evaluated the impact of adding semantic features to a Portuguese coreference resolution system. The semantic features are based on the identification of entity categories (person, place and organization) and on semantic relations of synonymy, hypernymy and hyponymy which are provided by the Portuguese semantic resource, Onto-PT. As a result, we show improvements in the Portuguese coreference resolution task. These semantic features allow the identification of coreferent pairs which share semantic similarity that are not realized through lexical similarity. Although we found apparently just a little improvement, it certainly adds quality on a new level to the output. As future work, we want to test our semantic model at the chains level, using the CoNLL scorer [21]. The source files used to generate the features are available to the community.[2]

Acknowledgments. The authors acknowledge the financial support of CNPq (Conselho Nacional de Desenvolvimento Científico e Tecnológico), CAPES (Coordenação de Aperfeiçoamento de Pessoal de Nível Superior) and FAPERGS (Fundação de Amparo à Pesquisa do Rio Grande do Sul).

References

1. Bick, E.: The parsing system "palavras": automatic grammatical analysis of Portuguese in a constraint grammar framework. Ph.D. thesis, Aarhus University, Aarhus University Press, Denmark (2000)
2. Bruckschen, M., Muniz, F., Souza, J., Fuchs, J., Infante, K., Muniz, M., Gonçalves, P., Vieira, R., Aluísio, S.: Anotação linguística em xml do corpus pln-br. Série de relatórios do NILC (2008)
3. Cardoso, N.: Rembrandt - a named-entity recognition framework. In: Proceedings of the Eighth International Conference on Language Resources and Evaluation - LREC, Istanbul, Turkey, pp. 1240–1243 (2012)
4. Carreras, X., Màrquez, L., Padró, L.: A simple named entity extractor using adaboost. In: Proceedings of the Seventh Conference on Natural Language Learning at HLT-NAACL, vol. 4, pp. 152–155. Association for Computational Linguistics (2003)
5. Collovini, S., Carbonel, T.I., Fuchs, J.T., Coelho, J.C., Rino, L., Vieira, R.: Summ-it: Um corpus anotado com informações discursivas visando a sumarização automática. In: Proceedings of V Workshop em Tecnologia da Informação e da Linguagem Humana, Rio de Janeiro, RJ, Brasil, pp. 1605–1614 (2007)
6. Coreixas, T.: Resolução de correferência e categorias de entidades nomeadas. Pontifícia Universidade Católica Do Rio Grande Do Sul, Dissertação de Mestrado (2010)

---

[2] http://www.inf.pucrs.br/linatural/scorref.html.

7. da Silva, F.J.V., Carvalho, A.M.B.R., Roman, N.T.: A comparative analysis of centering-based algorithms for pronoun resolution in Portuguese. In: Kuri-Morales, A., Simari, G.R. (eds.) IBERAMIA 2010. LNCS, vol. 6433, pp. 336–345. Springer, Heidelberg (2010)
8. de Souza, J.G.C., Gonçalves, P.N., Vieira, R.: Learning coreference resolution for Portuguese texts. In: Teixeira, A., de Lima, V.L.S., de Oliveira, L.C., Quaresma, P. (eds.) PROPOR 2008. LNCS (LNAI), vol. 5190, pp. 153–162. Springer, Heidelberg (2008)
9. do Amaral, D.O.F.: O reconhecimento de entidades nomeadas por meio de conditional random fields para a língua portuguesa. Dissertação de Mestrado, Pontifícia Universidade Católica do Rio Grande do Sul (2013)
10. Fonseca, E.B., Vieira, R., Vanin, A.: Dealing with imbalanced datasets for coreference resolution. In: Proceedings of The Twenty-Eighth International Flairs Conference - FLAIRS (2015)
11. Fonseca, E.B., Vieira, R., Vanin, A.A.: Coreference resolution in portuguese: detecting person, location and organization. J. Braz. Comput. Intell. Soc. **12**, 86–97 (2014)
12. Freitas, C., Mota, C., Santos, D., Oliveira, H.G., Carvalho, P.: Second HAREM: advancing the state of the art of named entity recognition in Portuguese. In: Proceedings of the International Conference on Language Resources and Evaluation, LREC, Valletta, Malta (2010)
13. Gabbard, R., Freedman, M., Weischedel, R.: Coreference for learning to extract relations: yes, virginia, coreference matters. In: Proceedings of the 49th Annual Meeting of the Association for Computational Linguistics: Human Language Technologies: short papers, vol. 2, pp. 288–293. Association for Computational Linguistics (2011)
14. Garcia, M., Gamallo, P.: An entity-centric coreference resolution system for person entities with rich linguistic information. In: Proceedings of 25th International Conference on Computational Linguistics, COLING, Dublin, Ireland, pp. 741–752 (2014)
15. Garcia, M., Gamallo, P.: Multilingual corpora with coreferential annotation of person entities. In: Proceedings of the 9th edn. of the Language Resources and Evaluation Conference - LREC, pp. 3229–3233 (2014)
16. Hou, Y., Markert, K., Strube, M.: A rule-based system for unrestricted bridging resolution: recognizing bridging anaphora and finding links to antecedents. In: Proceedings of Conference on Empirical Methods in Natural Language Processing - EMNLP, Doha, Qatar, pp. 2082–2093 (2014)
17. Lee, H., Chang, A., Peirsman, Y., Chambers, N., Surdeanu, M., Jurafsky, D.: Deterministic coreference resolution based on entity-centric, precision-ranked rules. Comput. Linguist. **39**, 885–916 (2013). MIT Press
18. Maziero, E.G., Pardo, T.A., Di Felippo, A., Dias da Silva, B.C.: A base de dados lexical e a interface web do tep 2.0: thesaurus eletrônico para o português do brasil. In: Proceedings of the XIV Brazilian Symposium on Multimedia and the Web, pp. 390–392. ACM (2008)
19. Miller, G.A.: WordNet: a lexical database for english. Commun. ACM **38**(11), 39–41 (1995)
20. Oliveira, H.G., Gomes, P.: ECO and Onto-PT: a flexible approach for creating a portuguese wordnet automatically. Lang. Resour. Eval. **48**(2), 373–393 (2014)

21. Pradhan, S., Ramshaw, L., Marcus, M., Palmer, M., Weischedel, R., Xue, N.: Conll-2011 shared task: Modeling unrestricted coreference in ontonotes. In: Proceedings of the Fifteenth Conference on Computational Natural Language Learning: Shared Task, pp. 1–27. Association for Computational Linguistics (2011)
22. Rahman, A., Ng, V.: Coreference resolution with world knowledge. In: Proceedings of the 49th Annual Meeting of the Association for Computational Linguistics: Human Language Technologies, Portland, Oregon, USA, pp. 814–824 (2011)
23. Sarmento, L., Pinto, A.S., Cabral, L.M.: REPENTINO – A wide-scope gazetteer for entity recognition in Portuguese. In: Vieira, R., Quaresma, P., Nunes, M.G.V., Mamede, N.J., Oliveira, C., Dias, M.C. (eds.) PROPOR 2006. LNCS (LNAI), vol. 3960, pp. 31–40. Springer, Heidelberg (2006)
24. da Silva, J.F.: Resolução de correferência em múltiplos documentos utilizando aprendizado não supervisionado. Dissertação de Mestrado, Universidade de São Paulo (2011)
25. Soon, W.M., Ng, H.T., Lim, C.Y.: A machine learning approach to coreference resolution of noun phrases. Comput. Linguist. **27**(4), 521–544 (2001)

Language Processing – Short Papers

Improving POS Tagging Across Portuguese Variants with Word Embeddings

Erick Rocha Fonseca$^{(\boxtimes)}$ and Sandra Maria Aluísio

ICMC – University of São Paulo, São Carlos, Brazil
erickrfonseca@gmail.com

Abstract. Brazilian Portuguese (BP) and European Portuguese (EP) have specific NLP resources and tools for many tasks. It is generally agreed upon that applying them to the variant other than their intended one results in a performance drop; however, very little research has measured it. We evaluated a POS tagger in a cross-variant setting under multiple combinations of word embeddings, train and test corpora, and found that (i) BP is easier than EP, (ii) word embeddings help increase tagger performance significantly, but not enough to close the accuracy gap in a cross-variant setting and (iii) embeddings generated from a corpus with both variants are useful in cross-variant scenarios. While we cannot generalize observations from POS tagging to any NLP task, this is an important first step for such evaluations.

1 Introduction

Brazilian Portuguese (BP) and European Portuguese (EP) have very distinct features in phonology, syntax, word choice and, less remarkably, orthography[1]. While mutually intelligible, the differences between both have motivated the development of specialized NLP resources and tools [4,11], and speakers of either variant usually agree that applying them to the variant other than their intended one hurts performance. While this is a reasonable assertion, there is very little published research exploring quantitatively this degradation.

In this study, we experimented with training a part-of-speech (POS) tagger on one variant of Portuguese and testing it on the other. We evaluate POS tagging for a number of reasons. First, speakers of one variant of Portuguese can tell the POS of words in sentences from another one, including words they do not know, based on their morphology and context. This is because the knowledge of the general properties of the language, even if acquired from one variant, allows the speaker to understand the structure of utterances in another variant.

Other reasons are more functional: the Bosque corpus [1] has texts from BP and EP annotated with the same tagset, making a direct comparison straightforward. The same is true for syntactic parsing, but we focus on the simpler task of POS tagging here, leaving syntactic parsing as a potential future work.

---

[1] The texts explored here are from before the Spelling Agreement of the Portuguese language taking place.

© Springer International Publishing Switzerland 2016
J. Silva et al. (Eds.): PROPOR 2016, LNAI 9727, pp. 227–232, 2016.
DOI: 10.1007/978-3-319-41552-9_22

We also fed a tagger with word embeddings (numeric vectors representing words) learned from texts in either variant. Specifically, we tried to answer the following questions: first, to what extent embeddings obtained from a variant v_1 can help a tagger trained on variant v_2 to improve its accuracy on v_1. This is analogous to using unlabeled data for domain adaptation. Second, as a generalization, whether embeddings obtained from a combination of texts from both variants could improve tagging performance on any cross-variant setting. Third, whether using embeddings learned from a sufficiently large amount of unlabeled texts from v_1 could make a tagger trained on v_1 perform on v_2 as well as a tagger trained on a labeled v_2 corpus.

The practical value of this study is to understand how to best use word embeddings to adapt NLP systems for a variant v_1 in cases when there is only annotated data in v_2. We cannot say to which extent observations in POS tagging apply to NLP in general, and therefore studies similar to this one focused on other tasks are needed to reinforce or contradict our findings.

In Sect. 2, we present some related work. Section 3 describes the experimental setup, Sect. 4 presents the results from experiments and some discussion. Finally, Sect. 5 summarizes our contribution.

2 Related Work

The interest in NLP focused on similar languages and variants has been growing in the last few years, as can be seen by the establishment of the VarDial workshops[2]. There have been recent studies in, for example, Mandarin Chinese [12], Occitan [13] and Arabic [8].

Most related to our research is the study from [7], which also compares POS taggers across variants of Portuguese. Unlike us, they do not exploit unsupervised data, but make use of variant-specific lexicons. The corpora in their experiments are different from ours: they use Mac-Morpho [2] for BP and the EP part of the Bosque corpus. Some tags of Mac-Morpho are subsumed by a single one from Bosque and vice-versa, and therefore they map both to a new one.

The authors also tested the taggers on a newly annotated corpus of web texts, which included texts in African Portuguese (namely, from Angola and Mozambique). The influence of the spelling agreement was measured, and the authors built new versions of their lexicons with post-agreement spellings. Their experiments showed that the tagger trained on the EP corpus achieved superior performance, and that lexicons with the new spellings had marginal effect.

3 Experimental Setup

We wanted to train our taggers for each variant with the same amount of data to ensure a fair comparison. Therefore, we truncated the EP part of Bosque, to

---

[2] More information about the workshops on http://ttg.uni-saarland.de/lt4vardial 2015/.

Table 1. Number of sentences and tokens in the Bosque corpus

Variety	Sentences	Tokens	Tokens/Sentence	Exclusive words
Brazilian	4,216	80,239	19.03	16.98 %
European	3,082	80,275	26.05	16.55 %

make it practically the same size as the BP part. Table 1 shows the token counts in the corpus and a tendency of EP to use longer sentences. The last column has the percentage of word tokens that only appear in one part, i.e., which are OOV for the other variant[3].

Concerning embedding generation, we needed the unlabeled corpora to be tokenized in the same way as Bosque, in order to have compatible vocabularies. This is not trivial, as Bosque was tokenized by Palavras [3], which groups many multiword expressions into a single token. Thus, we used CETENFolha[4] (BP) and CETEMPúblico (EP) [10], which were also analyzed by Palavras[5]. They differ significantly in size: the former has around 28 million tokens, while the latter has around 200 million.

We tried five embedding models, described in Table 2, generated by word2vec [9] (50 dimensions, minimum word frequency 20). The PT model was intended to be comparable with BR, while PT-FULL serves to answer the third question in the introduction.

The last column shows OOV counts considering the tagged corpora and the vocabulary of each model. As expected, all vocabularies have a lower OOV amount than the value shown in Table 1. Also, BR, PT and PT-FULL have less OOV words in their own variant, while BOTH has the lowest total count. RANDOM creates its vocabulary at (labeled) training time, and since we used 10-fold cross-validation, the exact OOV count varies.

Table 2. Embedding models and OOV counts with respect to the annotated corpora

Model	Origin	OOV – BP	OOV – EP
BR	CETENFolha	5.87 %	8.95 %
PT	Truncated CETEMPúblico	10.77 %	6.19 %
PT-FULL	Full CETEMPúblico	7.97 %	4.14 %
BOTH	CETENFolha + Truncated CETEMPúblico	5.02 %	4.27 %
RANDOM	Random initialization	varies (see text)	

---

[3] Many of these OOV words are common in both variants, but since the Bosque corpus is very small, they only appear in one of the halves.

[4] Available at http://www.linguateca.pt/cetempublico/.

[5] The Bosque corpus is composed of sentences from CETENFolha and CETEMPúblico. We removed all those sentences to avoid any overlap with the labeled corpus.

Table 3. Accuracies of POS taggers under all testing configurations

Embeddings	Trained on	Validation	Tested on	
			EP	BP
RANDOM	EP	93.51 %	93.41 %	92.82 %
	BP	94.40 %	92.20 %	94.36 %
BR	EP	95.86 %	95.81 %	95.73 %
	BP	96.72 %	94.83 %	96.70 %
PT	EP	96.18 %	96.20 %	95.23 %
	BP	96.54 %	95.19 %	96.46 %
PT-FULL	EP	96.50 %	96.42 %	95.71 %
	BP	96.82 %	95.42 %	96.79 %
BOTH	EP	96.33 %	96.29 %	95.84 %
	BP	96.92 %	95.40 %	96.85 %

4   Results

Here we present and analyze our experimental results. We use the abbreviations SV for "same variant" and XV for "cross-variant", referring to a test set from the same variant a tagger was trained on or not, respectively.

For SV, we performed 10-fold cross-validation; for XV, we ran all 10 models trained during the cross-validation on the whole corpus of the other variant. In both cases, we report average accuracy over the 10 runs. Statistical significance tests were paired t-tests with the accuracy of the 10 folds from two models, with a p-value of 0.05.

Experiments were performed with nlpnet [6], using indicators of capital letters, prefixes and suffixes up to length 5; word window 5, hidden layer of 100 units, learning rate of 0.01 adjusted by Adagrad [5]. Taggers initialized with RANDOM ran for 50 epochs; others ran for 30, since they converged faster.

Accuracies are shown in Table 3. In the BP SV setting, there was no statistically significant difference among BR, BOTH, PT-FULL; in XV, BOTH and PT-FULL were tied. In EP SV, PT-FULL was the isolated best model, and in XV, it was BOTH. Now we recall and answer our proposed research questions:

1. **How much can embeddings from a variant v_1 help a POS tagger trained on v_2 to improve its accuracy on texts from v_1?** Compared with RANDOM taggers, there is substantial improvement; however, these were very bad. In a fairer comparison with embeddings from the same variant as the training data, BOTH provided an absolute gain of 0.13 % for the EP tagger (vs. PT-FULL) and PT-FULL provided 0.52 % for the BP tagger (vs. BR). Especially for the latter, this is a very good improvement.
2. **Can embeddings obtained from a combination of v_1 and v_2 improve tagging performance on any cross-variant setting?** Yes. The model BOTH yielded a consistent performance boost on cross-variant scenarios.

3. **Can embeddings learned from a large corpus in v_1 help a tagger trained on v_1 perform in a v_2 as well as a tagger trained on v_2?** Not if the tagger trained on v_2 was also initialized with word embeddings.

Besides these questions, we can draw a few other interesting conclusions:

Our results differ greatly from the ones in [7]. Their tagger accuracies in EP were higher than in BP[6], even for the tagger trained with BP data. Additionally, their EP tagger SV accuracy was higher than any of ours, while their BP SV accuracy was lower than all our BP models but the one with RANDOM embeddings. This suggests their tagset mapping made the problem harder for BP and easier for EP.

BP is easier than EP. Validation and SV accuracy of BP taggers is always higher than EP ones with the same embeddings, and EP taggers are better on BP than the opposite. A supposedly richer EP vocabulary cannot explain it: OOV counts for both variants were shown to be well distributed among embedding models. Instead, it might be related to the usage of shorter sentences in BP.

There is a tradeoff between a small corpus of the target variant and large one of another variant. BR, BOTH and PT-FULL embeddings had comparable performance on BP, showing that both quantity and quality are important in the corpus that generates the embeddings. Remember we were limited by the tokenization of Bosque; when this is not an issue, gathering a huge unlabeled corpus is easy.

5   Conclusions and Future Work

We presented a quantitative analysis of the performance drop of an NLP tool used in a variant of Portuguese other than the one it was trained on. We evaluated changing labeled training corpora and supplying different word embeddings, and found that BP is easier to tag than EP; word embeddings improve cross-variant performance, but not enough to reach a tagger trained on the same variant as the test data; and generating embeddings from a corpus with both variants is a good idea. Still, we cannot assume that the cross-variant properties of POS tagging are representative of NLP in general; thus, it would be very interesting to perform similar investigations on other tasks.

Acknowledgments. This work was funded by grant#2013/22973-0 of the São Paulo Research Funding Agency (FAPESP).

References

1. Afonso, S., Bick, E., Haber, R., Santos, D.: Floresta sintá(c)tica: a treebank for Portuguese. In: Proceedings of the Third International Conference on Language Resources and Evaluation (LREC 2002). pp. 1698–1703 (2002)

---

[6] Remember that their BP and EP corpora are not the same as ours.

2. Aluísio, S.M., Pelizzoni, J.M., Marchi, A.R., de Oliveira, L., Manenti, R., Marquiafável, V.: An account of the challenge of tagging a reference corpus for brazilian portuguese. In: Mamede, N.J., Baptista, J., Trancoso, I., Nunes, M.G.V. (eds.) PROPOR 2003. LNCS, vol. 2721, pp. 110–117. Springer, Heidelberg (2003)

3. Bick, E.: The Parsing System PALAVRAS: Automatic Grammatical Analysis of Portuguese in a Constraint Grammar Framework. Ph.D. thesis, Aarhus University (2000)

4. Branco, A., Carvalheiro, C., Costa, F., Castro, S., Silva, J., Martins, C., Ramos, J.: DeepBankPT and companion portuguese treebanks in a multilingual collection of treebanks aligned with the penn treebank. In: Baptista, J., Mamede, N., Candeias, S., Paraboni, I., Pardo, T.A.S., Volpe Nunes, M.G. (eds.) PROPOR 2014. LNCS, vol. 8775, pp. 207–213. Springer, Heidelberg (2014)

5. Duchi, J., Hazan, E., Singer, Y.: Adaptive subgradient methods for online learning and stochastic optimization. J. Mach. Learn. Res. **12**, 2121–2159 (2011)

6. Fonseca, E.R., Rosa, J.L.G., Aluísio, S.M.: Evaluating word embeddings and a revised corpus for part-of-speech tagging in Portuguese. J. Braz. Comput. Soc. **21**(2), 1–14 (2015)

7. Garcia, M., Gamallo, P., Gayo, I., Cruz, M.A.P.: PoS-tagging the web in Portuguese. national varieties, text typologies and spelling systems. Procesamiento Lenguaje Nat. **53**, 95–101 (2014)

8. Hamdi, A., Nasr, A., Habash, N., Gala, N.: POS-tagging of tunisian dialect using standard arabic resources and tools. In: Proceedings of the Second Workshop on Arabic Natural Language Processing. pp. 59–68 (2015)

9. Mikolov, T., Chen, K., Corrado, G., Dean, J.: Efficient estimation of word representations in vector space. In: Proceedings of the ICLR Workshop (2013)

10. Rocha, P., Santos, D.: CETEMPúblico: Um corpus de grandes dimensões de linguagem jornalística portuguesa. In: Actas do V Encontro para o processamento computacional da língua portuguesa escrita e falada. pp. 131–140 (2000)

11. Scarton, C., Sanches Duran, M., Aluísio, S.M.: Using cross-linguistic knowledge to build verbnet-style lexicons: results for a (brazilian) portuguese verbnet. In: Baptista, J., Mamede, N., Candeias, S., Paraboni, I., Pardo, T.A.S., Volpe Nunes, M.G. (eds.) PROPOR 2014. LNCS, vol. 8775, pp. 149–160. Springer, Heidelberg (2014)

12. Tseng, H., Jurafsky, D., Manning, C.: Morphological features help POS tagging of unknown words across language varieties. In: Proceedings of the Fourth SIGHAN Workshop on Chinese Language Processing. pp. 32–39 (2005)

13. Vergez-Couret, M., Urieli, A.: Pos-tagging different varieties of Occitan with single-dialect resources. In: Proceedings of the First Workshop on Applying NLP Tools to Similar Languages, Varieties and Dialects. pp. 21–29 (2014)

Joining Forces for Multiword Expression Identification

Leonardo Zilio[1], Rodrigo Wilkens[1(✉)], Luís Möllmann[1], Eric Wehrli[2],
Silvio Cordeiro[1], and Aline Villavicencio[1]

[1] Institute of Informatics, Federal University of Rio Grande do Sul,
Porto Alegre, Brazil
{zilio,rodrigo.wilkens,luis.mollmann,srcordeiro,
avillavicencio}@inf.ufrgs.br
[2] Department of Linguistics, University of Geneva, Geneva, Switzerland
eric.wehrli@unige.ch

Abstract. Multiword Expressions (MWEs) display some kind of linguistic and statistical markedness that may influence the effectiveness of techniques that automatically identify them in texts. While parsing-based techniques for MWE identification are considered to be better at handling long-distance dependencies, passivization and internal modification, statistics-based techniques use association measures to detect statistical markedness regardless of syntactic form. In this paper we compare these two approaches focusing on nominal compounds in Portuguese. We compare the accuracy of each method and propose that combining the strengths of both for increased accuracy.

1 Introduction

Multiword Expressions (MWEs) are sequences of words that must be treated as a unit at some level of linguistic processing [2]. Since tasks like parsing and automatic translation depend on resources such as dictionaries and grammars, the automatic MWE identification can enhance the quality of such resources, especially in uncommon languages and specialized domains, by pointing out expressions that need to be treated as a single unit.

Various techniques for automatic identification of Multiword Expressions (MWEs) from corpora have been investigated, varying in the amount of linguistic or statistical knowledge they employ as basis for the identification [8,10]. Their effectiveness seems to depend on variables like the particular language, the MWE type and the size of the corpus in which the identification is performed [3,7,8]. As a consequence no consensus about the best technique that works for all languages and MWE types has been reached. On the one hand parsing-based techniques, where identification is based on syntactic structure, seem to produce more precise results and capture long-distance dependencies for MWEs and other grammatically well-formed expressions [10]. However, they depend on parsing coverage for a given language and there are other types of MWEs for which they cannot be used. For instance, they cannot easily retrieve

© Springer International Publishing Switzerland 2016
J. Silva et al. (Eds.): PROPOR 2016, LNAI 9727, pp. 233–238, 2016.
DOI: 10.1007/978-3-319-41552-9_23

compounds which are not grammatically well-formed, such as *by and large*, and do not work well for named entities. On the other hand, statistics-based techniques, while generally less precise and more recall-oriented, rely on association measures to capture the statistical markedness and can help to identify candidates whose components co-occur more often than chance, regardless of their syntactic structure. They may also be combined with linguistic patterns, for greater precision. One important advantage is that they are language and domain independent [8], as for many languages and domains a parser may not be available.

In this paper we compare a precision-oriented parsing-based technique with a language-independent recall-oriented statistics-based technique. The main goal here is to combine results from both techniques, observing how each technique complement the other. The focus lies on MWEs in Portuguese, more specifically on compound nouns, like *carro de polícia* (*police car*).

By examining the combination of these complementary techniques we aim to obtain the basis for a more precise MWE resource. This paper is structured as follows: we discuss the materials and methods used in this work, and the evaluation and results obtained respectively in Sects. 2 and 3. We wrap it all up by presenting final remarks and future work Sect. 4.

2   Corpus and Methodology

For the MWE identification process we used the Europarl corpus [4], which was pre-processed with the PALAVRAS parser [1]. The corpus contains 67 million tokens from which MWE candidates were extracted.[1] The selection of the Europarl corpus was driven by the possibility of future comparison with other languages, since it presents alignments between the EU member languages.

The corpus was then processed using both a parsing-based and a statistic-based approach for identificating MWE candidates. For the parsing-based identification we used the Fips "deep" linguistic parser [11,12] is a symbolic parser system which recognizes structures corresponding to grammatically well-formed phrases and uses statistical information for detecting recurrent sequences based on corpus frequency. For the statistics-based identification we used the mwe-toolkit [9], which allows the definition of linguistic (POS tag) patterns to extract an initial list of candidates and the use of statistical association measures to identify candidates whose components co-occur more often than chance. Measures implemented in the toolkit include Pointwise Mutual Information, Student's t-test score, Dice's coefficient and Maximum Likelihood Estimation.

For generating the MWE lexicon we selected predefined patterns for Portuguese, such as noun-preposition-noun (e.g. *política de desenvolvimento*, *condição de trabalho*), excluding proper nouns. An initial set of candidates was obtained, and it was automatically validated according to the agreement between

---

[1] Parsing increased the number of tokens from 49 to 67 million words as contractions like those involving preposition and determiner are expanded to *de + o* by the parser (e.g. *do* (of the)).

the syntactic and statistics-based methods, and, for the cases where they dis-
agree, a weighted voting scheme was used based on how confident they are about
each candidate. From the resulting lists of MWE candidates we selected a set
of the top 2,000 MWE candidates from each approach. We then created a third
set of MWE candidates based on the agreement from the two processes. Finally,
we evaluated the quality of the three sets automatically using BabelNet [5] and
Onto.PT [6]. A smaller portion of the results (100 top candidates from each
process) was also selected for manual validation[2].

3   Results

Table 1 presents the results from the automatic validation of the top 2,000 MWE
candidates identified by each of the process. This automatic evaluation process
was done by checking if at least one of the resources contained the MWE candidate. As we can see, the intersection of both processes present a more accurate
result, rising the accuracy of MWE identification to 17.9 %.

Table 1. Automatic validation of the top 2000 candidates identified by each method

Noun-Prep-Noun	Candidates	In resources	% MWEs
Parsing-based	2000	198	9.9 %
Statistics-based	2000	319	16.0 %
Intersection	797	143	17.9 %

As described in the methodology, the top 100 MWE candidates identified
with each approach were manually evaluated by a linguist that is a native-
speaker of Portuguese. As some of these candidates are already automatically
validated against Onto.PT and BabelNet, only the MWE candidates that were
not in these resources were manually evaluated. Results of this evaluation are
listed in Table 2.

Since the idea is to combine both MWE identification techniques to look
for improvements on the quality of the resource, we identified in Table 2 the
number of MWE candidates that are among the top 100 for both approaches
(their intersection), those that occur in only one of the lists (their differences)
and those that occur in at least one list (their union), in relation to those that
were judged to be valid. The results obtained are that they complement each
other, since most candidates (61 %) are unique to each technique, but 39 % are
common to both, and from these 79.5 % are valid MWEs.

This means that, again, by intersecting parsing-based with statistics-based
information, it is possible to obtain a more accurate list of MWE candidates, that

---

[2] Among the 100 top candidates from each process, some were already validated in the
automatic validation step. The manual validation was applied only to those MWE
candidates that were not prevalidated.

Table 2. Validation of the top 100 candidates in each system

Noun-Prep-Noun	Candidates	Valid candidates	% MWEs
Intersection	39	30	76.9 %
Parsing-based only	61	33	54.1 %
Statistics-based only	61	36	59.0 %
Union	161	99	61.5 %

presents the strength of both methods. The precision for the 100 candidates can be seen in Fig. 1. By uniting both results, while we have an increase in coverage (presenting a larger list of MWEs), there is also a decrease on precision when compared to the individual systems. Since we are looking for a more accurate resource, the intersection of both is the one we are focusing on right now.

A collateral result from the manual evaluation was the observation that the lexical resources we used for evaluating the MWE candidates lack coverage in terms of MWEs. Table 3 shows the number of automatically and manually validated MWE candidates, demonstrating the increase of correct MWEs from the automatic compared to the manual evaluation. It is important to stress that all MWEs that were manually validated were not present in the lexical resources, since the manual validation was done only on those MWE candidates among the top 100 that were not automatically validated.

Table 3. Automatic vs. manual validation of the MWE candidates in each system

Noun-Prep-Noun	Automatic validation	Manual validation
Parsing-based	30	31
Statistics-based	22	40

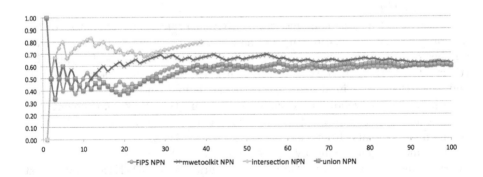

Fig. 1. Precision at 100

4   Conclusions and Future Work

In this paper we compared two approaches for the identification of MWEs, concentrating on nominal expressions in Portuguese. For MWE identification we examined a more precision-oriented with a more recall-oriented technique. We first looked at the first 2000 MWEs candidates and their automatic validation against lexical resources, and then we evaluated again the first 100 candidates, this time using the judgment of a native-speaker expert. The results showed that the combination of both techniques, by intersecting their results, leads to a more accurate result.

Since the corpus used in this work is a parallel corpus with alignment between language pairs, as future work, we plan on investigating how the transfer of information from one language can help identification of MWEs in the other.

Acknowledgments. This research was partially developed in the context of the project *Text Simplification of Complex Expressions*, sponsored by Samsung Eletrônica da Amazônia Ltda., in the terms of the Brazilian law n. 8.248/91. This work was also partly supported by CNPq (482520/2012-4, 312114/2015-0) and FAPERGS.

References

1. Bick, E.: The parsing system Palavras: Automatic grammatical analysis of Portuguese in a constraint grammar framework. Aarhus Universitetsforlag (2000)
2. Calzolari, N., Fillmore, C.J., Grishman, R., Ide, N., Lenci, A., MacLeod, C., Zampolli, A.: Towards best practice for multiword expressions in computational lexicons. In: Proceedings of the Third International Conference on Language Resources and Evaluation (LREC-2002), Las Palmas, Canary Islands, Spain. European Language Resources Association (ELRA), May 2002
3. Evert, S., Krenn, B.: Using small random samples for the manual evaluation of statistical association measures. Comput. Speech Lang. **19**(4), 450–466 (2005)
4. Koehn, P.: Europarl: a parallel corpus for statistical machine translation. In: Conference Proceedings: The Tenth Machine Translation Summit, pp. 79–86. AAMT, Phuket (2005)
5. Navigli, R., Ponzetto, S.P.: BabelNet: building a very large multilingual semantic network. In: Proceedings of the 48th Annual Meeting of the Association for Computational Linguistics, pp. 216–225. Association for Computational Linguistics (2010)
6. Oliveira, H.G., Gomes, P.: Onto.pt: automatic construction of a lexical ontology for portuguese. In: Stairs 2010: Proceedings of the Fifth Starting AI Researchers' Symposium, vol. 222, p. 199. IOS Press (2010)
7. Pecina, P.: Lexical association measures and collocation extraction. Lang. Resour. Eval. **44**(1–2), 137–158 (2010)
8. Ramisch, C., Schreiner, P., Idiart, M., Villavicencio, A.: An evaluation of methods for the extraction of multiword expressions. In: Proceedings of the LREC Workshop Towards a Shared Task for Multiword Expressions (MWE 2008), pp. 50–53 (2008)
9. Ramisch, C., Villavicencio, A., Boitet, C.: Multiword expressions in the wild?: the mwetoolkit comes in handy. In: Proceedings of the 23rd International Conference on Computational Linguistics: Demonstrations, pp. 57–60. Association for Computational Linguistics (2010)

10. Seretan, V.: Syntax-Based Collocation Extraction, Text, Speech and Language Technology, vol. 44, 1st edn, p. 212. Dordrecht, Netherlands (2011)
11. Wehrli, E., Nerima, L.: The fips multilingual parser. In: Gala, E., Rapp, R., Bel-Enguix, G. (eds.) Language Production, Cognition, and the Lexicon. Text, Speech and Language Technology, vol. 48, pp. 473–490. Springer, Switzerland (2015)
12. Wehrli, E., Seretan, V., Nerima, L.: Sentence analysis and collocation identification. In: 23rd International Conference on Computational Linguistics, COLING (2010)

A Construction Grammar Approach for Pronominal Clitics in European Portuguese

Tânia Marques[1]([✉]) and Katrien Beuls[2]

[1] School of Informatics, University of Edinburgh, Edinburgh EH8 9AB, UK
tmarques@inf.ed.ac.uk
[2] Artificial Intelligence Lab, Vrije Universiteit Brussel,
Pleinlaan 2, 1050 Brussels, Belgium
katrien@ai.vub.ac.be

Abstract. Cliticization in European Portuguese (EP) is unique amongst other Romance languages. While preverbal and postverbal placement of clitics is common, it is defined by the finiteness of the verb. In EP, however, the clitic placement does not depend on the verb, but on the context surrounding it. In this paper, we present an operational construction grammar model for parsing and production of proclitic contexts.

Keywords: Construction grammar · Pronominal clitics · European Portuguese · Language comprehension · Language production

1 Introduction

In European Portuguese (EP), as in other Romance languages, pronominal clitics can appear before the verb (proclisis) or after the verb (enclisis). However, in other languages, the type of cliticization is usually defined by the finiteness of the verb [1]. In EP, cliticization is more complex, because the same verb form can appear attached by the same pronominal clitic preverbally or postverbally, depending on the context. This is illustrated by sentence (1), where proclisis appears because of the conjunction 'porque' (because) before the verb in the second clause, while no trigger is present in the first clause, leading to an enclitic.

(1) Eu dei-te este livro, porque tu o querias
    I gave 2SG.DAT this book, because you 3SG.MASC.ACC wanted
    'I gave you this book, because you wanted it'

A solid theory for proclisis is essential for creating computational models that are able to produce sentences (i.e. generated from meaning) in EP. Existing parsers (e.g. LXParser [2], MSTParser [3], Palavras Parser [4]) have no problem identifying clitics which have more or less unique forms and are usually adjacent to the verb. But a system for producing sentences is dealing with far fewer hints about the placement of the clitics and is therefore likely to over-generate.

Luís and Otoguro [5] have identified proclitic contexts that could inform computational systems for language production. In their work, the position of the

J. Silva et al. (Eds.): PROPOR 2016, LNAI 9727, pp. 239–244, 2016.
DOI: 10.1007/978-3-319-41552-9_24

clitic and the triggers are defined through functional precedence relations, using Lexical-Functional Grammar (LFG) [6]. However, the argument structures and the constructions proposed have not been implemented in concrete operational-ized patterns, at least as far as we know. Inspired by Luís and Otoguro's the-ory [5], we present an initial computational model, where the full cycle of parsing and producing sentences is used as an evaluation mechanism. Fluid Construction Grammar (FCG) [7,8] is used instead of LFG, because of its partial mapping between syntactic and semantical arguments. This increases its flexibility when dealing with unknown arguments, in contrast with LFG which assumes that each lexical entry dictates all necessary linking information [7].

Our approach relates to semantic parsers, because it attempts to retrieve a meaning to enable production of the correct sentence back. Our initial grammar is based on syntactical phrasal templates obtained from the proclitic contexts found in the literature. Nevertheless, it could be further extended by introducing new constructions and lexicons. As in the work by Wang et al. [9], our model is fully functionality driven and could be easily adapted to a crowdsourcing model for automatic expansion. As stated by the authors, this has the advantages of not requiring annotation from error prone sources, and of enabling the system to learn a lot of the language variability with a small number of utterances.

2   Modeling a Fragment of Portuguese Grammar

Our grammar is implemented in Fluid Construction Grammar (FCG) [7,8]: a formalism that represents linguistic information by means of feature structures. Both for basic units (e.g. noun), or more complex structures, created by matching basic units. It works for strings (when parsing) and meaning networks (when producing). For more information, please consult Steels [7].

In our grammar, for simplification, we ignore all morphological changes that occur in enclisis and mesoclisis, and we do not accept auxiliary verbs.

Figure 1 gives an overview of our grammar[1]. There are six types of con-structions which apply in different orders in parsing and production. In parsing, morphological constructions are the first to apply by matching to the strings in the input, then semantic and syntactic features are added by the lexical construc-tions. If one of the words is a conjunction, then the sentence is divided into two clauses by the conjunction constructions. Argument linking constructions iden-tify the semantic connections between event participants and build grammatical dependencies. Trigger constructions find triggers to restrict the position of the clitic by cliticization constructions. Finally, word-order constructions take care of the remaining orderings between words by looking at the topic and/or focus of the clause. The output of parsing will be a fully connected meaning network that is used to generate the sentence back in production and a dependency graph with the connections amongst dependents. Production is similar to parsing, but lexical constructions are applied first by matching with the meanings, and the morphology of words is decided only after the argument linking constructions.

---

[1] A demo of the grammar can be found in http://fcg-net.org/demos/propor-2016/.

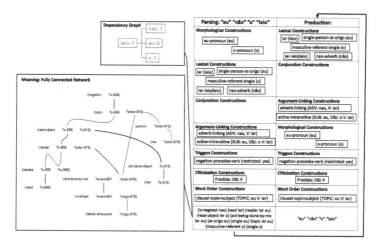

Fig. 1. Diagram with the order of application of the constructions in parsing (left), and in production (right). The meaning obtained in parsing is used as input in production.

3   Proclitic Contexts

Fronted Focus: Clitics appear preverbally if the focus comes before the verb (see (2) and (3), taken from Luís and Otoguro [10], which differ only in the focus). This happens when the first position of the clause (topic) is not the subject. The topic of a sentence is defined by a meaning predicate called 'topic' that is by default linked to the subject argument of the main verb. It is worth noting that the discourse topic is not the topic of the clause. In 'Esse livro, li-o eu', 'Esse livro' is the discourse topic, but not the topic of the second clause, which is empty. Furthermore, the focus 'eu' is not fronted, not leading to proclisis.

(2) Eu dei-te este livro   (3) Este livro te dei eu
     I gave 2SG.DAT this book   This book 2SG.DAT gave I
     'I gave you this book'   'I gave you this book'

This behaviour is achieved by two constructions: one that says that the clausal focus is not the subject, and adds a (restricted +) feature to the verb, activating proclisis; and the other one identifies the predicate focus (well-known concept in linguistics [11]), which moves the subject to the back of the sentence.

Wh Questions: Wh questions that are not the subject and precede the verb will automatically trigger proclisis, since they are fronted focus. For dealing with Wh questions that are subjects, an additional construction wh-fronted, was introduced that adds (restricted +) feature to the verb when a Wh question is identified before the verb. The wh-in-situ was also introduced, which adds a (restricted -) to the verb, allowing for Wh questions in the end of the sentence.

(4) Quem te deu este livro?
     Who 2SG.DAT gave this book
     'Who gave you this book?'

(5) Tu deste-o a quem?
     You gave 3SG.MASC.ACC to whom
     'You gave it to whom?'

Negation and Adverbs: Most adverbs can either precede the verb (7), or come in the end of the sentence (8). This is not true for negation (6), because it always comes before the verb. Furthermore, there is a class of adverbs that do not trigger proclisis even when preceding the verb as in (8). Luís and Otoguro [5] classify the proclisis triggering adverbs as operator-like. In our implementation, the constructions look for an adverb that precedes the verb and if it is operator-like or negation, which we use to restrict the verb. There is also a construction that accepts adverbs after the verb, if these adverbs can come postverbially.

(6) Ele não te deu este livro
     He not 2SG.DAT gave this book
     'He did not give you this book'
(7) Eu raramente o leio
     I rarely 3SG.MASC.ACC read
     'I rarely read it'

(8) Eu leio-o raramente
     I read 3SG.MASC.ACC rarely
     'I read it rarely'
(9) Eu ontem vi-te
     I saw 2SG.DAT
     'I saw you yesterday'

Quantified Subjects: Proclisis can also happen when specific quantified subjects precede the verb. This is not true for all quantifiers, though. The quantifier 'algumas' (some) is not a trigger (9), while 'poucas' (few) is (10). Crysmann [12] classifies the proclisis triggering quantifiers as 'downward entailing quantifiers', because they seem to have downward monotonicity. Following this idea, we added a feature *(downward +)* that is only present in lexical constructions for those quantifiers. Thus, the trigger construction looks for a 'downward entailing quantifier' that is quantifying an element related to the verb and precedes it.

(9) Algumas pessoas lêem-no
     Some people read 3SG.MASC.ACC
     'Some people read it'

(10) Poucas pessoas o lêem
     Few people 3SG.MASC.ACC read
     'Few people read it'

Subordinate Conjunctions: Complementarisers, relative clauses and subordinate conjunctions will also trigger proclisis. Here, we will focus on subordinate conjunctions, since they are all similar contexts. When a subordinating clause is connected to a subordinated clause, there are two different verbs in the sentence. So, our method for individualizing the clauses is by identifying the conjunction and the verb in each of them, and introducing the feature (restricted +) to the verb of the subclause.

(11) Eu dei-te este livro, porque tu o querias
     I gave 2SG.DAT this book, because you 3SG.MASC.ACC wanted
     'I gave you this book, because you wanted it'

4   Analysing the Model

All our examples and further sentences were tested to assure that they can be parsed and their meaning can produce exactly the same sentence, i.e. does not overgenerate. Yet, a more through evaluation of the system is needed. Usually, this is done with a well-known annotated corpus. In our case, such standard evaluation is unfeasible, because there are syntactic patterns that are unsupported, and because the required information differs from standard corpus. A solution would be to create our own corpus, but it would be inherently biased. Instead, we opted to describe the sentences that can be produced. Table 1 contains 18 templates. Each template was evaluated in terms of how many variations of a sentence can be generated (expected complexity column). The grammar requires grammatical agreement of gender and number. We consider 2 types of article (definite/indefinite with 4 forms each), 3 proximities in demonstratives, 8 nominative personal pronouns, 11 clitics (reflexives, accusatives and datives), 6 of which are datives. With the different orders that the elements can take, we expect our system to accept and generate $n_{sentences}$ sentences which could be obtained by automatically introducing morphological and lexical constructions.

Overgeneration of our model is limited by the grammatical agreement, but more importantly, by the meaning networks that directly steer the production process. The system mostly overgenerates in the word order. We could be more strict, but it would not capture the order flexibility in EP. Because of the lack of a grammaticality criterion, ungrammatical sentences are sometimes processed

Table 1. Templates of sentences that can be parsed and produced in our design, and expected complexity. (n - number of; cj_vb - conjugated verb tense; np_pr - person of noun phrase; n/op_adv - non/operator-type adverb; n/dqt - non/downward quantifier)

Structure	Examples	Expected Complexity
Noun Phrase		$np = \sum_{i=1}^{4} np_i$
NOUN	Livros	$np_1 = n_{noun}$
Art NOUN	O livro	$np_2 = 2 \times n_{nouns}$
DEM NOUN	Este livro	$np_3 = 3 \times n_{nouns}$
NPron	Eu	$np_4 = 8$
Clauses		$n_{clauses} = \sum_{i=1}^{18} cl_i$
1:NP1 V NP	Eu leio livros	$cl_1 = 2 \times np \times n_{cj\_vb \land pr = np1\_pr}$
2:NP1 V CL	Eu leio-o	$cl_2 = np \times n_{cj\_vb \land pr = np1\_pr} \times 11$
3:NP1 V CL NP	Eu leio-te este livro	$cl_3 = np \times n_{cj\_vb \land pr = np1\_pr} \times 6 \times np$
4:NP CL V NP1	Este livro te dei eu	$cl_4 = np \times n_{cj\_vb \land pr = np1\_pr} \times 6 \times np$
5:NP1 ADV CL V	Eu não te vejo	$cl_5 = np \times n_{op\_adv} \times n_{cj\_vb \land pr = np1\_pr} \times 11$
6:NP1 ADV CL V NP	Eu não te leio o livro	$cl_6 = 2 \times np \times n_{op\_adv} \times n_{cj\_vb \land pr = np1\_pr} \times 6$
7:NP1 ADV V CL	Eu ontem vi-te	$cl_7 = np \times n_{nop\_adv} \times n_{cj\_vb \land pr = np1\_pr} \times 11$
8:NP1 ADV V CL NP	Eu ontem li-te este livro	$cl_8 = 2 \times np \times n_{nop\_adv} \times n_{cj\_vb \land pr = np1\_pr} \times 6$
9:NP1 V CL ADV	Eu vejo-te raramente	$cl_9 = np \times n_{pos\_adv} \times n_{cj\_vb \land pr = np1\_pr} \times 11$
10:NP1 V CL NP ADV	Eu dou-te o livro docemente	$cl_{10} = 2 \times np \times n_{pos\_adv} \times n_{cj\_vb \land pr = np1\_pr} \times 6$
11:WH CL V NP	Quem te deu este livro?	$cl_{11} = n_{wh} \times np \times n_{cj\_vb \land pr = wh\_pr} \times 6$
12: WH CL V	Quem o viu?	$cl_{12} = n_{wh} \times n_{cj\_vb \land pr = wh\_pr} \times 11$
13: NP V CL WH	Tu deste-o a quem?	$cl_{13} = np \times n_{wh} \times n_{cj\_vb \land pr = np1\_pr} \times 11$
14: QT NP CL V	Poucas pessoas o lêem	$cl_{14} = n_{dqt} \times np \times n_{cj\_vb \land pr = np\_pr} \times 11$
15: QT NP1 CL V NP	Poucas pessoas te lêem livros	$cl_{15} = n_{dqt} \times 2 \times np \times n_{cj\_vb \land pr = np1\_pr} \times 6$
16: QT NP V CL	Algumas pessoas lêem-no	$cl_{16} = n_{ndqt} \times np \times n_{cj\_vb \land pr = np\_pr} \times 11$
17: QT NP1 V CL NP	Algumas pessoas lêem-te livros	$cl_{17} = n_{ndqt} \times 2 \times np \times n_{cj\_vb \land pr = np1\_pr} \times 6$
Complex Clauses		$n_{complex\_clauses} = ccl$
18: Clause CONJ Clause	Eu dei-te o livro, para tu o leres	$ccl = 2 \times n_{clauses} \times n_{conj}$
		$n_{sentences} = n_{complex\_clauses} + n_{clauses}$

coherently, meaning that the parsing process resulted in a fully connected meaning network and the produced utterance is actual a correction for the original input sentence. This robustness check is an advantage in error-prone discourse.

5   Conclusions and Further Research

In this paper, we presented an initial computational implementation to parse and produce sentences in EP with the correct placement of pronominal clitics. The sentences handled are still limited. In the future, we would also like to explore more proclitic contexts, the morphological aspects of the clitics, and automatic creation of constructions to expand the grammar with an annotated corpus.

Acknowledgments. The research presented in this paper has been funded by the European Community's Seventh Framework Programme (FP7/2007-2013) under grant agreement no. 607062 *ESSENCE: Evolution of Shared Semantics in Computational Environments* (http://www.essence-network.com/).

References

1. Spencer, A., Luís, A.R.: Clitics: An Introduction. Cambridge University Press, Cambridge (2012)
2. Silva, J., Branco, B., Gonalves, P.: Top-Performing Robust Constituency Parsing of Portuguese: Freely Available in as Many Ways as you Can Get it. LREC (2010)
3. McDonald, R.: Discriminative learning and spanning tree algorithms for dependency parsing. Dissertation, University of Pennsylvania (2006)
4. Bick, E.: The parsing system Palavras. Automatic Grammatical Analysis of Portuguese in a Constraint Grammar Framework (2000)
5. Luís, A.R., Otoguro, R.: Inflectional morphology and syntax in correspondence. Morphol. Interfaces **178**, 97–135 (2011)
6. Bresnan, J.: Lexical Functional Syntax. Blackwell Publisher, Oxford (2001)
7. Steels, L.: Design Patterns in Fluid Construction Grammar. John Benjamins, Amsterdam (2011)
8. Steels, L.: Basics of Fluid Construction Grammar. Constructions and Frames (in press)
9. Wang, Y., Berant, J. Liang, P.: Building a semantic parser overnight. Association for Computational Linguistics (ACL) (2015)
10. Luís, A.R., Otoguro, R.: Proclitic contexts in European Portuguese and their effect on clitic placement. In: Proceedings of the LFG 2004 Conference (2004)
11. Van Valin, R.D., La Polla, R.J.: Syntax: Structure, Meaning and Function. Cambridge University Press, Cambridge (1997)
12. Crysmann, B.: Constraint-based Coanalysis. Ph.D. thesis, Universität des Saarlandes and DFKI Gmbh (2002)

Investigating Machine Learning Approaches for Sentence Compression in Different Application Contexts for Portuguese

Fernando Antônio Asevedo Nóbrega[(✉)] and Thiago Alexandre Salgueiro Pardo

Interinstitutional Center for Computational Linguistics (NILC),
Institute of Mathematical and Computer Sciences,
University of São Paulo, São Carlos, SP, Brazil
{fasevedo,taspardo}@icmc.usp.br
http://www.nilc.icmc.usp.br

Abstract. Sentence compression aims to produce a shorter version of an input sentence and it is very useful for many Natural Language applications. However, investigations in this field are frequently task focused and for English language. In this paper, we report machine learning approaches to compress sentences in Portuguese. We analyze different application contexts and the available features. Our experiments produce good results, outperforming some previously investigated approaches.

1 Introduction

The sentence compression task aims to produce a shorter version of a sentence [9] and it may be useful for many Natural Language applications. As an example, in compressive summarization, the systems produce summaries with the most relevant content of one or more related texts and they compress some sentences before their inclusion in the summaries [2,6,8–11,15].

The current sentence compression methods usually delete some tokens or arcs in a syntactic tree of the input sentence [2,4,8,9,13,16,17], and these methods frequently only use information from the input sentence, although the situations and compression applications are varied and have the potential to provide different features and hints to the task. In this paper, we have investigated sentence compression approaches based on machine learning techniques, employing different types of features in order to analyze different application contexts. We have performed experiments for texts written in Portuguese and show that our methods outperform some previously investigated approaches.

The paper is organized as follows. We briefly introduce the main related work in Sect. 2. Our dataset is presented in Sect. 3. Our methods and the investigated features are presented in Sect. 4. In Sect. 5, we show the evaluation methodology based on different application contexts and the results of our methods. Some final remarks are presented in Sect. 6.

© Springer International Publishing Switzerland 2016
J. Silva et al. (Eds.): PROPOR 2016, LNAI 9727, pp. 245–250, 2016.
DOI: 10.1007/978-3-319-41552-9_25

2   Related Work

In one of the most used approaches in the area, [17] shows a sentence compression method based on deletions of segments in a syntactic tree using the Noisy-Channel framework. The authors report the problem of lack of datasets for the training process and improve their method by applying unsupervised approaches and some manually produced constraints.

[7] performs machine learning experiments for Portuguese by using a Decision Tree technique. They investigate syntactic and semantic features extracted by the PALAVRAS parser [3] and others based on the documents that the input sentence came from, as frequency of the token and position of the sentence.

[2] presents a compressive module for summarization based on Integer Linear Programming (ILP) in a similar way to [12]. Their system uses a bi-gram model as sentence representation and makes deletions over the arcs from the dependency tree. They use features based on the labels of the arcs in the tree, shallow features (frequency of the words, if the word is a stopword, its position in the sentence and document) and analysis of modifier tags (as negation, temporal words, and others). Furthermore, they use hard constraints, which avoid some arc deletions, in order to produce more grammatical sentences.

[16] investigates features based on two kinds of sentence representation, a list of tokens and a tree of syntactic dependencies, used with ILP. The authors say that this approach shows better results because these two kinds of knowledge complement each other.

[6] presents a sentence compression method based on token deletion using the deep learning framework. The authors defend the use of this approach because of the low performance of syntactic parsers. Their best results were achieved with embedding vectors obtained by a skip gram model [14] with 256 dimensions.

3   Dataset

We selected a dataset with 770 sentence pairs (25,966 tokens, including punctuation marks) of original sentences and their respective compressed versions from the Priberam Compressive Summarization Corpus (PCSC) [1]. In the PCSC corpus, there are 801 texts/documents organized into 80 clusters and 160 summaries (two for each cluster) that were manually made based on the compressive summarization approach.

When there were two or more compressed possibilities for the same source sentence in the corpus (there was a total of 78 cases like this), we have maintained only its shortest version. This way, we aim to produce a compression method that learns as many token deletions as possible. Aiming to maintain the data unbiased, we have also removed eventual duplicated pairs.

4   Our Methods

We handled the sentence compression task as a token deletion approach in a traditional classification problem of machine learning, in which we must to answer the question "Should this token be deleted?" for each token in the input sentence.

We have initially experimented three sceneries of features (Sentence, Document and Summary) based on the diversity of available information on different applications. For instance, in compressive summarization, we may use information of the input sentence, its source text and the output summary being produced. On the other hand, in text simplification, we only have the input sentence and its text. Furthermore, we also have used background features that may be added in any of these sceneries.

In the context of the **Sentence**, we have only used information from the input sentences, as simple shallow features and features derived from the syntactic and semantic analyses. As shallow features[1], we have used: if the token is inside parentheses; if the token occurs in the beginning (if it is one of the 20 % first tokens), ending (if it occurs after the 80 % first tokens) or in the middle of the sentence (otherwise); if the token is a stopword; the two previous and next tokens; and the token itself. It is important to say that, for the two last features, we have used a stemmer in order to reduce the dimensionality of the machine learning model. For the remaining features, we have used information extracted by the PALAVRAS parser [3], as: the POS (Part of Speech) of the token and of the two previous and next tokens; available syntactic functions of the token in the dependency tree; and semantic information (named entity and semantic class labels presented by PALAVRAS) of the token. Furthermore, PALAVRAS expands syntactic contractions ($do = de + o$; $dele = de + ele$; $no = em + o$). Thus, in order to produce compressed sentences with the same tokens of the input sentences, we contract these expansions with a simple set of manually developed rules.

In the scenery of **Document**, in addition to the Sentence features, we have also extracted information from the documents of the sentences, as follows: the position of the sentence in the document; if the token occurs in the most relevant sentence in the document (which is the sentence with the most frequent words); and the token frequency in the document (normalized by the log).

For the **Summary** scenery, in addition to the features above, we also have used information based on the summarization process, as follows: if the token was used in previous selected sentences in the summary; and the sentence position in the summary. It is important to say that we have used the available summaries in the PSCP corpus in order to extract these features.

We have also experimented **Background** features based on an embedding vector representation of words, in which each word is represented by a numeric array of n dimensions with values trained in a big text corpus. We have used vectors of 256 dimensions made by a Skip gram model [14], as performed by [6], over 1,008,353 texts from the G1 portal[2]. We have applied this vector representation in order to calculate lexical similarities among tokens and use them as features (the similarity of the token for the two previous and next tokens). Here, the idea is to identify pairs of terms that are very similar between each other and, therefore, may be simplified, as: names of companies (e.g., Microsoft

---

[1] Those that require limited linguistic processing.

[2] It is a famous news web portal in Brazil, at g1.globo.com.

Corporation may be simplified to Microsoft) and names of famous people (e.g., President Dilma Rousseff may be simplified to President Dilma or only Dilma), and others.

Finally, since the sentence compression process is interpreted as a sequence of token deletions from the sentence and, therefore, the decision if we will keep or remove a token probably affects the next decisions in the sentence, we have also included features that indicate if the two previous tokens in the **Context** of the target token were removed.

We have experimented 5 machine learning algorithms from different approaches, as follows: Decision Tree; Logistic Regression; MultiLayer Perceptron (MLP); Naïve Bayes; and Support Vector Machine (SVM). Furthermore, we have also experimented an Ensemble approach [5], in which a set of methods are used in order to classify the inputs by using a voting system (weighted or not). In this paper, our Ensemble method is simply composed of all the previously mentioned methods with a weighted voting strategy based on the evaluated f-measure values of the methods.

5   Evaluation

We used the ten-fold cross-validation strategy in order to perform our experiments. We report the traditional Precision ($\frac{|\text{correctly classified tokens}|}{|\text{compressed sentence}|}$), Recall ($\frac{|\text{correctly classified tokens}|}{|\text{original sentence}|}$)[3] and F-measure (F-1) metrics.

We contrast our methods with the system presented by Kawamoto and Pardo [7][4], which investigated the Decision Tree framework for Portuguese language, and the compression method used by Almeida and Martins [2]. Table 1 shows the evaluation results, in which we organize the methods on the rows and the scores on the columns. In the first column, we present the groups of features that were used. For instance, the Logistic Regression approach, using Sentence, Background, and Context features, presents a 0.887 f-measure.

One may see that Logistic Regression produced the best results, largely outperforming the baseline methods. We believe that Almeida and Martins method, which is based on the ILP approach, did not show good results probably because of the size of the dataset, since we train and evaluate this method with a cross-validation methodology. It is also interesting to see that, although there are differences in performance for the different application sceneries, the results are not very different. This may happen due to the fact that all the sceneries probably use the same main features (the ones related to the Sentence scenery).

Finally, a factor that we did not explore and that may be important is the compression rate. We simply adopted the compressed sentences in the corpus, without explicitly modeling the compression rate as a parameter. However, it is known that the number of deleted tokens is directly influenced by the desired size

---

[3] Where $| \bullet |$ is the size of $\bullet$.

[4] To the best of our knowledge, it was the first sentence compression investigation for Portuguese.

Table 1. Evaluation of the compressed sentences produced by our methods in different sceneries

Features		Method	Precision	Recall	F-measure
		Kawamoto and Pardo [7]	0.616	0.577	0.596
		Almeida and Martins [2]	0.491	0.460	0.475
Sentence	+	Decision Tree	0.870	0.885	0.878
Background	+	Ensemble	0.736	0.783	0.759
Context		Logistic Regression	0.882	0.892	**0.887**
		Naïve Bayes	0.658	0.457	0.558
		MLP	0.729	0.764	0.746
		SVM	0.869	0.872	0.870
Document	+	Decision Tree	0.857	0.875	0.866
Background	+	Ensemble	0.855	0.880	0.867
Context		Logistic Regression	0.881	0.892	**0.887**
		Naïve Bayes	0.693	0.471	0.582
		MLP	0.774	0.793	0.784
		SVM	0.869	0.872	0.870
Summary	+	Decision Tree	0.859	0.855	0.857
Background	+	Ensemble	0.881	0.891	0.886
Context		Logistic Regression	0.882	0.892	**0.887**
		Naïve Bayes	0.693	0.471	0.582
		MLP	0.776	0.808	0.792
		SVM	0.869	0.872	0.870

of the summary that contains the compressed sentences. Therefore, our machine learning is probably biased by this, but so far we have not analyzed its impact in our results.

6   Final Remarks

We have investigated 6 machine learning techniques for sentence compression in three different application sceneries for texts in Portuguese. For each application context, we have presented a different set of features. In general, we have produced good results that outperformed some previous approaches for Portuguese, but we believe that there is still room for improvements. Future work includes the investigation of some hard constraints to avoid the production of ungrammatical compressed sentences, as performed by many previous investigations [2,6,8,9,16], as well as the investigation of more sophisticated methods.

Acknowledgments. The authors are grateful to CAPES and FAPESP for supporting this work.

References

1. Almeida, M.B., Almeida, M.S.C., Martins, A.F.T., Figueira, H., Mendes, P., Pinto, C.: A new multi-document summarization corpus for european portuguese. In: Language Resources and Evaluation Conference (LREC 2014), pp. 1–7, Reykjavik, Iceland (2014)
2. Almeida, M.B., Martins, A.F.T.: Fast and robust compressive summarization with dual decomposition and multi-task learning. In: Proceedings of the Annual Meeting of the Association for Computational Linguistics, pp. 196–206 (2013)
3. Bick, E.: The Parsing System Palavras, Automatic Grammatical Analysis of Portuguese in a Constraint Grammar Framework. Aarhus University Press, Aarhus (2000)
4. Cohn, T., Lapata, M.: Sentence compression beyond word deletion. In: Proceedings of the International Conference on Computational Linguistics, pp. 137–144 (2008)
5. Dietterich, T.G.: Ensemble methods in machine learning. In: Kittler, J., Roli, F. (eds.) MCS 2000. LNCS, vol. 1857, pp. 1–15. Springer, Heidelberg (2000)
6. Filippova, K., Alfonseca, E., Colmenares, C., Kaiser, L., Vinyals, O.: Sentence compression by deletion with LSTMs. In: Proceedings of the 2015 Conference on Empirical Methods in Natural Language Processing, pp. 360–368 (2015)
7. Kawamoto, D., Pardo, T.A.S.: Learning sentence reduction rules for Brazilian Portuguese. In: Proceedings of the 7th International Workshop on Natural Language Processing and Cognitive Science - NLPCS, pp. 90–99 (2010)
8. Berg-Kirkpatrick, T., Gillick, D., Klein, D.: Jointly learning to extract and compress. In: Proceedings of the International Conference on Computational Linguistics, p. 10 (2011)
9. Knight, K., Marcu, D.: Statistics-based summarization - step one: sentence compression. In: Proceedings of the AAAI, pp. 703–711 (2000)
10. Li, C., Liu, F., Weng, F., Liu, Y.: Document summarization via guided sentence compression. In: Proceedings of the Conference on Empirical Methods in Natural Language Processing, pp. 490–500 (2013)
11. Madnani, N., Zajic, D., Dorr, B., Ayan, N.F., Lin, J.: Multiple alternative sentence compressions for automatic text summarization. In: Proceedings of the Document Understanding Conference, p. 8 (2007)
12. Martins, A.F.T., Smith, N.A.: Summarization with a joint model for sentence extraction and compression. In: Proceedings of the NAACL HLT Workshop on Integer Linear Programming for Natural Language Processing, pp. 1–9 (2009)
13. McDonald, R.: Discriminative sentence compression with soft syntactic evidence. In: Proceedings of the AAAI, pp. 297–304 (2006)
14. Mikolov, T., Sutskever, I., Chen, K., Corrado, G., Dean, J.: Distributed representations of words and phrases and their compositionality. In: Proceedings of the Advances in Neural Information Processing Systems, pp. 3111–3119 (2013)
15. Qian, X., Liu, Y.: Fast joint compression and summarization via graph cuts. In: Proceedings of the Conference on Empirical Methods in Natural Language Processing, pp. 1492–1502 (2013)
16. Thadani, K., McKeown, K.: Sentence compression with joint structural inference. In: Proceedings of the Seventeenth Conference on Computational Natural Language Learning, pp. 65–74 (2014)
17. Turner, J., Charniak, E.: Supervised and unsupervised learning for sentence compression. In: Proceedings of the 43rd Annual Meeting on Association for Computational, pp. 290–297 (2005)

A Model for Textual Entailment Based on Linguistic Rules

Evandro Metz Flores and Sandro José Rigo[✉]

Applied Computing Graduate Program (Pipca), University of Vale do Rio dos Sinos (Unisinos),
São Leopoldo, Brazil
evandrometzflores@gmail.com, rigo@unisinos.br

Abstract. This work proposes a model for recognition of textual entailment by presenting a new approach through the combined use of syntactic analysis, morphology, linguistic rules, detection of the bending voice, treatment of denial and the use of synonyms. A prototype was developed to evaluate the model proposed. The results, which are promising, allow the identification of textual linking of different textual samples accurately and with flexibility.

Keywords: Distance-learning · MOOC · Textual entailment

1 Introduction

The area of entailment recognition among textual samples develops a series of tools that can support textual analysis activities in education [9]. One of the most commonly used approach is the statistic, observed in a relevant number of works [4, 6]. Another approach is the syntactic analysis of texts [3], where the grammatical classification of each word and its position in sentences are identified and used to build a dependence tree, that is in turn used to identify the existence or not of entailment among textual samples [2, 5].

Works that aim to implement qualitative analysis resources in text excerpts through the combined use of syntactic analysis, morphology, linguistic rules and the use of synonyms are not observed in the literature. However, the growing use of the Distance Learning modality, Massive Online Open Courses and the widespread adoption of Distance Learning Environments as support in teaching and learning activities configure a scenario that can be beneficiated from using tools that allow the automatic treatment of textual excerpts, which can be a support element for the students and teachers.

The objective of this article is to present a model that makes possible to identify the entailment between a short textual message containing the correct answer to a question and other textual messages containing the answers to that same question. It is common knowledge that natural language possesses a complex and broad nature, having plenty of possibilities to answers and questions description. Despite this context, we consider that in controlled environments and with specific knowledge domains, it is viable to treat textual messages in natural language, to automatically evaluate the entailment between short sentences. The model was implemented in a prototype and the main differential is the textual entailment process based in a broad set of information involving syntactic information, morphological information, linguistic rules and the description of synonyms. This model allows the correct identification of short answers to questions about

© Springer International Publishing Switzerland 2016
J. Silva et al. (Eds.): PROPOR 2016, LNAI 9727, pp. 251–255, 2016.
DOI: 10.1007/978-3-319-41552-9_26

authorship and composition. The prototype was validated with case studies that demonstrated results with good accuracy and performance.

2 Related Work

The lexical approach can be expanded, for instance, as in the work of Litkowski [7], which consists in the use of comparisons between named entities, within the compared texts evaluations. Majumdar and Bhattacharyya [8] describe a system based in lexical representation, but uses references between the elements of the treated texts. The work of Tsuchida and Ishikawa [11] combines the use of a score generated with basis in the level of lexical representation and expanded with mechanisms based in machine learning. The work of Zhang [12] combines lexical information, syntactical information, and levels of semantic analysis with the aid of NLP tools. The work of Rios [10] apply semantics combined with an approach of semantic distance, where points are conceded to the candidate text according to lexical similarity and semantic similarity between the terms of both textual samples.

The analysis of studied works indicates diverse approaches for the representation of the elements and their treatment. The elements utilized for the representation are predominantly lexical, but there are also relevant works with the use of additional relations, such as synonyms and semantics annotation. The work presented in this article has as its main differentials the combination of information from distinct analysis (syntactic and semantic) and the use of linguistic rules.

3 Model and Evaluation

The main innovative aspect of the proposed model is the use of a broad set of resources to perform the task of textual entailment recognition: it uses morphological information, syntactical information, linguistic rules and synonyms. The objective of using this large set of resources is to evaluate the possibility of better results achievements. The evaluation or recognition of textual entailment occurs when a textual sample containing the hypothesis of a correct answer, that could have been constructed by a student, gets compared with a sample that contains the content representing all of the necessary knowledge to identify determined answer. The text in the answer would be normally developed by a teacher. These samples will be called candidate text and optimal text respectively.

The general vision of the proposed model is presented in Fig. 1. In this figure are presented the model components and the information flow among them. After selected, the messages containing both the optimal text and the candidate text are parsed through the morphosyntatic analyser. In the case of the implementation of this model the morphosyntatic parser "PALAVRAS" [1] was used, due to its precision. The parsed textual messages are manipulated with the treatment of denial and the voice flexion is verified. The verbs and synonyms are compared with an information base extracted from the repository WorldNet.Br. The linguistic rules are used to extract and label each piece of information, both in the optimal answer and in the candidates answer. Linguistic specialists using an editor available in the prototype manually generate the syntactic rules. The process of execution of the rules is about identifying which one of the present rules is able to represent the analysed text. Each rule is defined based on the information analysis of the used corpus, creating patterns composed of morphosyntatical information interspersed with the indication of elements that are important for the answer.

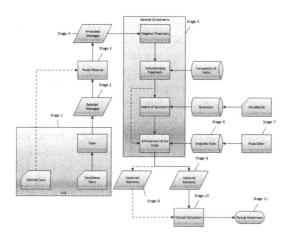

Fig. 1. General view of the proposed model

The evaluation performed was composed of three different case studies. All three evaluations realized counted with the participation of linguistic specialists. In the first evaluation, this linguistic specialists developed thirty questions created specifically for this evaluation, being twenty two in the "Which/Which One" format, five in the "Who" format, two in the "When" format and two in the "Of Which" format. Together with these questions were also developed one optimal answer and four candidate answers for each question, totaling one hundred and twenty answer groups to be evaluated by the prototype. These answers were also observed by the linguistic specialists for the construction of linguistic rules to be indexed in to the prototype, so that they would be available next to the linguistic rules repository. The rules present in the repository during the initial case study totals sixty nine rules. The prototype presented a 95,7 % success rate in the initial case study.

For the second case study the linguistic specialists developed eleven questions of different styles, being four related to the "Which/Which Ones" pattern, four to the "Who", two to the "What" and the last one to "How", as well as the optimal answers for each question. Eighteen people answered to these questions generating as such 198 candidate answers, analyzed by the linguistics specialist and also by the prototype. These answers, after being evaluated by the prototype, presented an accuracy of 97,22 %. The third evaluation was done with a larger set of questions and answers, in order to make possible to evaluate the behavior of the model in a context similar to the real situations of application of this resource in education context. This evaluation comprises a number of sixty questions with the same proportions of each of the types of questions. The results shows a consistent performance of the model, with a precision of 77,5 %, even to a greater set of very diverse questions.

4　Final Remarks

This article presented a model to identify textual entailment, which identifies as its main components the use of linguistic rules in combination with the morphosyntatic parser annotation, voice identification in the sentences, negation and synonym search at WordNet. A prototype was developed and evaluated. The prototype allowed us to demonstrate the positive results achieved with the combination of utilized resources. The prototype possesses functionality so that the linguistics specialists can add morphological patterns to the linguistic rules data base.

The area of recognition of textual entailment has been broadly explored for years but there are difficulties in the adaptation of other known works in its use in the Portuguese language. To bypass this characteristic, in this proposal it was presented the focus with the use of linguistic rules. This enabled the construction of a textual entailment system dedicated to the Portuguese language.

References

1. Bick, E.: The parsing system Palavras. Tese de Doutorado, Universidade de Aarhus (2000)
2. Chen, L., Liu, Y.: Automated scoring system using dependency-based weighted semantic similarity model. In: 2009 Second International Symposium on Knowledge Acquisition and Modeling, vol. 1, pp. 241–244 (2009)
3. Dagan, I.: Recognizing Textual Entailment – Models and Applications. Synthesis Lectures on Human Language Technologies. Claypool Publishers, Toronto (2013)
4. Fabricio, B.T.A., Behar, P.A., Reategui, E.B.: Análise das mensagens de fóruns de discussão através de um software para mineração de textos SBIE. n. 2004, pp. 20–29 (2011)
5. Lamjri, A.K., Kosseim, L., Radhakrishnan, T.: Comparing the contribution of syntactic and semantic features in closed versus open domain question answering. In: International Conference on Semantic Computing ICSC 2007, pp. 679–685 (2007)
6. Lin, F., Hsieh, L., Chuang, F.: Discovering genres of online discussion threads via text mining. Comput. Educ. **52**(2), 481–495 (2009)
7. Litkowski, K.: Overlap analysis in textual entailment recognition. In: Proceedings of Second Text Analysis Conference, TAC 2009, Gaithersburg, Maryland, USA (2009)
8. Majumdar, D., Bhattacharyya, P.: Lexical based text entailment system for summarization settings of RTE6. In: Proceedings of the Text Analysis Conference, TAC 2010, 15 November 2010
9. Rigo, S.J., Metz, E., Abech, M., Barbosa, J.L.V., Costa, C.A.: O papel do Processamento de Língua Natural e da Representação de Conhecimento na extração de informações em mensagens textuais na Educação a Distância. In: Simpósio Brasileiro de Informática na Educação (2013)

10. Rios, M., Gelbukh, A.: Recognizing textual entailment with a semantic edit distance Metric. In: 11th Mexican International Conference on Artificial Intelligence, MICAI 2012 (2012)
11. Tsuchida, M., Ishikawa, M., Ikoma, K.: A method for recognizing textual entailment using lexical-level and sentence structure-level features. In: TAC (2011)
12. Zhang, S., Wang, Y., Zhu, D., Shi, J.: Recognizing Textual Entailment with synthetic analysis based on SVM and feature value control. In: 2012 IEEE 3rd International Conference on Software Engineering and Service Science (ICSESS), Beijing (2012)

Language Resources – Full Papers

LX-DSemVectors: Distributional Semantics Models for Portuguese

João Rodrigues$^{(\boxtimes)}$, António Branco, Steven Neale, and João Silva

Department of Informatics,
Faculty of Sciences, University of Lisbon, Lisbon, Portugal
{joao.rodrigues,antonio.branco,steven.neale,jsilva}@di.fc.ul.pt

Abstract. In this article we describe the creation and distribution of the first publicly available word embeddings for Portuguese. Our embeddings are evaluated on their own and also compared with the original English models on a well-known analogy task. We gathered a large Portuguese corpus of 1.7 billion tokens, developed the first distributional semantic analogies test set for Portuguese, and proceeded with the first parametrization and evaluation of Portuguese word embeddings models.

Keywords: Distributional semantics · Word embeddings · Portuguese

1 Introduction

Current research trends focusing on distributional semantics are sparking interest in possible ways to enrich the resources and tools used for natural language processing (NLP) tasks. Researchers and practitioners are exploring possible improvements that can be achieved from integrating distributional vectors semantics, also known as word embeddings, in a range of syntactic and semantic tasks, including speech recognition [16], semantic similarity of words [15] part-of-speech (POS) tagging, named entity recognition, sentiment analysis [13] and logical semantics [4].

Experimenting with word embeddings in such tasks requires large data sets to extract word embeddings in a specific language. At the time of writing we have found no such freely available or evaluated data set for Portuguese to exist. There is therefore a need to create word embeddings in the Portuguese language that can be explored in the types of tasks mentioned above for the English language.

In this paper we describe our results in training, parameterizing and evaluating word embeddings for Portuguese – a computationally intensive and time consuming undertaking – as well as comparing them with an implementation for the English language. Our contribution is in making available a set of trained word embeddings for the computational processing of Portuguese, as well as a set of instructions for getting them running quickly and easily.

In Sect. 2 we briefly describe word embeddings and current methods for obtaining them, followed by a description of our own implementation of the

J. Silva et al. (Eds.): PROPOR 2016, LNAI 9727, pp. 259–270, 2016.
DOI: 10.1007/978-3-319-41552-9_27

models in Sect. 3 and the set of experiments we run to improve their accuracy. The resulting models are evaluated and analyzed in Sect. 4 against the English models, before we draw our conclusions and outline our plans for future work in Sect. 5.

2   Related Work

As concisely stated in [8], "distributional semantics is predicated on the assumption that linguistic units with certain semantic similarities also share certain similarities in the relevant environments". Addressing this so-called 'relevant environment' using distributional semantics methods is based on two key paradigms – count-based and prediction-based methods.

Both count and prediction-based methods generate a set of distributional vectors (also known as word embeddings or distributed word representations) that are able to reflect the semantic similarity between words, with the meaning of each word possibly characterized by a vector of real values. For example, cosine similarity can be used to find the similarity between two word vectors, and hence the meaning of the words they represent. Interestingly, it is possible to perform algebraic operations using the vectors, as demonstrated by the typical example of $vector(king) - vector(man) + vector(woman)$ resulting in a similar vector to $vector(queen)$, the distributed word representation of the word queen [15].

In this article we focus on the prediction-based methods, which – inspired by neural network-based probabilistic language models [3] – predict the co-occurrent context words for a word of interest using a sliding window of one or more (n-) words that continually moves along the corpus with each new word of interest. The co-occurring context words captured by this sliding window fill the cells of the vector for a given word of interest as we move through the corpus, in contrast to count-based methods that count all cases of co-occurrence with a word of interest across the entire corpus, often resulting in a huge and sparse vector space that typically grows with the quadratic size of the vocabulary.

An even simpler use of these methods was presented in [16], where a combination of an input and an output layer – each with a length corresponding to the size of the vocabulary – and one hidden layer with approximately 60 neurons are trained to estimate the probability distribution of the next word in a text given a previous word from the vocabulary, as in [3]. This work was later extended with two new models – the continuous bag of words (CBOW) and Skip-gram – to further simplify the neural probabilistic language models [15]. These two models represent a shallow use of neural networks – the CBOW model introduces a shared projection layer for all words, a weighted matrix between the first two layers, and a sliding co-occurrent context window for the training of the current word of interest; while the Skip-gram model creates a standalone vector representing the combination of contextual words and then predicts the context vector closest to the current word vector.

These new methods – designed to leverage maximum information from large data sets at minimum computational costs – began to be evaluated in a semantic textual similarity task. One of the comprehensive test sets for measuring both syntactic and semantic regularities using analogies that was made available [15] is the same test set we use for comparison in this paper[1]. As discussed in [12], no qualitative difference can be pinpointed to one or the other of the different approaches to word vectors, count or prediction-based ones. Any difference to be found among particular models is instead due to the optimization of various hyperparameters, which can yield better results using one approach or the other. Notwithstanding, both methods still resort to different computational means to create the shared semantic models and in a range of different sets of tasks the Skip-gram prediction-based model has been shown to be better, on average.

Both the CBOW and Skip-gram models are available with the standalone implementation 'word2vec'[2]. For the creation of the Portuguese vectors we used Gensim [19], a Python-based Skip-gram implementation. Gensim is a good choice for our work because it allows for different distributional semantic methods to be deployed within the same framework, models which can be ported to word2vec later if convenient.

Regarding related work concerning distributional semantics of Portuguese, Portuguese corpora are used for the creation of distributional semantic models in [10,20], the latter using the CharWNN deep neural network for boosting named entity recognition and the former applying a novel method that uses parallel data for document classification. Using long short-term memory (LSTM) neural networks, [14] constructs vector representations of words with a Portuguese model yielding state-of-the-art results in language modeling and part-of-speech (POS) tagging. [21] also uses distributed word representation in POS tagging for Portuguese by using a neural language model. They resort to word2vec and to the Portuguese Wikipedia, CETENFolha and CETEMPublico corpora, obtaining state-of-the-art results for POS tagging. Another article using Portuguese word embeddings is the Polyglot project [1], which uses a Portuguese Wikipedia corpus to support POS tagging. Finally, [7] also seeks to improve on the POS tagging of Portuguese using word embeddings.

Although all of these works used models for distributional semantics of Portuguese, none of them report an evaluation of parameter optimization or an assessment against current test sets. In this article, we seek to overcome this shortcoming by performing a comparison with state-of-the-art evaluation methods. The models trained in the related works above are also not available except the one from the Polyglot project, which has a unique model. A major contribution of this article is making these trained and tuned models of Portuguese word embeddings available as a freely available resource.

3 Implementation

For the creation of the Portuguese word embeddings we chose Skip-gram as the training algorithm, since it obtains the best accuracy, on average, from a range

---

[1] For a more complete description of the evaluation methods, see [22].
[2] http://code.google.com/p/word2vec/.

of test sets in the distributional semantics domain [12]. We installed the Gensim framework and developed the necessary scripts for the training and evaluation, which are made available at http://github.com/nlx-group.

The first step in the implementation process was the gathering of corpora, described below in Subsect. 3.1. For a reasonable comparison with the original (English) evaluation of Skip-gram, our evaluation should be performed with a similar test set. Since the original test set is in English, it was necessary to translate it – this process is described in Subsect. 3.2.

With Portuguese corpora and a test set in place, we could then design and undertake a set of experiments encompassing the training of different models with the objective of maximizing the accuracy of the models obtained. These experiments are described in Subsect. 3.3.

3.1   Acquisition of Corpora

For Portuguese (both Brazilian and European variants), a total of 1,723,693,241 tokens from 121,706,288 sentences were gathered. To the best of our knowledge, this is the largest raw text data set whose gathering was ever reported for the Portuguese language. Table 1 lists each of the gathered corpora used along with their respective token and sentence volumes (obtained after tokenization). We used a web crawler to gather news articles from Jornal Digital[3] and Observador[4]. The crawl gathered all public news articles available on November 20, 2015, including their titles, headlines and the articles themselves.

After the search, extraction and cleaning of the corpora, a tokenization process took place. No lowercasing was performed over the source texts and the original surface form of the word was used. For the tokenization, the LX-Tokenizer [5] was used, which has a reported f-score of 99.72 %.

3.2   Test Set

The test set described in [15] – a collection of word analogies – was used as the basis for the assessment of word embeddings. An example entry in this data set would read: 'Berlin Germany Lisbon Portugal'. With these four words relations – as in this example – one can test semantic analogies by using any of the possible combinations of three of the four word vectors in one entry and testing whether or not the resulting vector is similar to the (fourth) word vector missing from the combination being tested. In the example above, the completed analogy should read: 'Berlin is to Germany as Lisbon is to Portugal'.

The test set contains five types of semantic analogy: common capitals and countries, all capitals and countries, currency, cities and states, and family relations. Nine types of syntactic analogy are also represented: adjective to adverb, opposite, comparative, superlative, present participle, nationality (adjective),

---

[3] www.jornaldigital.com.

[4] www.observador.pt.

Table 1. Portuguese corpora used for training

Corpus	Tokens	Sentences	Description	Ref.
TCC	61,979	642	scientific corpus	[18]
QTLeap	56,255	4,000	q/a pairs in the IT domain	[9]
CRPC	133,497	5,061	oral communication of direct inquiries	[17]
Tanzil	178,225	9,377	Quran translation to Portuguese	[24]
CINTIL	707,444	30,344	International corpus of Portuguese	[2]
JDigital	3,891,407	110,227	news articles from Jornal Digital	
Ted2013	3,173,357	156,033	corpus from the TED talks	[6]
KDE4	3,123,310	230,178	KDE4 localization files	[23]
Observador	34,900,297	732,240	news articles from Observador	
EMEA	19,083,444	1,213,566	documentation from EMA	[23]
ECB	71,387,581	2,162,343	documentation from the ECB	[23]
Europarl	67,506,802	2,171,029	European Parliament sessions	[11]
DGT	73,788,835	3,153,654	translation memories from the Acquis	[23]
Stackoverflow	36,200,297	3,767,771	posts from Stackoverflow	
EUBookshop	203,762,634	7,310,336	documentation from the EU bookshop	[23]
Wikipedia	246,550,786	7,460,428	PT Wikipedia dump of 01/09/2015	
CETEMPublico	225,906,693	8,065,830	news articles from the Público	
OpenSubtitles	442,182,528	54,415,635	Portuguese subtitles until 2013	[23]
CETENFolha	291,097,870	30,707,594	news articles from Folha de S. Paulo	
Total	1,723,693,241	121,706,288		

past tense, plural nouns and plural verbs. The test set contains a total of 8869 semantic and 10675 syntactic entries.

For the evaluation of the Portuguese word embeddings, the original English test set was translated into Portuguese by skilled, native Portuguese-speaking language experts. The resulting translations, LX-4WAnalogies, and corresponding English terms are available at http://github.com/nlx-group.

There were some English words that could not be accurately translated into a unique Portuguese word. Given that the original evaluation does not support vector composition, if a single word from the original four words in an analogy could not be translated as a single word, then the analogy in question had to be dropped. Therefore, the resulting Portuguese test set kept only 17558 analogies from the original 19544 English analogies. The groups of analogies affected were:

- **family**: From the original 506 analogies, 462 were retained in the translation (506, 462, -44). For example, the single word *copwoman* is translated to the two word expression *mulher polícia*.
- **gram1-adjective-to-adverb** (992, 930, -62): For example, *most* is translated to *a maioria*.
- **gram2-opposite** (812, 756, -56): For example, *uncompetitive* is translated to *não competitivo*.
- **gram3-comparative** (1332, 30, -1302): The Portuguese language needs, in most of the cases, a separate word to mark the comparison. This is accomplished with adverbs that quantify the adjective. For example *brighter* is translatable to *mais brilhante*.
- **gram4-superlative** (1122, 600, -522): This group is affected by the same linguistic phenomena as in the comparative group. For example, *tastiest* is translated to *o mais saboroso*.

3.3   Experiments

A total of five experiments were ran, with the objective of narrowing the choice of corpora and parameters to arrive at the most accurate word embeddings. The evaluation was performed in two ways: with the original restriction, where only analogies in which the frequencies of all four words are in the top 30000 most frequent words overall, and a second evaluation where this restriction is not applied. The unrestricted evaluation is useful for grasping the achievable generalization of a model.

- **First experiment** – Firstly, we use the vanilla parameters of Gensim to evaluate each of the gathered corpora separately (both with and without restriction). Secondly, we evaluate each of the gathered corpora incrementally – that is, one of the other gathered corpora is added to the whole in each incrementation step.
- **Second experiment** – We take the largest resulting data sets from the incremental phase of the first experiment and evaluate them with larger vector dimensions. The reason for performing this experiment is to test whether or not the proportionality of data and vector dimensions influences the result, as expected.
- **Third experiment** – We use only the data sets that in the first experiment yielded an improvement in accuracy when they were incremented with other corpora (using the vanilla parameters).

– **Fourth experiment** – We compare the best model obtained in the third experiment with the model obtained by assembling together: (a) corpora that improved over each other during the incremental phase of experiment 1; (b) Europarl (which improved over the previous two increments although they had not improved over the highest incremental score at that point); and (c) CETEMPublico and CETENFolha (which were shown to yield the best scores without restriction in experiment 2). Both models were evaluated along a range of vector dimensions.

– **Fifth experiment** – We evaluated the effect of additional parameterization (besides vector dimensions) for the most accurate model obtained in the fourth experiment, including: (a) sliding window size (value 5 or 10); (b) initial learning rate, which linearly drops to zero as the training progresses (0.025 or 0.05); (c) the threshold for configuring which higher-frequency words are randomly down sampled (0 or 0.00005); (d) hierarchical sampling (0-off or 1-on); and (e) negative sampling (5 or 15).

4 Evaluation

4.1 Experiments with Portuguese Embeddings

The first experiment (see Table 2) shows that – as expected – better accuracy is obtained with larger data sets. Wikipedia excels here both because of its higher quality and its relevance to the analogies test set. When incrementing the data set in a step wise fashion, a large increase in accuracy can be seen when Wikipedia is added to all of the previous corpora incremented up to that point (incr_15, Table 3). Beyond this (incr_16 to 18) the accuracy drops, possibly due to the fixed vector dimension.

The second experiment (Table 4) tests for the vector dimension with the four largest corpora (incr_15 to 18 in Table 3). Increasing the vector dimension here leads to increased accuracy, with the strongest results appearing around vectors with dimension 400. The OpenSubtitles corpus seems to introduce some noise, which reduces the accuracy. Although the incrementing with CETEMPublico and CETENFolha did not improve over the top accuracy score with the typical restriction, an increase can be seen without such restriction that reaches the top value in that setting.

The third experiment (Table 5) reveals that by using only those corpora that induced improved accuracy better generalization is obtained as indicated in the score 28.5 for accuracy without restriction (against 26.3 from the first experiment).

The fourth experiment (Table 6) reveals that although the assembled corpora in the third experiment (chosen_incr_9 in Table 5, labeled as model 2 in Table 6) permit to obtain the best result in the restricted evaluation, that accuracy can be surpassed by a non-restricted evaluation with the selected larger corpus as the vector dimension is increased to 400.

Because model 1 in fourth experiment yielded higher accuracy than model 2 in almost all non-restricted evaluations (Table 6), model 1 was chosen – with a

Table 2. First experiment – accuracy (with/without restriction) obtained by training on the different corpora

Corpus	Accuracy %
TCC	0.0/0.0
QTLeap	0.0/0.0
CRPC	0.0/0.0
Tanzil	5.0/5.0
CINTIL	0.8/0.8
JDigital	2.4/2.4
Ted2013	3.2/3.2
KDE4	3.7/3.7
Observador	9.8/5.9
EMEA	4.0/3.6
ECB	6.5/2.6
Europarl	12.7/6.4
DGT	8.4/3.8
Stackoverflow	6.9/3.1
EUBookshop	17.3/5.9
Wikipedia	**36.3/26.1**
CETEMPublico	27.6/20.1
OpenSubtitles	25.4/18.5
CETENFolha	19.0/13.8

Table 3. First experiment – accuracy (with/without restriction) obtained from incrementally adding each corpus to the previous (for example, incre_ 3 consists of the TCC+QTLeap+CRPC+Tanzil corpora)

Corpus	Accuracy %
incr_0 (TCC)	0.0/0.0
incr_1 (+QTLeap)	0.0/0.0
incr_2 (+CRPC)	0.0/0.0
incr_3 (+Tanzil)	1.8/1.8
incr_4 (+CINTIL)	2.9/2.9
incr_5 (+JDigital)	2.0/1.9
incr_6 (+Ted2013)	6.3/4.9
incr_7 (+KDE4)	6.9/4.9
incr_8 (+Observador)	13.2/8.3
incr_9 (+EMEA)	5.9/3.4
incr_10 (+ECB)	3.9/2.2
incr_11 (+Europarl)	8.4/4.5
incr_12 (+DGT)	8.3/3.5
incr_13 (+Stackoverflow)	7.2/3.5
incr_14 (+EUBookshop)	16.9/6.6
incr_15 (+Wikipedia)	**38.2/26.3**
incr_16 (+CETEMPublico)	32.8/23.9
incr_17 (+OpenSubtitles)	30.3/20.5
incr_18 (+CETENFolha)	30.5/21.4

vector dimension of 400 – to be used in the fifth and final experiment (Table 7). In the fifth experiment model 1 of the fourth experiment is evaluated against different settings of parameters. The results of this experiment clearly shows that all models using hierarchical sampling induce a reduced accuracy, while increasing the negative sampling from 5 to 15 units increases accuracy. The best obtained score was in p_17 with an accuracy of 52.8 % in the restricted evaluation and 37.7 % without restriction.

4.2   Comparison with English Models

In the original evaluation of the English word embeddings [15] – trained with a vector dimensionality of 300 and a corpus of 783 million tokens – accuracy of 50.4 % was obtained with restriction. In a second experiment – where the word embeddings were trained with a vector dimensionality of 1000 and a corpus of 6 billion tokens – accuracy of 65.6 % was obtained without restriction.

Our best trained word embeddings for Portuguese – using a corpus of approximately 1 billion tokens – obtained an accuracy of 52.8 % with restriction (compared to 50.4 % in English, with 783 million tokens) and 37.7 % without restriction (compared to 65 % in English, with 6 billion tokens).

Table 4. Second experiment – different ranges of vector dimensions evaluated on the incremental corpora with the highest accuracy (incr_15 to 18) from experiment 1 (with/without restriction)

corpus	vector dimension					
	100	200	300	400	500	600
incr_15	38.2/26.3	43.4/29.2	44.7/29.6	**44.9**/30.6	43.4/29.5	40.4/26.8
incr_16	32.8/23.9	40.0/29.8	43.6/**32.8**	43.6/32.1	42.4/32.2	43.5/32.0
incr_17	30.3/20.5	35.2/26.2	35.0/25.4	35.3/25.8	37.2/28.3	36.7/27.2
incr_18	30.5/21.4	36.3/27.0	37.3/27.2	35.3/27.0	35.2/27.1	35.3/26.1

The results obtained for the Portuguese embeddings seem to be in line with those obtained for English when using data sets of similar size. When analyzing this kind of comparison, the variety of different training conditions – including differences in parameterization, vector dimensionalities and the larger English corpora – must be taken into account. We believe that our work reported here suggests room for further improvements in the Portuguese models, and that the work described in this paper paves the way for further exploration, specially if larger data sets are used.

Table 5. Third experiment – selected incremental corpora, accuracy with/without restriction (each new corpus is added to the existing with each incrementation – for example, incr_3 consists of the TCC+QTLeap+CRPC+Tanzil corpora)

Corpus	Accuracy %
chosen_incr_0 (TCC)	0.0/0.0
chosen_incr_1 (+QTLeap)	0.0/0.0
chosen_incr_2 (+CRPC)	0.0/0.0
chosen_incr_3 (+Tanzil)	1.8/1.8
chosen_incr_4 (+CINTIL)	2.8/2.8
chosen_incr_5 (+Ted2013)	4.1/3.9
chosen_incr_6 (+KDE4)	5.1/4.1
chosen_incr_7 (+Observador)	13.8/9.1
chosen_incr_8 (+EUBookshop)	15.5/6.4
chosen_incr_9 (+Wikipedia)	**37.3/28.5**
chosen_incr_10 (+CETEMPublico)	31.4/25.2
chosen_incr_11 (+CETENFolha)	33.9/26.4

Table 6. Fourth experiment – comparing model 1 (corpora: TCC, QTLeap, CRPC, Tanzil, CINTIL, Ted2013, KDE4, Observador, Europarl, DGT, EUBookshop, Wikipedia, CETEMPublico and CETENfolha) with model 2 (corpora: chosen_incr_9 from experiment 3, accuracy with/without restriction)

	vector dimension					
corpus	100	200	300	400	500	600
model_1	34.4/26.3	40.3/30.7	41.4/32.3	42.6/**33.1**	43.0/32.2	41.8/32.0
model_2	37.3/28.5	40.9/30.0	42.4/31.5	**43.1**/31.6	42.6/30.4	42.4/30.7

Table 7. Fifth experiment – accuracy (acc. with/without restriction) of model 1 from experiment 4 with additional parameterization including: sliding window size (win), learning rate (lrate), threshold for configuring which higher-frequency words are randomly downsampled (hf), hierarchical sampling (hs), and negative sampling (ns). The training of these models took a week and a half using a server consisting of 30 processors (Intel(R) Xeon® 8C CPU E5-2640 V2 @ 2.00 GHz, 20 M Cache, RAM 16x 16 GB RDIMM, 1600 MHz)

	win	lrate	hf	hs	ns	acc. %		win	lrate	hf	hs	ns	acc. %
p_0	5	0.025	0	0	5	51.6/35.4	p_16	10	0.025	0	0	5	52.0/37.0
p_1	5	0.025	0	0	15	49.3/34.9	p_17	10	0.025	0	0	15	**52.8/37.7**
p_2	5	0.025	0	1	5	45.4/36.3	p_18	10	0.025	0	1	5	48.0/36.5
p_3	5	0.025	0	1	15	47.2/36.1	p_19	10	0.025	0	1	15	48.6/36.6
p_4	5	0.025	1e-05	0	5	50.7/31.4	p_20	10	0.025	1e-05	0	5	50.0/30.5
p_5	5	0.025	1e-05	0	15	52.1/32.6	p_21	10	0.025	1e-05	0	15	51.3/32.0
p_6	5	0.025	1e-05	1	5	45.2/33.7	p_22	10	0.025	1e-05	1	5	44.4/33.0
p_7	5	0.025	1e-05	1	15	47.1/35.0	p_23	10	0.025	1e-05	1	15	44.4/32.3
p_8	5	0.05	0	0	5	50.2/36.4	p_24	10	0.05	0	0	5	50.7/36.4
p_9	5	0.05	0	0	15	50.5/36.7	p_25	10	0.05	0	0	15	51.0/36.8
p_10	5	0.05	0	1	5	45.8/34.7	p_26	10	0.05	0	1	5	44.6/32.1
p_11	5	0.05	0	1	15	44.8/34.6	p_27	10	0.05	0	1	15	46.1/33.2
p_12	5	0.05	1e-05	0	5	50.6/30.5	p_28	10	0.05	1e-05	0	5	46.4/28.0
p_13	5	0.05	1e-05	0	15	52.5/34.3	p_29	10	0.05	1e-05	0	15	49.7/31.8
p_14	5	0.05	1e-05	1	5	43.9/30.9	p_30	10	0.05	1e-05	1	5	41.2/28.5
p_15	5	0.05	1e-05	1	15	44.3/31.6	p_31	10	0.05	1e-05	1	15	40.5/28.6

5   Conclusion

In this paper we described the creation, parameterization and evaluation of the first publicly available distributional semantic models for Portuguese, which perform in line with the original state-of-the-art models for English. All the models from the fifth experiment are made available from http://github.com/nlx-group.

In future work we plan to account for missing analogies in our test set by using phrases instead of words. While introducing lowercasing and lemmatization steps and making use of richer linguistic knowledge are also promising directions, acquiring a larger Portuguese corpora to train on remains the most important step as we seek to improve the accuracy of our models.

Acknowledgements. The results reported in this paper were partially supported by the Portuguese Government's P2020 program under the grant 08/SI/2015/3279: ASSET-Intelligent Assistance for Everyone Everywhere, and by the EC's FP7 program under the grant number 610516: QTLeap-Quality Translation by Deep Language Engineering Approaches.

References

1. Al-Rfou, R., Perozzi, B., Skiena, S.: Polyglot: distributed word representations for multilingual NLP. In: Proceedings of the Seventeenth Conference on Computational Natural Language Learning. pp. 183–192. Association for Computational Linguistics, Sofia, August 2013
2. Barreto, F., Branco, A., Ferreira, E., Mendes, A., Nascimento, M.F., Nunes, F., Silva, J.: Open resources and tools for the shallow processing of portuguese: the tagshare project. In: Proceedings of LREC 2006. Citeseer (2006)
3. Bengio, Y., Ducharme, R., Vincent, P., Janvin, C.: A neural probabilistic language model. J. Mach. Learn. Res. **3**, 1137–1155 (2003)
4. Bowman, S.R., Potts, C., Manning, C.D.: Recursive neural networks can learn logical semantics. In: ACL-IJCNLP, p. 12 (2015)
5. Branco, A., Silva, J.: Evaluating solutions for the rapid development of state-of-the-art pos taggers for portuguese. In: LREC (2004)
6. Cettolo, M., Girardi, C., Federico, M.: Wit3: Web inventory of transcribed and translated talks. In: Proceedings of the 16th Conference of the European Association for Machine Translation (EAMT), pp. 261–268 (2012)
7. Fonseca, E.R., Rosa, J.L.G., Aluísio, S.M.: Evaluating word embeddings and a revised corpus for part-of-speech tagging in portuguese. J. Braz. Comput. Soc. **21**(1), 1–14 (2015)
8. Garvin, P.L.: Computer participation in linguistic research. Language **38**, 385–389 (1962)
9. Gaudio, R.D., Burchardt, A., Branco, A.: Evaluating machine translation in a usage scenario. In: Proceedings of LREC (to appear in print, 2016)
10. Hermann, K.M., Blunsom, P.: Multilingual models for compositional distributed semantics. arXiv preprint arXiv:1404.4641 (2014)
11. Koehn, P.: Europarl: a parallel corpus for statistical machine translation. In: MT Summit, vol. 5, pp. 79–86. Citeseer (2005)
12. Levy, O., Goldberg, Y., Dagan, I.: Improving distributional similarity with lessons learned from word embeddings. Trans. Assoc. Comput. Linguist. **3**, 211–225 (2015)
13. Li, J., Jurafsky, D.: Do multi-sense embeddings improve natural language understanding? arXiv preprint arXiv:1506.01070 (2015)
14. Ling, W., Luís, T., Marujo, L., Astudillo, R.F., Amir, S., Dyer, C., Black, A.W., Trancoso, I.: Finding function in form: Compositional character models for open vocabulary word representation. arXiv preprint arXiv:1508.02096 (2015)
15. Mikolov, T., Chen, K., Corrado, G., Dean, J.: Efficient estimation of word representations in vector space. arXiv preprint arXiv:1301.3781 (2013)

16. Mikolov, T., Kopecký, J., Burget, L., Glembek, O., Černocký, J.H.: Neural network based language models for highly inflective languages. In: IEEE International Conference on Acoustics, Speech and Signal Processing, pp. 4725–4728. IEEE (2009)

17. do Nascimento, M.F.B., Pereira, L., Saramago, J.: Portuguese corpora at CLUL. PRAXIS **2**(2.1/759), 95 (2000)

18. Pardo, T.A.S., Nunes, M.d.G.V.: A construção de um corpus de textos científicos em português do brasil e sua marcação retórica. Tech. rep. (2003)

19. Rehurek, R., Sojka, P.: Gensim–python framework for vector space modelling. NLP Centre, Faculty of Informatics, Masaryk University, Brno, Czech Republic (2011)

20. dos Santos, C., Guimaraes, V., Niterói, R., de Janeiro, R.: Boosting named entity recognition with neural character embeddings. In: Proceedings of NEWS 2015 The Fifth Named Entities Workshop, p. 25 (2015)

21. Santos, C.D., Zadrozny, B.: Learning character-level representations for part-of-speech tagging. In: Proceedings of the 31st International Conference on Machine Learning (ICML), pp. 1818–1826 (2014)

22. Schnabel, T., Labutov, I., Mimno, D., Joachims, T.: Evaluation methods for unsupervised word embeddings. In: Proceedings of EMNLP (2015)

23. Tiedemann, J.: News from OPUS - A collection of multilingual parallel corpora with tools and interfaces. In: Nicolov, N., Bontcheva, K., Angelova, G., Mitkov, R. (eds.) Recent Advances in Natural Language Processing, vol. V, pp. 237–248. John Benjamins, Amsterdam/Philadelphia (2009)

24. Tiedemann, J.: Parallel data, tools and interfaces in OPUS. In: Chair, N.C.C., Choukri, K., Declerck, T., Dogan, M.U., Maegaard, B., Mariani, J., Odijk, J., Piperidis, S. (eds.) Proceedings of the Eight International Conference on Language Resources and Evaluation (LREC). European Language Resources Association (ELRA), Istanbul, May 2012

Making Virtue of Necessity: A Verb Lexicon

Valeria de Paiva[1]([✉]), Fabricio Chalub[2],
Livy Real[4], and Alexandre Rademaker[3]

[1] Nuance Communications, Sunnyvale, USA
valeria.depaiva@nuance.com
[2] IBM Research, Rio de Janeiro, Brazil
[3] IBM Research and FGV/EMAp, Rio de Janeiro, Brazil
[4] IBM Research, São Paulo, Brazil
{fchalub,livym,alexrad}@br.ibm.com

Abstract. We describe the verb lexicon of OpenWordNet-PT, a wordnet-like resource for (mostly Brazilian) Portuguese and a series of experiments that we designed to extend its coverage. These experiments include checking online lists of most common verbs, checking corpora freely available such as the Bosque-UD (the Bosque corpus annotated with Universal Dependencies) and especially checking a dictionary of Brazilian politicians' biographies (the DHBB) that we consider an ideal corpus for the kind of information extraction we are after. We certainly succeeded into extending the coverage of the verb lexicon, however it remains to be seen whether this new coverage is enough for the original application.

1 Introduction

Verbs, together with nouns, are usually the main bearers of meaning in sentences. We could not agree more with [8] when they say

> Verbs are the primary vehicle for describing events and expressing relations between entities. Hence, verb semantics could help in many natural language processing (NLP) tasks that deal with events or relations between entities. For tasks which require canonicalization of natural language statements or derivation of (plausible) inferences from such statements, a particularly valuable resource is one which (i) relates verbs to one another and (ii) provides broad coverage of the verbs in the target language.

Portuguese is the 6th most spoken language in the world, according to Etnologue [11], but lexical resources for Portuguese are still not very well-developed. Despite some recent work on Portuguese verbs, such as VerbNet.BR [19,20], Viper [3], and the catalog of Brazilian Portuguese Verbs [6], there are still no freely available, comprehensive resources that provide human users and automated programs with access to Portuguese verbs, their meanings and information about their subcategorization frames.

Given the essential role played by verbs in sentence understanding we decided to improve the state of the verb lexicon in the basic resource OpenWordNet-PT [14]. OpenWordnet-PT already provides some of the functionality desired,

© Springer International Publishing Switzerland 2016
J. Silva et al. (Eds.): PROPOR 2016, LNAI 9727, pp. 271–282, 2016.
DOI: 10.1007/978-3-319-41552-9_28

as it has 5902 verbal synsets in Portuguese, with as many as 4511 verbal lemmas. It also has 7865 synsets in English that are empty in Portuguese and for many of these we know there are Portuguese words that fit them perfectly, but they are not there, yet[1]. An example is the verb *popularize*: the verb *popularizar* exists in Portuguese with the same sense as popularize has in English. We only need to add it to the appropriate synset, but our problem is to find out within the 'soup' of these 7865 empty synsets, which ones are easy cases, where a corresponding verb exists with the same meaning in English, like this one. We also need to find out which ones are the hard verbs to translate or even the impossible ones. For an example of a Portuguese verb that is impossible to translate as a single word in English, we are using *apaulistar* (to make similar to what the natives of São Paulo, *paulistas*, do). We do not expect English to have such a word, nor many others like this that correspond to particular facets of Brazilian reality, but these should also be part of a truly useful Portuguese (verb) lexicon.

For the verbs OpenWordnet-PT already knows about, we can provide some indication of meaning, by giving other words that the given verb is related to, as well as its placing in both the OpenMultilingual Wordnet (OMW) [5] and in the SUMO ontology [16]. We can also provide some possible subcategorization frames, inherited from OMW, but not checked for Portuguese, so far. This, as well as making sure that all Portuguese synsets have verified glosses in Portuguese, is left to future work. For the moment we would like to acknowledge the helpful work of Alberto Simões [22] in producing automatically translated glosses, which are extremely helpful for the work described here.

2   OpenWordNet-PT

OpenWordnet-PT is a lexical-semantic resource describing (mostly Brazilian) Portuguese words and their relationships. It is modelled after and fully interoperable with the original Princeton WordNet [9] for English (henceforth PWN), relying on the same identifiers as Princeton WordNet 3.0. This means that one can easily find Portuguese equivalents for some specific English word senses and conversely. This also means that OpenWordnet-PT is part of a large ecosystem of compatible resources, including domain identifiers and mappings to Wikipedia, DBpedia and Wikidata, amongst others. OpenWordnet-PT is encoded and distributed in RDF/OWL [2].

2.1   Related Work

As indicated above there are several works on Portuguese verbs, some more linguistic, some more computational. The more linguistic work seems not to be available online or tends not to have meanings associated to the verbs. The computational work on VerbNet.BR is very encompassing, but it has not been verified for consistency or accuracy. We discuss the golden subset of VerbNet.BR

---

[1] For up to date numbers check http://wnpt.brlcloud.com/wn/stats.

in the following section. Moreover, we would not like to take the syntactic classes as so fundamental in our own work. The work on Viper is not open source, at the moment. The work on TeP [12] has unclear licensing status and its definitive version is, apparently, not available yet.

3 Extending the Verb Lexicon

It is always easier to check whether one has coverage of a lexical resource than accuracy of the same, so we decided to check the coverage of our verb lexicon, using resources available online. From this perspective this work is a continuation of [13]. A series of experiments, with different collections of verbs, from corpora and otherwise was devised. We describe some of these experiments here. Data sources used in our experiments can be found in our GitHub repository[2].

3.1 Golden VerbNet.BR

Since there is an available VerbNet.BR 1.0, with a manually verified golden subset, we first decided to investigate whether we had all the verbs in this golden subset. Exactly 50 verbs were found to be missing from OpenWordNet-PT from the 604 verbs in the golden subset of VerbNet.BR. These verbs were added to OpenWordNet-PT, in their respective places, with the exception of two verbs *entreabrir, rebolar* (meaning, respectively 'to partially open' and 'to move your hips in a rolling way', both used literally and metaphorically in Brazilian Portuguese) that we did not find perfect placements for.

Adding these verbs was not difficult, but showed us some of the problems and issues we have to deal with. First there are typos and misspellings everywhere. Even the (short) list of golden verbs in VerbNet.BR[3] has a typo *captura*, instead of *capturar* (to capture). Then the different ways of writing in Portugal and Brazil sometimes duplicate entries. For instance the verb *adjectivar* (to add, perhaps too many, adjetives to your sentences), is not really different from *adjetivar*, the Brazilian spelling. This verb apparently does not exist in English (or at least in English as considered by PWN). This orthographic difference is well-known, but there are many entries like this. While there is an official agreement between Portuguese speaking countries that has 'settled' these orthographic differences, it seems absurd to ignore how the language is really written at the moment. Thirdly, as expected, many English verbs 'pack in' an adverb or two, when in Portuguese we only have the basic verb. For example the verb *to jog* is to run slowly or walk fast, hence between *correr* and *andar* in Portuguese, for the fun of it. In Portuguese we have no verb between running and walking, we need the adverbs *slowly, quickly* and we need to indicate that the purpose is fun. But of course this process also happens in the opposite direction and this is much harder for us to ascertain. We have a huge English lexicon

---

[2] https://github.com/own-pt/cl-wnbrowser/tree/master/corpora.
[3] To be found at http://www.nilc.icmc.usp.br/verbnetbr/.

(117K synsets in PWN) and no guarantees that the humans trying to fit meanings into it, know the whole lexicon. Fourthly, the different kinds of affixes used both in English and Portuguese make some comparisons difficult. In particular a negating prefix, such as *mis-* does not exist in Portuguese, as such, while the Portuguese prefix *auto-* corresponding to doing something to yourself, *self-*, seems much more used in Portuguese than in English. The English PWN lists only one verb with prefix *self*, *self-destroy*, while the Bosque corpus (not a very large one) lists at least four verbs with the corresponding prefix *autodenominar/self-denominate, auto-excluir/self-exclude, autoparodiar/self-parody, autopunir/self-punish* in Portuguese. Finally, one of the main problems has to do with the frequency and popularity of lexical items. We have no reliable frequency data and particularly with two very different vocabularies, corresponding to Portugal and Brazil, for daily things and actions, it is hard to decide on the level of coverage that is required.

As said, we only had problems to fit in two verbs from the golden subset of VerbNet-BR: *entreabrir, rebolar*. The first one *entreabrir*, meaning 'open partially' shows the phenomenon described above: a kind of conceptualization that seems to be done via an adverb in English, as it is a *partial* opening. The second one *rebolar* we can find an approximated sense in the English verb 'to roll'. However, while in English this seems to indicate a particular gait, a way of walking, in Portuguese there is no need to cover any ground while you *ondulate your hips*. This small exercise made us wonder whether there were too many other verbs missing from our resource and we decided to investigate other resources, described below.

To find where to fit in the PWN network the 'missing' Portuguese verbs from the golden VerbNet.BR we established a modus operandi: we translate the desired Portuguese verbs using machine translation and then we manually verify the translation. A list of words in Portuguese and corresponding words in English is then fed to an algorithm that looks for strict matches both of Portuguese and English words, in synsets and in glosses and then suggests these synsets to the human annotators. Finally at least two human annotators have to agree on the appropriateness of the word sense and its placement into the network to make it part of the official resource. The suggesting and voting processes of OpenWordNet-PT are described in [17].

3.2   Basic Coverage

First we used a list of the thousand most common Portuguese verbs as collected by the 'Corpus do Português'[4]. The list in that website actually has 999 verbs instead of a thousand ones and we have all of them in OpenWordNet-PT.

Then we investigated a Swadesh list of the most important Portuguese words[5]. American linguist Morris Swadesh used vocabulary lists to try to understand not only change of languages over time but also the relationships between

---

[4] http://www.corpusdoportugues.org/.
[5] https://en.wiktionary.org/wiki/Appendix:Portuguese_Swadesh_list.

extant languages. He based his lists on meanings he presumed would be available in as many cultures as possible, so we are using his list here as a basic sanity check. There are several variations of Swadesh lists, and we used the one coming from the archives of the Open Language Archives Community (OLAC) of the University of Pennsylvania. The file in their link to the Project Rosetta[6] was easy to deal with, but seems more about European Portuguese than the list at Wiktionary itself. The whole Swadesh list has 298 items, but many are pronouns and demonstratives that are not part of a traditional wordnet. From this list we found two verbs that we did not have (*fender/'to split', desamolar/'blunt'*), which we added in, but that are not that common in Brazilian Portuguese.

3.3   VerbOcean Translated

A different source of verbs to extend our lexicon was VerbOcean[7]. Work on textual entailment of the traditional kind, using logical forms, could be helped considerably if the algorithms doing the matching of assumptions and conclusions had access to relations of entailment and causation between verbs. One of Princeton's Wordnet's weaknesses is that not many of these relations are recorded in the database. Chklovski and Pantel's work [8] in VerbOcean was meant to address this problem. Given our avowed disposition to do logical reasoning with our representations, as soon as possible, it made sense for us to discuss the collection of verbs in VerbOcean.

Previous work in [13] describes a first attempt to clean up and improve the extant verb lexicon of OWN-PT, using a constructed manual translation of the verbs in VerbOcean. We have checked that all these translated VerbOcean verbs are included in OpenWordnet-PT. Out of the original 2119 verbs in VerbOcean, we already had in OWN-PT 1182 verbs. Now we also have in suggestions 930 verbs. Altogether there were only six verbs still missing: *escantear, gazetear, prototipar, reconfigurar, subempregar,* and *desinstalar.* These show that, even if morphologically related, sometimes words can have very different meanings, the so-called *semantic drifting.* While the verb *gazette* in English means *to publish in a gazette,* in Portuguese the verb *gazetear* means *to play truant.* The verbs *prototipar, 'to prototype', desinstalar, 'to uninstall',* and *reconfigurar, 'to reconfigure'* seem to arise from technology and hence should exist in English, but in PWN one does not have these verbs. Maybe one should. The verb *subempregar* shows a different social reality. In English one says *underpay* for the practice of paying less than customary to workers, but in Portuguese we prefer to say *subempregar,* or 'under-employ'. Finally the verb *escantear* shows the issues with different national sports as being represented in lexicons. There is a whole collection of verbs in PWN having to do with baseball, American football, golf and basketball that have no direct correspondents in Portuguese (e.g. *to tee* in golf). By contrast in Brazilian Portuguese we have many verbs and especially verbal expressions derived from soccer, the national sport, as *escantear.*

---

[6] http://dla.library.upenn.edu/dla/olac/record.html?id=rosettaproject_org_rosettaproject_por_swadesh-1.

[7] http://demo.patrickpantel.com/demos/verbocean/.

3.4 'Bosque' Universal Dependencies

The corpus Bosque [1] is a paradigmatic corpus of Portuguese. It has been used in the CoNLL-X Shared Task in dependency parsing (2006); and very recently it has been converted to Universal Dependencies [21]. The corpus consists of texts in Portuguese (both from Brazil and Portugal) annotated (and analyzed) automatically by the syntactic parser PALAVRAS [4] and reviewed by trained, native speaker linguists. The data comes from news sources. For many reasons, it would be reasonable to expect to have all verbs in this corpus already in OpenWordNet-PT. Nonetheless we found out that a massive number of verbs were not available in OpenWordNet-PT, in any of their senses.

Despite being initially surprised by this finding, we believe that this shows the beginnings of the maturity of OpenWordnet-PT. While subscribing to the view that meaning can be translated from language to language, it seems also clear that different languages will conceptualize different realities, so while an English speaker may not need verbs such as *abrasileirar, aportuguesar, apaulistar, argentinizar, africanizar* (to make or to make more Brazilian, Portuguese, native of São Paulo, Argentinian or African), a Brazilian speaker does need them. These are very easy to explain. Then there are misspellings: despite the fact that the corpus was hand-checked, apparently there was a theoretical decision not to touch the contents of the texts themselves, only hand-correct the processing. Hence we have 'verbs' that are instead typos such as *abanadonar, apretechar, assessoriar, assitir, atinjir*. These are all clearly typos from *abandonar, apetrechar, assessorar, assistir, atingir* ('*to abandon*', '*to equip*', '*to be a consultant*', '*to assist*', '*to reach*'), which are all present in the lexicon.

Before removing typos and deciding on what to do about prefixes we have 1981 verbs in Bosque-UD. We had already in OWN-PT 1043 of these. We have managed to add suggestions to 831 synsets. But there are still some 107 missing, which are mostly cases of prefixes and typos. There are six verbs where the prefix *recém* (meaning *recently*) was added to an existing verb, 13 with the prefix *des-*, and 20 typos. A few true Portuguese verbs, arising from nouns (e.g. *vampirizar*, '*vampirize*') and adjectives (*minorar*, '*to make minor*') which do not seem to be conceptualized as verbs in English were added to the list of candidate Portuguese-only synsets, to be dealt with later on.

However, all in all, easy typos and clear-cut cases of a different social reality are rare. Most of the cases of the verbs missing from OpenWordNet-PT seem to be either differences in prefixes used and cases of adjectives and nouns that are made into verbs in Portuguese, but not in English. The prefixes *des-, di-, re-, in-* are used extensively in both Portuguese and English, but they apply to different verbs. For instance, in Portuguese we have *independer, indeterminar* for *to be independent*, and for *not determining* something. These are not treated as verbs in English, or so it seems to us. In Portuguese we use the suffix *-ar* to make verbs out of nouns and adjectives and many in our list of candidates to truly missing Portuguese synsets correspond to these, e.g. *bacharelar, biografar, conveniar, desertificar*, respectively, 'to obtain a Bachelor's degree', 'to write a biography', 'to get a *convênio* (a contract for health insurance or such like) set up' or 'to make a place a desert'.

Of course, deciding that there is no verb in English that expresses exactly the same idea of a Portuguese word is a much harder task then deciding which words in Portuguese fit a given synset. Given this state-of-affairs and the difficult task of deciding which new Portuguese synsets we need to create, we have decided to collect these "possibly Portuguese-only" candidate synsets into a spreadsheet to see if others would be able to find appropriate PWN synsets for these meanings. A comma separated file[8] is available in GitHub with these proposed new synsets. The number of proposed extensions coming from Bosque is not so big, around a hundred, but these still need another checking and devising of principles to add them in.

3.5   Diário Gaúcho

The structured corpus of the Diário Gaúcho (here called DG) is one of the products of PorPopular project[9] that aims to describe and study patterns of written popular Portuguese. Diário Gaúcho is a popular newspaper from the south of Brazil and we have chosen to work with this corpus hoping to find colloquial verbs that were not in OpenWordnet-PT, yet. The DG corpus has approximately 5 millions of tokens and the news were extracted from newspaper issues from 2008. Since OpenWordnet-PT comes from bilingual dictionaries and Wikipedia links, as well as some translated lists, we were worried that we might lack some popular or colloquial verbs.

But our worries were mostly unfounded. Many of the colloquialisms brought in by the DG corpus had synsets that was were good fits. Actually out of all the 2042 verbs in the corpus, 1044 were in OWN-PT and 937 were already in suggestions. Most of the missing 61 verbs are actually typos and processing errors. However, from this work the question of how to deal with different kinds of orthography resurfaced. Differently from some other languages, Portuguese has an official formally approved lexicon that dictates which words are sanctioned as Portuguese words, which ones are not.

There is also the 'new' Portuguese Language Orthographic Agreement, an agreement from 1990, revised several times since then, signed by nine Portuguese speaking countries, which is trying to achieve an unified way to spell Portuguese. The Brazilian Government has announced that the agreement will be fully implemented in 2016.

Mostly the lists of verbs we have been using so far already follow the proposed new Orthographic Agreement. However in the DG corpus, we have found many words that do not follow the new rules, since the corpus was extracted from 2008's news; for example *argüir, 'to argue'*, officially spelled nowadays as *arguir* and *sub-alugar*, officially spelled as *subalugar, 'sublet'*. Since our resource also works as a dictionary for human users, these cases of old spelling rules made us

---

[8] http://wnpt.brlcloud.com/wn/prototypes/corpora#candidates.
[9] http://www.ufrgs.br/textecc/porlexbras/porpopular/.

worry that populating the base with all the ways of spelling the same word is not the best course of action. Those old ways are now considered wrong and might lead users astray. On the other hand, to be used as a tool for the analysis of texts written before the agreement (the largest part of documents in Portuguese), we should, perhaps, have those old forms in the lexicon. We decided to include in our base mainly the forms that follow the agreement, but also to report some old spellings that appear in the chosen corpora. Thus we include e.g. *argüir* and *sub-alugar* in the same synsets that have *arguir* and *subalugar* but do not insist on making it consistently for all variants in the base.

3.6   Verbs from Other Sources

One of the applications envisaged for our lexical resources is their use on Digital Humanities studies, in particular to help with information extraction from unstructured text. In our case we work with a small, but very interesting corpus of biographies of historical figures in Brazil, from the 1930's onwards, which we abbreviate as DHBB ('Dicionário Histórico Biográfico Brasileiro', in Portuguese, or "Dictionary of Brazilian Historical Biographies"). What makes this corpus particularly nice for information extraction is that the writers of the entries were asked to follow a set of guidelines with respect to the information that these entries about the historical figures should contain. Hence there is a sense in which the corpus is 'semantically contained'.

For the purposes of our work here it means that we expect to find a relatively limited collection of verbs that would appear, about people being born, marrying, campaigning, being elected, writing laws, approving them and such like. The first attempt at analyzing the results from a shallow processing of this corpus have been described in [15]. We quickly realized that for processing this corpus we needed to deal with *named entities* (NE) and their recognition. Our processing of NEs is not up to the levels expected, yet. However, even processing that is not as precise and accurate as one would like it to be, can be enough to provide useful pointers when it comes to verifying the coverage, precision and recall of your lexical resource. Thus we use the list of verbs of the DHBB, as processed by Freeling [7], as an evaluation measure for the coverage of our system.

From the whole DHBB corpus we choose to check verbs with more than ten occurrences. We still have 51 such verbs missing. Amongst these, there are no typos, but there are some specific items from the politics domain (e.g. the verb *subsecretariar,* 'to act as a subsecretary') and some oddities that need investigation (e.g. verbs *pedrar, extremar* and *bondar*). If we add to these, the list of candidate synsets already extracted from Bosque-UD and the other corpora we have some a hundred and fifty verbs that we think deserve new Portuguese synsets. Providing synsets for these these, we do have all verbs in both Bosque-UD and the DHBB corpora covered. It is interesting to note that indeed the DHBB corpus shows some interesting social differences: we have several different verbs in Portuguese for graduating from college *bacharelar, graduar, formar, doutorar, mestrar*, while there is simply *graduate* in PWN. We also have a collection of verbs related to *enrol,* both in schools and in political parties, such as *ingressar, afiliar, matricular, juntar-se.* And we have at least three different ways of expressing the meaning of *separate from your spouse* in Portuguese, with different legal status, *descasar, desquitar, divorciar,* of which only the last one exists as such in PWN.

The work in PropBank-BR [18] says it has identified, automatically, 5688 verbs as candidate members of VerbNet.Br, distributed within 257 classes, inherited from VerbNet. They suggest that a human verification of their results would be highly valuable. We have not been able to verify all of those verbs, which means that there is no resource with semantics for them, at the moment. Hopefully our methods will produce similar results in terms of subcategorization frames, when we get to these and to semantic role labelling.

Finally to see how much work we would still need to do to claim comprehensive coverage, we compared our number of verbs with the ones in the Portal da Língua Portuguesa[10]. Of course we are missing even more verbs, but the Portal is an inclusive resource, covering many variations of Portuguese and not concentrating in high frequency items, like us. José Pedro Ferreira, from the Portal da Língua Portuguesa, was kind enough to extract from their database only verbs mainly used in Brazil and of higher frequency. The construction of the lexical resources in the Portal website are described in [10]. This list has 3918 verbs, of which we have in OWN-PT 1822 verbs, and in suggestions 1290. We still miss 806 verbs, at the moment.

4 What's Next?

Like VerbNet-BR [20], our goal is a domain independent lexicon that provides semantic and syntactic information about Portuguese verbs. Like VerbNet-BR, we would like to have a Levin-style classification of Portuguese verbs, together with a comprehensive listing of the subcategorization frames for these verbs in our lexicon. This is because subcategorization frames provide information about the syntactic realization of verbs as well as diathesis alternations, which can be used as features for machine learning of semantic roles, our eventual goal.

Thus far we are completing our translations from English to Portuguese, using as criteria of evaluation of coverage open source corpora such as Bosque-UD and PropBank-BR, which is also the same corpus Bosque, but under different processing. We would like to bootstrap a comprehensive lexicon of subcategorization frames from both the minimal frames already present in Princeton WordNet and the annotated corpora available.

Princeton WordNet has 13767 verbal synsets. More than half of these synsets have no words in Portuguese. How many of these really constitute synsets that should not exist in a Portuguese wordnet? And how many new synsets do we need to add to have a resource as useful for Portuguese as PWN is for English? Surely we must have coverage to deal with basic news text, such as the ones in Bosque, Diário Gaúcho or the DHBB. But we do not have, as yet, an worked-out measure for accuracy or adequacy of our resource.

---

[10] http://www.portaldalinguaportuguesa.org/.

5   Conclusions

This work describes some of our effort to complete and to extend the verb lexicon of OpenWordnet-PT, our Portuguese wordnet. As OpenWordnet-PT is a semi-automatic construction from the translation of Princeton Wordnet 3.0, we do need some strategies to complete the synsets where the automated process was not refined enough. Thus we have used a few different verb lists complete the English synsets with no Portuguese words. Since our work was also based in (recent) corpora, we have found some verbs that should, perhaps, be in Princeton Wordnet and are not, and many truly 'Brazilian verbs', that should be in a Portuguese wordnet, but have no reason to appear in an English one.

There is still plenty of work to do, both in terms of coverage and accuracy of our verb lexicon. But we reckon we now have a much more substantial verb lexicon. We verified that we do have all the verbs in the golden standard subset of VerbNet.BR, except two. We verified that we have all the verbs in the translation of VerbOcean, with the exception of six verbs, that we need to add to OpenWordNet-PT. We verified the list of the thousand most frequent verbs in the Corpus do Português and only four verbs needed to be added. We verified all the verbs in the Bosque-UD corpus and DG corpus, adding the ones that we need to create to a jointly curated list, of approximately a hundred 'missing' items. Similarly we verified all the verbs that occur in the DHBB ten or more times and all of these have been either added to OpenWordNet-PT or added to the list of new candidate synsets. With all of these in place, we believe that the verb lexicon is comprehensive enough for our immediate goals.

We need new ways of making sure that the empty synsets in English are there because there are no good translations. Most importantly, we need to come up with principled ways of extending OpenWordNet-PT in the directions that we are clear that it needs to be extended. Frequency in corpora that we deem relevant is one of the tools to be used. Another idea we are pursuing is using the data from the common orthographic Vocabulary of the Portuguese Language (VOC) [10] to improve OpenWordnet-PT, given that we have very little in terms of morphology and frequency of use information, which they do have. On a different direction, we would like to find ways of verifying the Portuguese glosses. We would like to tackle also the issue of the nominalizations, that VOC has many more than we do. Continuing the theme of incorporating morphology together with semantics, we would like to finish the inclusion of morpho-semantic links in OpenWordnet-PT.

References

1. Afonso, S., Bick, E., Haber, R., Santos, D.: Floresta sintá(c)tica: a treebank for Portuguese. In: Proceedings of LREC 2002 (2002)
2. van Assem, M., Gangemi, A., Schreiber, G.: RDF/OWL representation of Word-Net. W3C Working Draft, World Wide Web Consortium, June 2006. http://www.w3.org/TR/2006/WD-wordnet-rdf-20060619/
3. Baptista, J.: Viper: a lexicon-grammar of european portuguese verbs. In: 31e Colloque International sur le Lexique et la Grammaire (2012)

4. Bick, E.: The Parsing System Palavras - Automatic Grammatical Analysis of Portuguese in a Constraint Grammar Famework. Ph.D. thesis, Aarhus University (2000)

5. Bond, F., Paik, K.: A survey of wordnets and their licenses. In: Proceedings of the 6th Global WordNet Conference (GWC 2012), Matsue, pp. 64–71 (2012)

6. Cançado, M., Godoy, L., Amaral, L.: The construction of a catalog of Brazilian Portuguese verbs. In: Jancsary, J. (ed.) Empirical Methods in Natural Language Processing: Proceedings of the Conference on Natural Language Processing (2012)

7. Carreras, X., Chao, I., Padró, L., Padró, M.: Freeling: an open-source suite of language analyzers. In: Proceedings of the 4th International Conference on Language Resources and Evaluation (LREC 2004) (2004)

8. Chklovski, T., Pantel, P.: VerbOcean: mining the web for fine-grained semantic verb relations. In: Proceedings of EMNLP (2004)

9. Fellbaum, C. (ed.): WordNet: An Electronic Lexical Database (Language, Speech, and Communication). The MIT Press, Cambridge (1998)

10. Ferreira, J.P., Janssen, M., Correia, M., De Oliveira, G.M.: The common orthographic vocabulary of the Portuguese language: a set of open lexical resources for a pluricentric language. In: Proceedings of LREC (2012)

11. Lewis, M.P., Simons, G.F., Fennig, C.D. (eds.): Ethnologue: Languages of the World, 18th edn. SIL International, Dallas (2015). http://www.ethnologue.com

12. da-Silva, B.C.D., de Moraes, H.R.: A construção de um thesaurus para o português do Brasil. Alfa **47**(2), 101–115 (2003)

13. de Paiva, V., Freitas, C., Real, L., Rademaker, A.: Improving the verb lexicon of OpenWordnet-PT. In: Alemany, L.A., Padró, M., Rademaker, A., Villavicencio, A. (eds.) Proceedings of Workshop on Tools and Resources for Automatically Processing Portuguese and Spanish (ToRPorEsp). Biblioteca Digital Brasileira de Computação, UFMG, Brazil, São Carlos, Brazil, October 2014. http://www.lbd. dcc.ufmg.br/bdbcomp/servlet/Evento?id=755

14. de Paiva, V., Rademaker, A., de Melo, G.: Openwordnet-pt: an open Brazilian Wordnet for reasoning. In: Proceedings of COLING 2012: Demonstration Papers, pp. 353–360. The COLING 2012 Organizing Committee, Mumbai, December 2012. http://www.aclweb.org/anthology/C12-3044, Published also as Techreport. http://hdl.handle.net/10438/10274

15. Paiva, V.D., Oliveira, D., Higuchi, S., Rademaker, A., Melo, G.D.: Exploratory information extraction from a historical dictionary. In: IEEE 10th International Conference on e-Science (e-Science). vol. 2, pp. 11–18. IEEE, October 2014

16. Pease, A.: Ontology: A Practical Guide. Articulate Software Press, Angwin (2011)

17. Real, L., Chalub, F., de Paiva, V., Freitas, C., Rademaker, A.: Seeing is correcting: curating lexical resources using social interfaces. In: Proceedings of 53rd Annual Meeting of the Association for Computational Linguistics and The 7th International Joint Conference on Natural Language Processing of Asian Federation of Natural Language Processing - Fourth Workshop on Linked Data in Linguistics: Resources and Applications (LDL 2015), Beijing, China, July 2015

18. Scarton, C., Aluísio, S.: Towards a cross-linguistic Verbnet-style lexicon for Brazilian Portuguese. In: Workshop on Creating Cross-language Resources for Disconnected Languages and Styles Workshop Programme, p. 11 (2012)

19. Scarton, C., Sun, L., Kipper-Schuler, K., Duran, M.S., Palmer, M., Korhonen, A.: Verb clustering for brazilian portuguese. In: Gelbukh, A. (ed.) CICLing 2014, Part I. LNCS, vol. 8403, pp. 25–39. Springer, Heidelberg (2014)

20. Scarton, C.E.: VerbNet.BR: construção semiautomática de um léxico computacional de verbos para o português do brasil. In: 8th Brazilian Symposium in Information and Human Language Technology, pp. 20–29 (2011)

21. Silveira, N., Dozat, T., de Marneffe, M.C., Bowman, S., Connor, M., Bauer, J., Manning, C.D.: A gold standard dependency corpus for English. In: Proceedings of the Ninth International Conference on Language Resources and Evaluation (LREC 2014) (2014)

22. Simões, A., Guinovart, X.G.: Bootstrapping a Portuguese wordnet from Galician, Spanish and English wordnets. In: Navarro Mesa, J.L., Ortega, A., Teixeira, A., Hernández Pérez, E., Quintana Morales, P., Ravelo García, A., Guerra Moreno, I., Toledano, D.T. (eds.) IberSPEECH 2014. LNCS, vol. 8854, pp. 239–248. Springer, Heidelberg (2014). http://dx.doi.org/10.1007/978-3-319-13623-3_25

CONTO.PT: Groundwork for the Automatic Creation of a Fuzzy Portuguese Wordnet

Hugo Gonçalo Oliveira[✉]

CISUC, Department of Informatics Engineering,
University of Coimbra, Coimbra, Portugal
hroliv@dei.uc.pt

Abstract. There are several lexical resources available for the computational processing of Portuguese, organised differently and created by different people with different approaches and limitations. This paper presents the first experiments towards the exploitation of seven of those resources in the automatic creation of a large wordnet, where numerical scores are assigned to the inclusion of words in synsets and to the connection of synsets by semantic relations. Experiments confirm that a large wordnet can indeed be created and, to some extent, computed scores can be used as a confidence measure, which will enable the users to select only a portion of the resource, depending on the needs of their application on quantity and quality of lexical-semantic knowledge.

Keywords: Wordnet · Semantic relations · Confidence · Redundancy · Fuzzy

1 Introduction

Wordnets are lexical-semantic knowledge bases, modelled after Princeton WordNet (PWN) [1]. They group synonyms in synsets, which represent concepts by their possible lexicalisations. Together with the synset glosses, different types of semantic relation, including hypernym and meronymy, are established between synsets and help to describe their meaning. As the same meaning might be transmitted by different words, the same word might be in more than one synset, one for each of its senses.

Due to its machine-friendly structure, wordnet became the standard model of a lexical knowledge base. We have seen the creation of wordnets for many languages, including Portuguese [2], though none is as consensual as PWN is for English. Given the overwhelming task of populating a wordnet from scratch, the open Portuguese wordnets are created automatically or semi-automatically, and rely heavily on the contents of other lexical resources, including wordnets of other languages. On the one hand, automatic processes enable a faster creation but, at the same time, existing noise leads to less reliable resources.

In order to tackle existing limitations, we aim go further on leveraging the advantages of automatic approaches, and to give the users some control on coverage and reliability, depending on their needs. We believe in the potential of

© Springer International Publishing Switzerland 2016
J. Silva et al. (Eds.): PROPOR 2016, LNAI 9727, pp. 283–295, 2016.
DOI: 10.1007/978-3-319-41552-9_29

redundant information across open Portuguese lexical-semantic resources, which should enable the creation of a new broad-coverage wordnet where confidence degrees are assigned to the decisions taken, including the membership of words in synsets or the connection of two synsets by a semantic relation. This should enable users to select their own confidence cut-points, which will set either large but less reliable or smaller and more reliable wordnets. The result can be seen as a fuzzy wordnet, an idea that is not completely new (see [3]), but has not been much explored. Moreover, the fuzzy representation is less artificial, as we know that word senses are not discrete [4], but complex and overlapping structures, so their representation as crisp objects does not reflect the human language.

This paper presents the first experiments towards the creation of a fuzzy Portuguese wordnet. Next section overviews the current Portuguese wordnet initiatives. Resources exploited in this work are then enumerated, and their contents and redundancy analysed. After that, the proposed approach for discovering fuzzy synsets and fuzzy semantic connections is described, together with some results and their evaluation. It follows the steps of ECO [5] – extraction, clustering and ontolosiging –, an abstract model tailored for the automatic creation of Onto.PT, one of the open Portuguese wordnets, but flexible enough to the creation of other resources of the same kind. This is also why this new wordnet is baptised as CONTO.PT – as in *Confidence-enriched* Onto.PT. The paper ends with the first conclusions of this approach and some lines for further work.

2   Portuguese Wordnets

There are at least six Portuguese lexical-semantic knowledge bases structured according to the wordnet model [2], created by independent teams, following different approaches, and with different licenses and usage restrictions. WordNet.PT Global [6] is the most recent instantiation of the first Portuguese wordnet, in development since 1998. It is essentially handcrafted and created from scratch, for Portuguese, it can be browsed online, but it is not available for download. WordNet.Br is a wordnet project for Brazilian Portuguese where synsets and antonymy relations were first manually produced, based on dictionaries and corpora, and released under the name TeP [7]. Synsets were then manually aligned with PWN and semantic relations between Portuguese synsets with English equivalents were inherited [8]. To our knowledge, this part is not publicly available. MultiWordNet.PT[1] is a Portuguese wordnet with synsets derived from the translation of PWN synsets. It can be browsed online and used under the payment of a license.

Besides the previous, there are four open Portuguese wordnets. Onto.PT [5] is created in a completely automatic fashion – both synset boundaries and the attachment of semantic relations are learned from the exploitation of available lexical semantic resources, without any human supervision. Its development follows ECO, a three-step approach to integrate words and relations from different sources: (i) relation extraction between words; (ii) synset discovery from

---

[1] See http://mwnpt.di.fc.ul.pt/.

the synonymy relations; (iii) mapping of words in remaining relations to discovered synsets. OpenWordNet-PT [9] was originally developed as a syntactic projection of the Universal WordNet [10] for Portuguese. Its development is thus based on the translation of lexical information in PWN, across multiple languages of Wikipedia, open dictionaries, and also some information from corpora. It is aligned to PWN and a manual curation process is currently undergoing. PULO [11] is based on the probabilistic translation of open wordnets of other languages, with special focus to those included in the MCR project [12], where wordnets of the Iberian languages are aligned to PWN. UfesWN [13] is another Portuguese wordnet, based on the automatic translation of PWN.

With more than 168 k lexical items, 248 k word senses, 117 k synsets, and 340 k relation instances, Onto.PT is the largest Portuguese wordnet [2], which additionally covers a broad range of relation types. On the other hand, it is not aligned to PWN nor any other wordnet and it is far from being 100 % reliable. In a manual evaluation [5], 74 % of synsets were labelled as correct, in 18 % there was no agreement between two judges, and the remaining had at least one incorrect word. Moreover, considering that relations between incorrect synsets are also wrong, between 78 %–82 % were labelled as correct. This highlights the need for incorporating confidence information in large automatically-created wordnets, such as Onto.PT, which may allow users to, depending on their needs, define their coverage *vs* reliability trade-off.

3 Redundancy in Portuguese Lexical-Semantic Resources

This section overviews the contents of the lexical-semantic resources exploited in the reported work and analyses their redundancy, which can be useful for the computation of confidence measures, as shown in the following section.

3.1 Open Portuguese Lexical-Semantic Resources Used

Seven Portuguese lexical-semantic resources are exploited. All of them, listed here, are freely available for download:

- Semantic relation instances of the network PAPEL [14], extracted automatically from a commercial Portuguese dictionary;
- Additional semantic relation instances extracted from **two** dictionaries – Dicionário Aberto (DA) [15] and Wiktionary.PT[2] (Wikt.PT) – using the same grammars as PAPEL, and included in the network CARTÃO [16];
- Synonymy and antonymy instances from **two** handcrafted synset-based thesauri: TeP 2.0 [17] and OpenThesaurus.PT[3] (OT.PT);
- Semantic relation instances acquired from **two** open Portuguese wordnets: OpenWordNet-PT (OWN.PT) [9] and PULO [11].

---

[2] http://pt.wiktionary.org.
[3] http://paginas.fe.up.pt/~arocha/AED1/0607/trabalhos/thesaurus.txt.

All the obtained lexical-semantic information was converted to a suitable input format for the second and third steps of ECO – term-based triples *(a related-to b)*, where words *a* and *b* are connected by a predicate *(related-to)* that is the name of a semantic relation. For that purpose, thesauri and word-nets synsets had to be deconstructed. For instance, a part-of relation between the synsets {*porta, portão*} and {*automóvel, carro, viatura*} would result in the triples: (*porta* synonym-of *portão*), (*automóvel* synonym-of *carro*), (*automóvel* synonym-of *viatura*), (*carro* synonym-of *viatura*), (*portão* part-of *automóvel*), (*porta* part-of *carro*), (*porta* part-of *viatura*), (*porto* part-of *automóvel*), (*portão* part-of *carro*), (*portã* part-of *viatura*). Relation types used were those covered by PAPEL, with a minor extension to include wordnet relations not extracted from dictionaries, such as hypernymy between verbs (`hiperonimoAccaoDe`) or entailment (`accaoQueCausaAccao`). Other wordnet relation names were adapted to the equivalent names in PAPEL. For instance, `hypernymOf` became `hiperonimoDe` and `substanceHolonymOf` became `materialDe`.

From all the resources, a lexical-semantic network was established with 355,026 lexical items and 1,139,243 triples (excluding inverse relations in the wordnets) respectively distributed according to Table 1.

3.2   Redundancy

As expected, although most triples in the network occurred in only one resource, about 109 k were in more than one, and 192 in all the seven. Table 2 distributes the triples of covered types according to the number of resources they occur at.

A key intuition behind this work is that the more resources a triple is in, the more likely it is to transmit a consensual and useful relation, which is confirmed by selected examples in Table 3. On the other hand, triples that only occur in one resource are more likely to either be incorrect, resulting from noise on the automatic process, or to involve very specific meanings, though less useful.

4   Computing Confidence from Redundancy

We aim at exploiting the potential of redundancy for computing confidence towards the creation of a fuzzy Portuguese wordnet. For this purpose, triples acquired from the seven resources might be the input of a new implementation of the second and third steps of the ECO [5] that should encompass the assignment of scores that transmit confidence. In the second step, fuzzy synsets are discovered from synonymy triples and, in the third, they are connected by different semantic relations, based on the exploitation of all available triples.

4.1   Discovering Fuzzy Synsets

Though not very explored, the idea of fuzzy synsets is not new. Fuzzy memberships of words to synsets have been obtained from manual judgements [18] or from the structure of synonymy networks [19]. In order to integrate domain

Table 1. Number of lexical items and triples used from each exploited resource.

Lexical items

POS	PAPEL	DA	Wikt.PT	TeP	OT.PT	OWN.PT	PULO
Nouns	56,660	61,334	30,170	17,149	6,110	32,509	5,149
Verbs	21,585	16,429	8,918	8,280	2,856	3,626	1,573
Adjectives	22,561	18,892	9,536	14,568	3,747	4,401	1,316
Adverbs	1,376	3,160	610	1,095	143	1,120	153
Total	102,182	99,815	49,234	41,092	12,856	41,656	8,191

Relations

Type	PAPEL	DA	Wikt.PT	TeP	OT.PT	OWN.PT	PULO
Synonymy	83,432	52,278	35,330	388,698	51,410	35,597	9,189
Antonymy	388	440	1,263	92,234	–	5,774	2,818
Hypernymy	49,210	46,079	22,931	–	–	78,854	26,596
Part	5,491	4,367	1,574	–	–	14,275	1,146
Member	6,585	1,057	1,578	–	–	5,153	259
Material	336	518	192	–	–	958	67
Contains	391	263	120	–	–	–	–
Cause	7,700	7,211	3,278	–	–	295	291
Producer	1,336	913	500	–	–	–	–
Purpose	9,144	5,220	4,227	–	–	–	–
Property	23,354	15,732	7,020	–	–	10,825	3,327
State	394	237	79	–	–	–	505
Quality	1,636	1,221	381	–	–	–	–
Manner	1,268	3,381	439	–	–	–	–
Place	832	487	1,159	–	–	–	–
Total	191,497	139,404	80,071	480,932	51,410	151,731	44,198

knowledge, PWN has been extended with fuzzy memberships of words to synsets, as well as fuzzy semantic relations [3]. Fuzzy sets of highly related words have also been discovered from text, to represent word senses [20].

Despite its similarities with word sense disambiguation [21], this part of the work can be seen as a kind of word sense induction [22] because, instead of assigning words to senses in an inventory, word senses are drawn from scratch, based on the structure of the synonymy network.

Method: We have recently proposed an alternative approach for discovering fuzzy synsets from synonymy networks, in two steps [23]: (i) centroid discovery; (ii) fuzzy memberships computation. It is applied to a weighted synonymy network $N = (W, P)$, where W is a set of words and P a set of weighted synonym pairs, with a weight reflecting the number of times a synonym pair, $P(W_i, W_j)$,

Table 2. Occurrences of the same triples in different resources, per type.

Relation	1	2	3	4	5	6	7
Synonymy	262,325	38,495	14,945	6,035	2,301	792	192
Antonymy	48,444	1,257	345	96	22	7	–
Hypernymy	165,484	23,320	3,188	413	66	–	–
Part	22,620	1,883	146	6	1	–	–
Member	13,200	638	48	3	–	–	–
Material	1,735	159	6	–	–	–	–
Contains	635	65	3	–	–	–	–
Cause	9,286	3,354	927	–	–	–	–
Producer	2,225	217	33	–	–	–	–
Purpose	15,657	1,272	130	–	–	–	–
Property	45,431	6,057	798	76	3	–	–
State	1,031	81	6	1	–	–	–
Quality	1,760	631	72	–	–	–	–
Manner	3,845	551	47	–	–	–	–
Place	1,609	286	99	–	–	–	–
Total	595,287 (85 %)	78,266 (11 %)	20,793 (3 %)	6,630 (0.9 %)	2,393 (0.3 %)	799 (0.1 %)	192 (0.0 %)

Table 3. Examples of redundant triples.

#	Relation triples
7	*gruta* sinonimoNDe *caverna, vulgar* sinonimoAdjDe *ordinário, agarrar* sinonimoVDe *pegar porventura* sinonimoAdvDe *talvez*
6	*público* antonimoAdjDe *privado, fácil* antonimoAdjDe *difícil, parcial* antonimoAdjDe *imparcial*
5	*pessoa* hiperonimoDe *artista, mudança* hiperonimoDe *mutação, degrau* parteDe *escada, convencional* dizSeSobre *convenção, sexual* dizSeSobre *sexo, humanitário* dizSeSobre *humanidade*
4	*feliz* devidoAEstado *felicidade, gendarme* membroDe *gendarmaria, carta* membroDe *baralho, letra* membroDe *alfabeto, decisivo* dizSeDoQue *decidir*

occurs in the exploited sources. In the first step, Chinese Whispers [24] (CW), an efficient graph clustering algorithm, is run in the network. This results in a set of hard words clusters, used as centroids. In the second step, the membership degree of each word W_i to each centroid C_k is computed by Eq. 1, which considers the number of synonym pairs between W_i and each word in C_k.

$$\mu(W_i, C_k) = \frac{\sum_{j=0}^{|C_k|} \#(W_i \text{ synonym-of } [C_k]_j)}{|C_k|} \tag{1}$$

Example: The synset discovery approach is illustrated in Fig. 1, with the help of a weighted graph where two senses of the Portuguese word *canudo* arise: a tube/pipe, or, more informally, a diploma. If CW identifies the hard clusters C_A and C_B, to compute the membership of *canudo* to the fuzzy cluster C'_A, the weights of the connections between this word and words in C_A are summed and divided by the size of C_A. Since #(*canudo* synonym-of *diploma*) = 2,

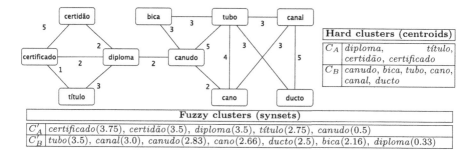

Fig. 1. Weighted lexical network, resulting hard clusters, and fuzzy synsets.

$\mu(canudo, C'_A) = \frac{2}{4} = 0.5$. For the membership of *canudo* to C'_B, the three connections between this word and words in C_B are considered, plus the word *canudo* itself, which belongs to C_B and has the maximum weight (7, if seven sources are exploited). So $\mu(canudo, C'_B) = \frac{3+5+2+7}{6} = \frac{17}{6} = 2.83$

Results: A total of 20,315 fuzzy synsets (13,735 noun, 4,827 adjective, 1,126 verb, 627 adverbs) were discovered from the synonymy network obtained from the seven exploited resources. On average, noun synsets had 9.4 words, adjectives 11.9 and verbs 59.3, because their network has more connections, which can be interpreted as a higher ambiguity and/or more synonyms for the Portuguese verbs. The resulting fuzzy thesaurus was baptised as CLIP 2.1 [23].

Evaluation: To assess the quality of the fuzzy synsets and computed memberships, random pairs of words from the same synset (240 nouns, 150 verbs, 150 adjectives), organised in sets of ten, were uploaded to the Crowdflower platform[4], where Portuguese-speaking volunteer contributors, living in Portuguese-speaking countries, manually labelled each pair either as possible synonyms or not[5]. In the end, 59 % of the noun pairs, 46 % verb and 55 % adjective pairs were labelled as correct. Each pair was labelled by two judges, respectively with an agreement (IAA) of 87 %, 85 % and 75 %. At first, quality does not look very promising. However, it improves for increasing membership degrees. Figure 2 plots the evolution of the proportion of correct pairs for different cut-points – if the membership of one of the words in the pair is below the cut-point, the pair is ignored – and confirms that the computed memberships behave as a confidence measure, because they are positively correlated with the quality. For instance, for a cut-point of 1.0, the proportion of correct noun and adjective pairs is 85 % and for verbs 89 %. Moreover, there is a point after which all the pairs are correct. Also in Fig. 2, the total number of words and their average number of senses is presented for each cut-point.

---

[4] https://crowdflower.com/.
[5] The same contributor was not allowed to label more than two sets of pairs.

Cut	Correct pairs			Size	
	N	V	Adj	#words	$\overline{senses}$
0.00	59%	46%	55%	94,835	2.74±3.93
0.25	64%	52%	63%	73,958	1.32±0.85
0.50	79%	67%	78%	63,116	1.13±0.45
0.75	87%	65%	84%	50,897	1.08±0.31
1.00	89%	85%	89%	45,163	1.05±0.24
1.25	89%	89%	96%	21,401	1.04±0.20
1.50	87%	90%	95%	16,949	1.02±0.15
1.75	85%	91%	100%	7,581	1.01±0.11
2.00	83%	94%	100%	5,389	1.01±0.08
2.25	90%	100%	100%	2,378	1.00±0.06
2.50	100%	100%	100%	1,546	1.00±0.04
IAA	87%	85%	75%		

Fig. 2. Evolution of the correct synonymy pairs while increasing the cut-point. (Color figure online)

4.2 Discovering Fuzzy Synset Connections

After discovering the fuzzy synsets, some of them may be automatically connected by semantic relations. Possible attachment points can be discovered by exploiting the non-synonymy triples, which is done in this step.

Method: Each pair of synsets, S_a and S_b, is analysed to set attachment points with a fuzzy score, computed by Eq. 2. For each relation type R, this equation considers the: (i) number of triples of type R between a word from each synset, a_i and b_j; (ii) number of resources where each of the previous triples occurs, $\#(a_i, R, b_j)$; (iii) membership of each word in the previous triples to their synset, $\mu(a_i, S_a)$ and $\mu(b_j, S_b)$.

$$c(S_a, R, S_b) = \frac{\sum_{i=0,j=0}^{|S_a|,|S_b|} \left(\#(a_i, R, b_j) \times (\mu(a_i, S_a) + \mu(b_j, S_b)) \right)}{|S_a| + |S_b|} : a_i \in S_a, b_j \in S_b \tag{2}$$

Example: Figure 3 illustrates the computation of the proposed measure in two synsets with several hypernymy triples between their words. Hypernymy triples used are represented in a graph, where the only redundant triple has weight 3.

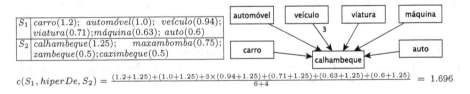

$$c(S_1, hiperDe, S_2) = \frac{(1.2+1.25)+(1.0+1.25)+3\times(0.94+1.25)+(0.71+1.25)+(0.63+1.25)+(0.6+1.25)}{6+4} = 1.696$$

Fig. 3. Computing the confidence of the connection S_1 hiperonimoDe S_2.

condição(0.97);disposição(0.92);situação(0.88) hiperonimoDe crispação(0.8);tensão(0.73);contração(0.6)	**Confidence:** 0.82 **Rendering:** *crispação é um tipo/género de condição*
origem(1.10);princípio(0.81);começo(0.70) antonimoNDe término(1.0)	**Confidence:** 2.28 **Rendering:** *origem é o contrário de término*
pressentir(1.73);prognosticar(1.73);prever(1.61) accaoQueCausa prognóstico(2.0);presságio(1.77);vaticínio(1.74)	**Confidence:** 0.45 **Rendering:** *pressentir pode levar a prognóstico*
educativo(1.75);doutrinal(1.75);educacional(1.25) dizSeDoQue ensinar(2.24);instruir(1.91);doutrinar(1.44)	**Confidence:** 0.23 **Rendering:** *pode ser educativo por ensinar*
desordenar(1.92);destemperar(1.73);desarranjar(1.67) hiperonimoAccaoDe contrabalançar(3.75);compensar(3.6);equilibrar(3.0)	**Confidence:** 0.25 **Rendering:** *desordenar é uma forma de contrabalançar*
pitoresco(2.25);pictórico(1.875);pinturesco(1.25) dizSeSobre novidade(3.33);nova(3.0);notícia(2.75)	**Confidence:** 0.24 **Rendering:** *pitoresco pode qualificar novidade*
incumbir(3.08);encarregar(2.85);confiar(2.54) finalidadeDe mística(2.5);misticismo(2.5);misticidade(0.67)	**Confidence:** 0.32 **Rendering:** *mística pode servir para incumbir*
lado(1.08);ilharga(1.08);flanco(1.0) parteDeAlgoComPropriedade trilateral(1.5);trilátero(1.5)	**Confidence:** 0.52 **Rendering:** *lado pode fazer parte de algo que é trilateral*
enfeite(1.0);adorno(0.98);ornato(0.80) fazSeCom jarro(1.71);jarra(1.29);vaso(0.63)	**Confidence:** 0.42 **Rendering:** *enfeite pode fazer-se com jarro*
loureiro(2.33);louro(2.0);papagaio(0.75) membroDe lauráceas(1.0)	**Confidence:** 1.11 **Rendering:** *loureiro pode ser um membro de lauráceas*

Fig. 4. Examples of discovered synset connections, their computed confidence, and their rendering, used in the crowdsourced evaluation.

Results: The previous measure was computed between all pairs of discovered fuzzy synsets, with a cut-point of 0.1, for relation triples of any type that were in at least two resources. A total of 52,504 synset connections were discovered, with a score higher than 0. As those did not include triples between words without synonyms, and thus not in the discovered fuzzy synsets, in a second step, when a word w involved in a triple was not in any synset, a new synset S_w containing just that w was created, with $\mu(w, S_w) = 1.0$. In the end, 406,751 additional synset connections were made, with at least one synset with a single word. Moreover, 13,542 new single-word synsets were added to the 20,315 multiword synsets discovered earlier.

Evaluation: To assess the quality of the discovered synset connections and the suitability of their computed confidence, we relied once again on Crowd-flower, where a random selection of 930 synset connections were uploaded. These included only connections where at least one synset had more than one word. To make labelling faster for the contributors, the following was done before upload-ing: (i) only the first word of each synset was used, as we noticed that they are often the most representative for the underlying concept; (ii) each triple was ren-dered to a natural language sentence, depending on the relation type. Contribu-tors could label each rendering as either: (i) correct; (ii) incorrect; or (iii) unsure.

Figure 4 illustrates, at the same time, the output of the fuzzy attachments and of the evaluation samples. It includes the first three words and respective memberships of several synset connections in the sample, their computed confidence, and the textual rendering shown to the contributors.

Figure 5 shows the results of the crowdsourced evaluation and the evolution of the correct connections for increasing cut-points. It also presents the proportion of answers where the contributors were unsure and insights on the size of the fuzzy wordnet for the same cut-points, namely the number of synsets and connections between them. Once again, the initial quality is far from impressive: 49.5 % renderings were labelled as correct and 44.3 % as incorrect. Agreement was also lower, 70 %. It should still be noted that connections between two single-word synsets, with a higher chance of being correct, were not used. Not to mention that, in some cases, the used renderings might be too limitative and they show just one word per synset. Moreover, though less consistently than for the synonyms evaluation, quality still increased for higher cut-points, which indicates that the computed score behaves as a confidence measure. At the same time, the number of connections is drastically reduced each time the cut-point increases, especially from 0 to 0.25.

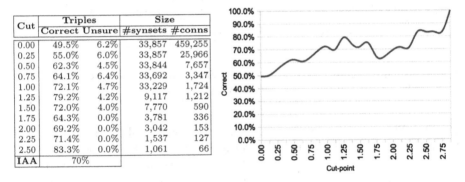

Cut	Triples		Size	
	Correct	Unsure	#synsets	#conns
0.00	49.5%	6.2%	33,857	459,255
0.25	55.0%	6.0%	33,857	25,966
0.50	62.3%	4.5%	33,844	7,657
0.75	64.1%	6.4%	33,692	3,347
1.00	72.1%	4.7%	33,229	1,724
1.25	79.2%	4.2%	9,117	1,212
1.50	72.0%	4.0%	7,770	590
1.75	64.3%	0.0%	3,781	336
2.00	69.2%	0.0%	3,042	153
2.25	71.4%	0.0%	1,537	127
2.50	83.3%	0.0%	1,061	66
IAA	70%			

Fig. 5. Evolution of the correct triples while increasing the cut-point.

After a shallow error analysis, we noticed that there were several renderings that should have been labelled as correct, but were not. Those included connections with confidence higher than 1.8, such as (*origem* antonimoDe *término*), (*dicéfalo* dizSeDoQue *ter_duas_cabeças*), or (*planta* hiperonimoDe *bisnaga*). Although we asked the contributors to confirm their answers in electronic dictionaries and check for less known senses, or to mark unknown answers as unsure, most of them were probably less experienced or have answered the questions too fast, thus not following the instructions strictly.

5   Conclusion and Further Work

The first experiments towards the automatic creation of a fuzzy Portuguese wordnet, through the exploitation of redundancy in available lexical-semantic resources, were presented. The projected wordnet combines the advantages of an automatic creation approach, including lower creation effort for a broad-coverage resource, with the option of controlling the quantity-quality trade-off, with a confidence cut-point. Synsets, discovered from synonymy networks, have words with variable memberships, and they can be connected, by semantic relations of different types, to other synsets, also with variable degrees.

A preliminary version of the resulting wordnet is available, in a non-standard format, from http://ontopt.dei.uc.pt, under the option CONTO.PT. We are still studying alternatives for representing CONTO.PT with standard formats, such as RDF/OWL.

Besides dealing with the previous issue, there is additional work to do. Alternative ways of computing confidence from redundancy should be explored, especially on the synset attachment, where the current measure seems to be biased towards smaller synsets. In order to measure progress, we can use the annotated data collected from crowdsourcing or, given the limitations of the previous, a more controlled evaluation might be performed by more experienced and trustful judges. It should also be analysed whether the synset memberships can be adjusted when connecting synsets. For instance, if several words of the same synset share a relation with another word, their memberships may increase.

It should be added that, although applied to Portuguese, this approach can be used to create fuzzy wordnets in other languages, as long as there are available computational lexical resources, whether they are dictionaries, thesauri, wordnets or even relations extracted from corpora.

References

1. Fellbaum, C. (ed.): WordNet: An Electronic Lexical Database (Language, Speech, and Communication). The MIT Press, Cambridge (1998)
2. Gonçalo Oliveira, H., de Paiva, V., Freitas, C., Rademaker, A., Real, L., Simões, A.: As wordnets do Português. In: Simões, A., Barreiro, A., Santos, D., Sousa-Silva, R., Tagnin, S.E.O. (eds.) Linguística, Informática e Tradução: Mundos que se Cruzam, pp. 397–424. OSLa: Oslo Studies in Language, University of Oslo (2015)
3. Araúz, P.L., Gómez-Romero, J., Bobillo, F.: A fuzzy ontology extension of WordNet and EuroWordnet for specialized knowledge. In: Proceedings of Terminology and Knowledge Engineering Conference, TKE 2012, Madrid, Spain, June 2012
4. Kilgarriff, A.: Word senses are not bona fide objects: implications for cognitive science, formal semantics, NLP. In: Proceedings of 5th International Conference on the Cognitive Science of Natural Language Processing, pp. 193–200 (1996)
5. Gonçalo Oliveira, H., Gomes, P.: ECO and Onto.PT: a flexible approach for creating a Portuguese wordnet automatically. Lang. Resour. Eval. **48**(2), 373–393 (2014)
6. Marrafa, P., Amaro, R., Mendes, S.: WordNet.PT global - extending WordNet.PT to Portuguese varieties. In: Proceedings of 1st Workshop on Algorithms and Resources for Modelling of Dialects and Language Varieties, Edinburgh, Scotland, pp. 70–74. ACL Press (2011)
7. Dias-da-Silva, B.C., de Oliveira, M.F., de Moraes, H.R.: Groundwork for the development of the Brazilian Portuguese wordnet. In: Ranchhod, E., Mamede, N.J. (eds.) PorTAL 2002. LNCS (LNAI), vol. 2389, pp. 189–196. Springer, Heidelberg (2002)

8. Dias-da-Silva, B.C.: Wordnet.Br: an exercise of human language technology research. In: Proceedings of 3rd International WordNet Conference (GWC), GWC 2006, South Jeju Island, Korea, pp. 301–303, January 2006

9. de Paiva, V., Rademaker, A., de Melo, G.: OpenWordNet-PT: an open Brazilian wordnet for reasoning. In: Proceedings of 24th International Conference on Computational Linguistics, COLING (Demo Paper) (2012)

10. de Melo, G., Weikum, G.: Towards a universal wordnet by learning from combined evidence. In: Proceedings of the 18th ACM Conference on Information and Knowledge Management (CIKM 2009), pp. 513–522. ACM, New York (2009)

11. Simões, A., Guinovart, X.G.: Bootstrapping a Portuguese wordnet from Galician, Spanish and English wordnets. In: Navarro Mesa, J.L., Ortega, A., Teixeira, A., Hernández Pérez, E., Quintana Morales, P., Ravelo García, A., Guerra Moreno, I., Toledano, D.T. (eds.) IberSPEECH 2014. LNCS, vol. 8854, pp. 239–248. Springer, Heidelberg (2014)

12. Gonzalez-Agirre, A., Laparra, E., Rigau, G.: Multilingual central repository version 3.0. In: Proceedings of the 8th International Conference on Language Resources and Evaluation (LREC 2012), pp. 2525–2529. ELRA (2012)

13. Gomes, M.M., Beltrame, W., Cury, D.: Automatic construction of Brazilian Portuguese WordNet. In: Proceedings of X National Meeting on Artificial and Computational Intelligence, ENIAC 2013 (2013)

14. Gonçalo Oliveira, H., Santos, D., Gomes, P., Seco, N.: PAPEL: a dictionary-based lexical ontology for Portuguese. In: Teixeira, A., de Lima, V.L.S., de Oliveira, L.C., Quaresma, P. (eds.) PROPOR 2008. LNCS (LNAI), vol. 5190, pp. 31–40. Springer, Heidelberg (2008)

15. Simões, A., Sanromán, Á.I., Almeida, J.J.: Dicionário-Aberto: a source of resources for the Portuguese language processing. In: Caseli, H., Villavicencio, A., Teixeira, A., Perdigão, F. (eds.) PROPOR 2012. LNCS, vol. 7243, pp. 121–127. Springer, Heidelberg (2012)

16. Gonçalo Oliveira, H., Antón Pérez, L., Costa, H., Gomes, P.: Uma rede léxico-semântica de grandes dimensões para o português, extraída a partir de dicionários electrónicos. Linguamática 3(2) 23–38, 2011

17. Maziero, E.G., Pardo, T.A.S., Felippo, A.D., Dias-da-Silva, B.C.: A Base de Dados Lexical e a Interface Web do TeP 2.0 - Thesaurus Eletrônico para o Português do Brasil. In: VI Workshop em Tecnologia da Informação e da Linguagem Humana (TIL), pp. 390–392 (2008)

18. Borin, L., Forsberg, M.: From the people's synonym dictionary to fuzzy synsets - first steps. In: Proceedings of LREC 2010 Workshop on Semantic Relations. Theory and Applications, La Valleta, Malta, pp. 18–25 (2010)

19. Gonçalo Oliveira, H., Gomes, P.: Automatic discovery of fuzzy synsets from dictionary definitions. In: Proceedings of 22nd International Joint Conference on Artificial Intelligence, IJCAI 2011, Barcelona, Spain, pp. 1801–1806. IJCAI/AAAI, July 2011

20. Velldal, E.: A fuzzy clustering approach to word sense discrimination. In: Proceedings of 7th International Conference on Terminology and Knowledge Engineering, Copenhagen, Denmark (2005)

21. Navigli, R.: Word sense disambiguation: a survey. ACM Comput. Surv. 41(2), 1–69 (2009)

22. Nasiruddin, M.: A state of the art of word sense induction: a way towards word sense disambiguation for under resourced languages. In: Proceedings of Traitement Automatique des Langues Naturelles and Rencontres des Étudiants Chercheurs en Informatique pour le Traitement Automatique des Langues, TALN/RECITAL 2013 (2013)

23. Gonçalo Oliveira, H., Santos, F.: Discovering fuzzy synsets from the redundancy in different lexical-semantic resources. In: Proceedings of 10th Language Resources and Evaluation Conference, LREC 2016, Portorož, Slovenia. ELRA, May 2016

24. Biemann, C.: Chinese whispers: an efficient graph clustering algorithm and its application to natural language processing problems. In: Proceedings of 1st Workshop on Graph Based Methods for Natural Language Processing, TextGraphs-1, New York City, pp. 73–80. ACL Press (2006)

Lexical Semantics Annotation for Enriched Portuguese Corpora

Steven Neale, Rita Valadas Pereira, João Silva[(⊠)], and António Branco

NLX - Natural Language and Speech Group, Department of Informatics,
Faculty of Sciences, University of Lisbon, Lisbon, Portugal
{steven.neale,ana.pereira,jsilva,antonio.branco}@di.fc.ul.pt

Abstract. The semantic annotation of corpora has an important role to play in ensuring that sentences occurring in natural language texts are correctly understood based on their intended context. Two examples of lexical semantic units that contribute to this knowledge are word senses – which allow words with multiple meanings to be understood based on the context in which they are used – and named entities – which can be disambiguated and linked back to the specific encyclopedic resources that describe them.

In this paper, we describe the construction of lexical semantically-annotated corpora for Portuguese, annotated with both word senses linked to senses in a Portuguese wordnet and named entities linked to Portuguese Wikipedia entries using DBpedia. The result is a gold-standard lexical semantically-annotated resource that is useful in supporting the training and evaluation of tools for the disambiguation of these lexical units in Portuguese.

Keywords: Annotated corpora · Lexical semantics · Word senses · Named entities · Portuguese

1 Introduction

The representation of complex semantic linguistic features and information in the annotation of corpora has resulted in the availability of increasingly sophisticated resources for natural language processing (NLP) tasks. Word sense disambiguation (WSD) – annotating words that have more than one meaning in the lexicon with a suitable label that refers to correct sense of that word in the context in which it was used – and named entity disambiguation (NED) – determining the identity of the entities mentioned in a text and then linking them to the corresponding information in existing knowledge bases – are two such semantic tasks whose representation in corpora can be vital for the training and evaluation of NLP tools.

An example of a simple case for WSD would be the English word 'bank' – this might refer either to the financial institution or to the slope of land by the side of a river, and the chosen sense would usually be represented by an entry from a

© Springer International Publishing Switzerland 2016
J. Silva et al. (Eds.): PROPOR 2016, LNAI 9727, pp. 296–305, 2016.
DOI: 10.1007/978-3-319-41552-9_30

stand-alone knowledge base or ontology such as WordNet [6], where nouns, verbs, adjectives and adverbs are stored as sets of synonyms or 'synsets' and linked by their semantic relations. Similarly, a simple case for NED would be a phrase such as 'The President lives in the White House', in which 'President' would be recognized as an entity and annotated with a tag denoting 'person', and 'White House' with a tag denoting 'location' and would usually be represented by a link to a descriptive entry in a database such as DBpedia [8], a large-scale multilingual knowledge base extracted from 111 language editions of Wikipedia.[1] However, most of the available annotated corpora for Portuguese contain other types of semantic information, such as the semantic role of phrases within sentences [2] or the semantic type of named entities [1] – specific information about word senses and named entities senses is not yet represented.

In this paper, we describe the creation of new Portuguese corpora annotated with lexical semantic units, CINTIL-WordSenses and CINTIL-NamedEntities. The new corpora are built upon the CINTIL International Corpus of Portuguese [1] and are annotated with synset identifiers selected from the Portuguese Multi-WordNet [9] (word senses) and with links to appropriate Portuguese Wikipedia entries extracted from DBpedia (named entities). Our contribution is a pair of gold-standard annotated datasets for Portuguese that can support the development of WSD, named entity recognition and classification (NERC) and NED tools, either as dedicated training materials or as a baseline against which Portuguese tools can be evaluated.

We first describe some related work (Sect. 2) before outlining the construction of our corpora (Sect. 3) – focusing on the CINTIL corpus we build upon, its enrichment with word senses, linking disambiguated named entities to it, and the resulting corpora statistics. Next, we describe some of the issues encountered and the future work necessary to develop the corpora (Sect. 4), before offering our concluding remarks (Sect. 5).

2   Related Work

While there are various semantically-annotated corpora in existence in other languages, there are relatively few examples in Portuguese. In the case of NED, recent work by Santos et al. [13] describes their efforts to resolve the linking of named entities in Spanish and Portuguese texts to Wikipedia pages, and outlines that their approach is based on extracted dumps of the Spanish and Portuguese versions of Wikipedia and XLEL-21, a dataset developed to support the training and evaluation of cross-language named entity linking systems in twenty-one languages other than English. However, our understanding is that these datasets are based not on spontaneously occurring natural language in context, but rather on lists of singular entities mapped to the links of corresponding Wikipedia pages.

In their work on WSD for Portuguese, Nóbrega and Pardo [11] describe evaluating their work against a manually-annotated subsection of the CSTNews corpus [12], the original version of which contains 140 news texts grouped by

---

[1] Wikipedia, the free encyclopedia: http://en.wikipedia.org.

topic – 2,088 sentences amounting to 47,240 words. Due to the difficulty and time constraints of their annotation task, they annotated only a small portion of the original corpus – the most frequent 10 % of nouns in each of the 50 clusters that they divided the texts into – resulting in 4,366 annotated words in total. An additional caveat of their work is that their approach relies on translating ambiguous terms to and from English and using the English WordNet for the disambiguation and annotation tasks.

Both of these examples highlight the importance – for both the training and the evaluation of tools – of having a large, dedicated corpora of Portuguese, accurately annotated with word senses and with named entities. For WSD, annotating ambiguous terms with senses from a Portuguese-specific lexicon is important, while for NED annotating named entities as they occur in natural language – as opposed to relying solely on datasets with stand-alone entities mapped to Wikipedia or DBpedia – will be beneficial for the training of NED and also NERC tools. Finally, the example provided by previous work on WSD for Portuguese highlights how useful it would be to have corpora of a much larger size, to better account for the variety found in natural language.

3   Constructing the Corpora

This section describes the CINTIL International Corpus of Portuguese, the lexical semantically-annotated corpora that have been built upon it - CINTIL-WordSenses and CINTIL-NamedEntities - and the tools and processes used to annotate these new resources.

3.1   The CINTIL International Corpus of Portuguese

The CINTIL International Corpus of Portuguese [1] is a linguistic resource of 1 million tokens containing data from both written sources and transcriptions of spoken Portuguese – the written part, sourced mainly from newspaper articles and short novels, comprises approximately 700,000 tokens. After first being manually annotated with (a) accurate sentence, paragraph and token boundaries; part-of-speech and morphosyntactic categories; inflectional features; and named entities boundaries and semantic types, the corpus was then used to train (b) the sentence chunker; tokenizer; POS tagger; lemmatizer; conjugator; nominal inflector; and named entity recognizer that collectively form the LX-Suite [5]. For the existing version of CINTIL on top of which our lexical semantically-annotated corpora are built, this progressive process of manual annotation, verification and subsequent re-training of the auxiliary annotation tools has ensured that all of the tokens in the existing corpus have been hand annotated and verified, and that the whole of the corpus has been used to train the LX-Suite.

The CINTIL corpus continues to be extended and developed following its original construction. CINTIL-DeepGramBank [4] includes deep grammatical representations, with the output of LX-Gram [3] – a dedicated deep linguistic grammar for Portuguese – having been manually verified by Portuguese linguististic experts to extend CINTIL with representations of deep linguistic treebanks. This was followed by CINTIL-PropBank [2], whereby syntactic constituency trees from CINTIL-DeepGramBank have been leveraged and enriched with semantic role tags to construct a complete PropBank with both syntactic and semantic levels of annotation.

3.2 Enriching CINTIL with Word Senses

CINTIL-WordSenses is the result of the manual assignment of appropriate sense or meaning labels to words that are lexically ambiguous, taking into account the context in which they appear in a given sentence. For example, given a phrase such as 'John deposited Mary's money in the bank', words such as 'deposit' and 'money' give us enough context to determine that the word 'bank' refers to the financial institution, and not to the slope of land at the side of a river.

The Word Sense Annotation Tool. CINTIL-WordSenses was annotated using our own graphical user interface tool for assigning synset identifers from WordNet-style lexicons to pre-tagged input texts, LX-SenseAnnotator [10]. The tool was developed to provide a more user-friendly way to annotate texts with word sense information, as part of our research into WSD for Portuguese. The initial version of the tool was developed specifically to provide us with a flexible way to complete the word sense annotation task, after deciding that a gold-standard corpus for use in our WSD tasks was needed.

LX-SenseAnnotator requires that input texts have already been processed using the LX-Suite [5], resulting in tokens already being lemmatized, POS-tagged and morphologically analyzed – the POS-tagging in particular makes it very straightforward to separate the input text according to tokens that can (i.e. open-class words) and cannot be annotated. Loaded text is displayed in a text panel, with potential candidates for annotation marked in red and those words that have already been annotated marked in green. To the right of the interface, a second text panel displays the available senses of a given word to the annotator.

Annotators are able to choose from senses (synsets) extracted directly from the Portuguese MultiWordNet. On highlighting an ambiguous word, the main lemma, POS and the eight-digit synset identifier for each possible sense that could be assigned are displayed by LX-SenseAnnotator, as well as the available synonyms of the synset and a selection of other semantically-related words (hypernyms, hyponyms, holonyms etc.) that offer additional context. As will be described in the next sections, annotators were only able to select from the words and synsets present in the Portuguese MultiWordNet, and as expected not all of the open-class words in the corpus were annotated.

The Word Sense Annotation Task. Sections of the CINTIL corpus – divided into short segments of around 50 sentences each – were given to a team of linguistic annotators whose task was to select the correct sense for words in a given sentence, taking into account the discursive context in which the word is used. The available senses (synsets) from which annotators can choose come directly from the Portuguese MultiWordNet, which currently stands at around 19,700 verified synsets. For example, given a simple phrase such as:

"Produção nacional e qualidade são os objectivos."

A rough translation of which would be:

"National production and quality are the goals."

We might wish to add the following synset identifiers to the open-class words, linking them with their correct sense (synset) in the wordnet:

"Produção (00600686) nacional e qualidade (00765551) são os objectivos (00884793)."

Within the Portuguese MultiWordNet – as with the English WordNet [6] – the open-class words that account for most of the occurrences of semantic ambiguity in natural language are represented by groups of terms or 'synsets', each of which represents a specific meaning or concept for which multiple words may be appropriate. Each synset is linked to others by semantic relations – such as synonymy, hyponymy, hypernymy etc. – and is labeled using an eight-digit number that acts as a unique identifier and can be used to represent the meaning of a word that a human annotator might assign it to (see the above example).

3.3   Linking Disambiguated Named Entities in CINTIL

CINTIL-NamedEntities has been created by manually linking pre-recognized named entities to appropriate Portuguese Wikipedia URIs via their entries in DBpedia. For example, the word 'Portugal' on being recognized as a named entity of semantic type 'LOC' (location) would be annotated with a link to the DBpedia entry for the country of Portugal in its geographical sense. Prior to the disambiguation and annotation task, the corpus was automatically tagged using LX-NER, a hybrid rule-based and statistical NERC tool for Portuguese [7]. For the task this paper describes we have focused particularly on the annotation of entities in the names category, which includes types for locations (LOC), organizations (ORG), persons (PER), events (EVT), works (WRK) and an additional miscellaneous (MSC) category.

Brat. The disambiguation and annotation task was completed using version 1.3 of the brat annotation tool [14],[2] a web-based annotation system with several features that make it a good choice for our task. b Brat runs directly in the browser – meaning that there is no installation of specific software required for the human annotator – and has an intuitive user interface that allows annotators to define the span of a named entity simply by dragging a selection over text and then selecting the tag to be assigned from a pre-defined drop-down list. Recent

---

[2] Available from: http://brat.nlplab.org.

versions of brat also have built-in support for normalization [15], which allow annotations to be associated with external resources. In the context of our task, this allows annotators to associate each disambiguated named entity with its corresponding Portuguese Wikipedia page entry in DBpedia.

Disambiguating and Annotating Named Entities. The task of disambiguating and annotating named entities was performed on a version of the corpus that had already been pre-processed using our in-house NERC tool, LX-NER. As well as helping the linguistic annotators to identify the ambiguous named entities within the text, having the entities recognized and classified prior to the start of the task allowed for entities to be fed to a pre-annotation normalization script that could associate them with the Portuguese Wikipedia pages in DBpedia's database [8]. Therefore, the linguistic annotators' task was not concerned with the recognition or coarse classification of the entities themselves, but instead on first verifying that the normalization was correct, and then on choosing the appropriate DBpedia entry for the named entity in question.

To complete the disambiguation task, annotators had to take into account both the tag assigned to the entity by LX-NER and the discursive context in which the entity occurs. Often, the same word could be associated with different Wikipedia entries, depending on the context in which it is used – for example, the word 'Portugal' with the LOC tag should link to the Wikipedia page for the country of Portugal in the geographical sense, while the same word with the ORG tag would be better linked to the Wikipedia page for the Portuguese government.

3.4 Semantically-Annotated Corpora Statistics

The final result of the annotation of the corpora is a language resource for Portuguese of 23,825 word sense-annotated and 30,493 named-entity annotated sentences[3] (see Table 1). For word senses, from a total of 508,717 tokens there are 193,443 open-class (potentially ambiguous) words, of which 45,502 (23.52 %) were manually disambiguated and annotated with synset identifiers from the Portuguese MultiWordNet. For named entities, from a total of 684,467 tokens there were 26,371 entities recognized by LX-NER during pre-processing, of which 16,120 (61.13 %) were manually disambiguated and annotated with links to their appropriate DBpedia entries. Both corpora – CINTIL-WordSenses and CINTIL-NamedEntities – are available via META-SHARE.[4]

To create these first versions of both CINTIL-WordSenses and CINTIL-NamedEntities, the sections into which the corpus was divided for each task were each assigned to a single annotator. In the future, we plan to annotate each section of the corpus again with a different annotator (for inter-annotator agreement)

---

[3] In this first version of the word sense annotation task, fewer sentences were distributed to annotators than in the named entity disambiguation task. These gaps will be addressed in future versions of the word sense annotation task.

[4] Accessible from: http://www.meta-share.eu/.

Table 1. Composition of the lexical semantically-annotated corpora for Portuguese, CINTIL-WordSenses and CINTIL-NamedEntities.

	WordSenses	NamedEntities
Sentences	23,825	30,492
Tokens	508,717	684,467
Senses:		
Potentially Ambiguous	193,443	—
Manually Annotated	45,502 (23.52 %)	—
Entities:		
Recognized (LX-NER)	—	26,371
Manually Annotated	—	16,120 (61.13 %)

and to introduce a process of adjudication, such that the reliability of both resources can be properly quantified.

4   Future Work and Development of the Corpora

Following the creation of CINTIL-WordSenses and CINTIL-NamedEntities, there are a number of possible improvements – affecting both the word sense and named entity annotation tasks – that can be considered going forward in order to continue the development of the corpora.

4.1   Gaps in the Resources Essential to Disambiguation

For both annotation tasks, some gaps are apparent in the lexical resources used to disambiguate word senses and to normalize named entities. In the case of CINTIL-WordSenses, we encountered some examples where the task would be improved by extending the Portuguese MultiWordNet used by the human annotators. The usual case is of words not having an entry present in the wordnet, but there were other cases in which while synsets may currently exist that contain the word in question, these synsets represent alternative meanings to the one most appropriate for the given context. For example, there is a well-formed synset for the word 'corredor' ('athlete' or 'runner' in English) in the wordnet, but currently no synset representing the alternative meaning of the word 'corredor' ('hallway' in English). The homonymous Portuguese words 'corredor' ('hallway') and 'corredor' ('runner') – with one of these senses being well-represented and another not available at present – highlight the positive impact that growing the Portuguese wordnet will have on future versions of CINTIL-WordSenses.

Similarly, in the case of CINTIL-NamedEntities, it was not possible to provide normalized links for all of the named entities discovered by LX-NER tool, as many of them had no suitable entries in DBpedia – in these cases, annotators marked the entities in question as 'Not found' during the annotation.

There were also cases in which the entity in question is more specific than the available DBpedia entry (or vice-versa), cases that annotators currently account for by adopting a 'generalized' annotation approach. For example, given a more abstract or ambiguous entity (to use an English example, 'President of the United States'), the annotator will link the entity to the most specific appropriate entry in DBpedia (i.e. 'Barack Obama') if it can be identified by context, or otherwise to the DBpedia entry of the more general concept (i.e. 'President of the United States'). This generalized annotation approach is also used when a DBpedia entry for a very specific named entity cannot be found. For example, many of the journalistic texts in the original corpus come from one of the numerous supplementary sections of the Portuguese newspaper 'Diário de Notícias', such as 'DN - Internacional'. These specific sections of the newspaper often appear as named entities in the corpus, but because DBpedia does not contain entries for the individual sections the entities must instead be linked to the less-specific page of the newspaper, 'DN - Diário de Notícias'.

4.2   Problems Inherited from Prior Tagging and Annotation

The prior annotation of the original corpus has also introduced some considerations for the development of both the word sense and named entity annotation tasks, usually as a result of discrepancies in the existing POS tagging or incorrectly recognized named entities. For example, previous annotation schemes allowed for nominal multi-word expressions (MWEs) to be compositionally annotated – for example, 'sala de estar' ('living room' in English), adverbial MWEs such as 'em princípio' ('in principle' in English), or prepositional MWEs such as 'por volta de' ('around' in English). The initial version of the word sense annotation tool used to annotate CINTIL-WordSenses reads input texts as individual tokens, and so currently only allows for nominal tokens (i.e. 'sala', 'princípio', 'volta') to be annotated with synset identifiers. Given that the disambiguation of these expressions relies on the role that each word plays in the compositional sense formed with the other words in the expression, future versions of the task – and the resulting corpus – could be greatly improved by updating the word sense annotation tool to allow for compositionally annotated expressions to be understood, and to link to compositional terms representing these expressions (as opposed to nominal tokens only) in the Portuguese MultiWordNet.

Considering also CINTIL-NamedEntities, a number of entities have been incorrectly tagged prior to the task by LX-NER – consider the sentence:

"Estas declarações não escondem as divergências entre Paris (LOC) e Washington (LOC)"

Which could be roughly translated as:

"These statements make clear the differences of opinion between Paris (LOC) and Washington (LOC)"

In this example, 'Paris' and 'Washington' would both have been better tagged as ORG than LOC – however, we have chosen not to disambiguate such occurrences at this stage. Instead, we intend to improve CINTIL-NamedEntities in future versions by first correcting the erroneous output from the LX-NER tool, and then by retraining our tools on the corrected output.

5 Conclusions

We have described the creation of two new lexical semantically-annotated corpora for Portuguese – CINTIL-WordSenses and CINTIL-NamedEntities – manually disambiguated and annotated with word senses and with named entities linked to appropriate Portuguese Wikipedia entries using DBpedia, respectively. Our work contributes gold-standard lexical semantically-annotated resources that can be used in the development of WSD, NERC and NED tools for Portuguese, either as dedicated data for the training of such tools or as a baseline against which their output can be evaluated.

We are now continuing with the development of the corpora, for which our immediate attention is on implementing inter-annotator agreement and adjudication steps in the next version to properly quantify the accuracy and reliability of the resource. Following this, we will then focus on addressing some of the issues encountered in Sect. 4 – manually correcting some of the existing errors inherited from pre-processing (concerning MWEs and recognized entities in particular) will allow us to retrain our existing tools on the corrected output, and to continue developing the corpora in the knowledge that it becomes a more reliable resource with each version.

Acknowledgements. The results reported in this paper were partially supported by the Portuguese Government's P2020 program under the grant 08/SI/2015/3279: ASSET-Intelligent Assistance for Everyone Everywhere, by FCT-Fundao para a Cinciao e Tecnologia under the grant PTDC/EEI-SII/1940/2012: DP4LT-Deep Language Processing for Language Technology, and by the ECs FP7 program under the grant number 610516: QTLeap-Quality Translation by Deep Language Engineering Approaches.

References

1. Barreto, F., Branco, A., Ferreira, E., Mendes, A., Nascimento, M.F.B., Nunes, F., Silva, J.: Open resources and tools for the shallow processing of Portuguese: the TagShare Project. In: Proceedings of the 5th International Conference on Language Resources and Evaluation, LREC 2006, pp. 1438–1443 (2006)
2. Branco, A., Carvalheiro, C., Pereira, S., Silveira, S., Silva, J., Castro, S., Graça, J.: A PropBank for Portuguese: the CINTIL-PropBank. In: Proceedings of the Eight International Conference on Language Resources and Evaluation (LREC 2012). European Language Resources Association (ELRA), Istanbul (2012)
3. Costa, F., Branco, A.: LXGram: a deep linguistic processing grammar for Portuguese. In: Pardo, T.A.S., Branco, A., Klautau, A., Vieira, R., de Lima, V.L.S. (eds.) PROPOR 2010. LNCS, vol. 6001, pp. 86–89. Springer, Heidelberg (2010)

4. Branco, A., Costa, F., Silva, J., Silveira, S., Castro, S., Avelãs, M., Pinto, C., Graça, J.: Developing a deep linguistic databank supporting a collection of tree-banks: the CINTIL deepgrambank. In: Proceedings of the Seventh International Conference on Language Resources and Evaluation (LREC 2010). European Language Resources Association (ELRA), Valletta (2010)
5. Branco, A., Silva, J.: A suite of shallow processing tools for Portuguese: LX-suite. In: Proceedings of the 11th Conference of the European Chapter of the Association for Computational Linguistics: Posters and Demonstrations, EACL 2006, pp. 179–182. Association for Computational Linguistics, Trento (2006)
6. Fellbaum, C.: WordNet: An Electronic Lexical Database. MIT Press, Cambridge (1998)
7. Ferreira, E., Balsa, J., Branco, A.: Combining rule-based and statistical methods for named entity recognition in Portuguese. In: V Workshop em Tecnologia da Informação e da Linguagem Humana, TIL 2007, pp. 1615–1624 (2007)
8. Lehmann, J., Isele, R., Jakob, M., Jentzsch, A., Kontokostas, D., Mendes, P., Hellmann, S., Morsey, M., van Kleef, P., Auer, S., Bizer, C.: DBpedia - a large-scale, multilingual knowledge base extracted from wikipedia. Semant. Web J. **6**(2), 167–195 (2012)
9. MultiWordNet: The MultiWordNet project. http://multiwordnet.fbk.eu/english/home.php (nd). Accessed 13 Jan 2015
10. Neale, S., Silva, J., Branco, A.: A flexible interface tool for manual word sense annotation. In: Proceedings of the 11th Joint ACL-ISO Workshop on Interoperable Semantic Annotation, ISA-11, pp. 67–71. Association for Computational Linguistics, London (2015)
11. Nóbrega, F.A.A., Pardo, T.A.S.: General purpose word sense disambiguation methods for nouns in Portuguese. In: Baptista, J., Mamede, N., Candeias, S., Paraboni, I., Pardo, T.A.S., Volpe Nunes, M.G. (eds.) PROPOR 2014. LNCS, vol. 8775, pp. 94–101. Springer, Heidelberg (2014)
12. Cardoso, P.C.F., Maziero, E.G., Jorge, M.L.R.C., Seno, E.M.R., di Felippo, A., Rino, L.H.M., das Nunes, M.G.V., Pardo, T.A.S.: CSTNews - a discourse-annotated corpus forsingle and multi-document summarization of news texts in Brazilian Portuguese. In: Proceedings of the Third Annual RST and Text Studies Workshop, pp. 88–105 (2011)
13. Santos, J., Anastacio, I., Martins, B.: Named entity disambiguation over texts written in the Portuguese or Spanish languages. Lat. Am. Trans. IEEE (Rev. IEEE Am. Lat.) **13**(3), 856–862 (2015)
14. Stenetorp, P., Pyysalo, S., Topić, G., Ananiadou, S., Aizawa, A.: Normalisation with the BRAT rapid annotation tool. In: Proceedings of the 5th International Symposium on Semantic Mining in Biomedicine, Zürich, Switzerland (2012)
15. Stenetorp, P., Pyysalo, S., Topić, G., Ohta, T., Ananiadou, S., Tsujii, J.: Brat: a web-based tool for nlp-assisted text annotation. In: Proceedings of the Demonstrations at the 13th Conference of the European Chapter of the Association for Computational Linguistics, pp. 102–107. Association for Computational Linguistics, Avignon (2012)

Crawling by Readability Level

Jorge A. Wagner Filho[1], Rodrigo Wilkens[1(✉)], Leonardo Zilio[1],
Marco Idiart[2], and Aline Villavicencio[1]

[1] Institute of Informatics,
Federal University of Rio Grande do Sul, Porto Alegre, Brazil
{jawfilho,rodrigo.wilkens,lzilio,avillavicencio}@inf.ufrgs.br
[2] Institute of Physics, Federal University of Rio Grande do Sul, Porto Alegre, Brazil
idiart@if.ufrgs.br

Abstract. The availability of annotated corpora for research in the area
of Readability Assessment is still very limited. On the other hand, the Web
is increasingly being used by researchers as a source of written content to
build very large and rich corpora, in the Web as Corpus (WaC) initiative.
This paper proposes a framework for automatic generation of large corpora classified by readability. It adopts a supervised learning method to
incorporate a readability filter based in features with low computational
cost to a crawler, to collect texts targeted at a specific reading level. We
evaluate this framework by comparing a readability-assessed web crawled
corpus to a reference corpus (Both corpora are available in http://www.
inf.ufrgs.br/pln/resource/CrawlingByReadabilityLevel.zip.). The results
obtained indicate that these features are good at separating texts from
level 1 (initial grades) from other levels. As a result of this work two
Portuguese corpora were constructed: the Wikilivros Readability Corpus,
classified by grade level, and a crawled WaC classified by readability level.

Keywords: Readability assessment · Web as a corpus · Focused crawling

1 Introduction

Readability assessment has been a popular and important research topic for
many decades, and by the 1980s more than a thousand papers had already
been published discussing more than 200 different proposed readability formulas
[6]. This is in part due to the fact that determining the reading level of a given
document is a very subjective task, and many different semantic (e.g. word usage)
and syntactic (e.g. sentence length) metrics can be used to offer an automatic
complexity evaluation. It is also a consequence of the importance of readability
level assessment in practice, which aims to, for example, support educators in
selecting appropriate reading materials for students [3,25] or for people with
intellectual disabilities [7].

With advances in Natural Language Processing and Machine Learning, this
problem has often been viewed as a classification task and more complex features have been used to determine if a given text belongs in a predetermined

© Springer International Publishing Switzerland 2016
J. Silva et al. (Eds.): PROPOR 2016, LNAI 9727, pp. 306–318, 2016.
DOI: 10.1007/978-3-319-41552-9_31

reading level, such as those derived from n-gram language models [21,23,28]. The Coh-Metrix system [13,17], for example, analyzes more than 200 features to determine text cohesion and readability. Nonetheless, these features generally incur in a high computational cost, often relying, for example, in parsing and annotation of the entire corpus while simpler features have been shown to be strong predictors of text readability [11]. Moreover, the availability of annotated corpora for research on this task is limited [21], frequently consisting of manually adapted content.

In this context, we propose a framework for the automatic generation of readability-assessed corpora, which adopts a supervised learning method to incorporate a readability filter with various low-cost complexity features in a crawler. As a consequence the framework can be used to collect suitable texts targeted at a pre-selected reading level. As a case study we focus on Brazilian Portuguese, but the framework could be straightforwardly adapted to other languages. Evaluation was performed by analyzing the correlation between a web crawled corpus classified by readability and a reference corpus. The results indicate that these low-cost features are good predictors of level 1 (initial grades) texts. While levels 2 (high school) and 3 (college) do differ in content, they seem to have no clear lexical or syntactic differences that could be measured by these features. As a result of this work two corpora were constructed: the Wikilivros Readability Corpus, classified by age, and a crawled WaC classified by readability.

This paper is structured as follows. In Sect. 2, we discuss some relevant work, while Sect. 3 presents the methodology and materials used in the experiments. Section 4 describes the evaluation method and results. We finish with conclusions and ideas for future work.

2   Related Work

Readability assessment has for a long time been a topic of interest, generating influential works like those by Flesch [10], Coleman and Liau [5], and Stenner [26]. Each proposes a set of measures for calculating the readability level of a given text. For instance, Flesch created the famous index of the same name which calculates readability based on the number of syllables per word and the number of words per sentence. The Flesch index is still broadly used today, being included in popular text editing tools such as Microsoft Word. Although originally designed for assessing English texts, it was adapted for Portuguese by Martins [16], by observing that Portuguese texts scored in average 42 points less than their English counterparts, due to the fact that Portuguese words present a higher average number of syllables because of its Graeco-Latin origins. The Coleman Index, on the other hand, is based on the average number of letters and sentences per hundred words [5], while the Lexile framework [26] combines word frequency counts and sentence length. The Dale-Chall formula combines sentence length and percentage of words not found on a list of 3000 easy words [4]. The open version of the Coh-Metrix system [18] analyses text cohesion and readability based in 108 different features, such as the incidence of connectives and pronouns.

More recently, readability assessment has been viewed as a classification task, with machine learning algorithms being trained with features that include some of these measures. For instance, Petersen and Ostendorf [21] propose the use of Support Vector Machines to combine features from language models, parsers and classic readability indexes to automate the task of selecting appropriate materials for second language learners. In their work they employ text classification and feature selection. The SVM models are trained on texts for children with reading level indicated by, for instance, the Weekly Reader, an educational newspaper with versions targeted at different grade levels, and are contrasted with other corpora consisting of articles for adults. They also discuss the large variability observed in the assessments of multiple human annotators and the poor agreement of those assessments with the reference corpora, showing that a well-trained system can achieve better results considering the desired conventions.

Feng et al. [8] also treat readability assessment as a classification task, evaluating how accurately features used to train these classifiers can predict if a given text is suited for a particular age group. Their best combination of features results in a 72 % accuracy. On similar lines Vajjala and Meurers [28] apply readability features and machine learning to classify a corpus of subtitles in terms of target audience age group. In relation to the features, François and Miltsakaki [11] compare the contribution of classical vs non-classical features and the effects of different machine learning algorithms. They focus on French and observe that the classical features are strong single predictors of text readability.

Scarton et al. [24] experimented with different features, machine learning algorithms and feature selection strategies for classifying Portuguese texts as simple or complex and obtained good results using Support Vector Machines. Automatic reading level assessment can be combined with simplification as an evaluation of the outcome of simplification, determining whether more simplification is needed or the desired reading level was reached [12].

With the increasing availability of language materials in the World Wide Web, repositories of texts not only include carefully curated collections, but also data from the web. Indeed, initiatives for treating the Web as Corpus include the WaCky (*Web-As-Corpus Kool Ynitiative*) framework which has been used to produce very large corpora for different languages [1], including Portuguese [2]. Ferraresi and Bernardini [9] also explore this idea of a focused Web as Corpus, developing acWaC-EU, a large corpus of non-native English academic pages from European universities to study the differences in language usage. Given this rich and ever growing source of texts, it is important to understand how they can be better leveraged, especially considering their heterogeneity and the ubiquitous presence of noise. For instance, regarding the application of readability models to texts from the web, Vajjala and Meurers [27] achieve good classification performance across different corpora, consisting of different genres of texts and different targeted age groups.

In this paper, we build on these works and propose a framework for the dynamic collection of texts from the web assessed according to readability features as a way of obtaining large amounts of text content that is suitable for particular reading levels.

3   Materials and Methods

The readability-focused Web-as-Corpus construction framework that we present consists of a focused crawler equipped with a readability assessment module. It adopts the pipeline proposed by Baroni et al. [1], which consists of four steps (1–2 and 4–5), and adds an intermediate step (3) for readability assessment:

1. identification of an appropriate set of seed URLs,
2. post-crawl cleaning,
3. readability assessment,
4. near-duplicate detection and removal, and
5. annotation.

For the first step, seed selection, we followed the same procedure applied in the construction of the brWaC [2]. We selected random pairs of medium frequency words (between a hundred and ten thousand occurrences) from the Linguateca[1] word frequency list[2] after the removal of stopwords. This list of bigrams was used as input to a search engine API (Microsoft Bing)[3], and the top ten results for each bigram were selected. This procedure aims at increasing corpus variety while avoiding undesirable pages such as word definitions.

For the second and third steps we used the Web as a Corpus Toolkit [29], a toolkit in Perl based on the principles of Web as corpus construction, which was chosen due to its modular and easily extensible architecture.[4] In the post-crawl cleaning, the toolkit applies several filters, removing non-HTML content, very small or large pages and boilerplate based on HTML tag density. We also introduced a stopword density filter to remove texts with less than 25 % of stopwords, which are unlikely to be content texts [22]. This filter also helps to eliminate any possible non-Portuguese texts resultant from the crawling phase.

In the third step, the readability assessment module eliminates all the documents that are not suitable to the specified target level, using the features described in Sect. 3.1. This reduces the amount of data effectively processed by the subsequent modules, minimizing the annotation cost of the relevant target readability level.

In the near-duplicate detection and removal stage all documents with more than 60 percent of duplicated sentences were discarded by the toolkit. This is important to avoid duplicated content in the corpus, since many search engine results can point to similar texts and this would make the corpus size a bad metric for content variation.

In the last step, the resulting corpus was compiled as a vertical file and annotated. Figure 1 summarizes the operation of the complete pipeline. We also

---

[1] http://www.linguateca.pt/ACDC/.

[2] http://dinis2.linguateca.pt/acesso/tokens/formas.totalbr.txt.

[3] http://www.bing.com/toolbox/bingsearchapi.

[4] The toolkit is divided in a web crawling module, several combinable filter modules, a deduplication module and a post-processing module responsible for the annotation and compilation of the corpus.

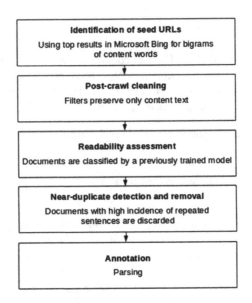

Fig. 1. The adapted Web-as-Corpus pipeline

extended the toolkit to include all the text complexity features calculated as part of the document header in the vertical corpus file. Then, depending on whether the classification filter is enabled or not, the document level classification can also be kept as a document annotation.

3.1 Readability Assessment

The readability assessment module is responsible for calculating several readability features for each document, that are subsequently used as input to a machine learning classification model. The features used in this module were selected based on efficiency, given the potentially very large sizes of the collected corpora, and on the information available for each document at this stage:

Type Token Ration (TTR): is a measure of lexical diversity that calculates how often the different types are repeated in a given corpus.

Flesch index (Flesch): this classic lexical and syntactic complexity measure [10] is based on the number of syllables per word and the number of words per sentence and is commonly included in readability models. We used the Portuguese version, adapted by Martins [16], calculated as Eq. 1:

$$Flesch = 248.835 - 84.6 \times SPW - 1.015 \times WPS \tag{1}$$

where SPW is the number of syllables per word and WPS is the number of words per sentence. This formula produces a value from 0 to 100, which is generally interpreted in a four-level scale of increasing ease of readability: very difficult (0 to 25), difficult (25 to 50), easy (50 to 75) or very easy (75 to 100).

In order to compute the number of syllables of each word, we used a rule-based syllabification tool, which achieved a performance of 99 % correctly syllabified words [20].

Coleman-Liau index (Coleman): this measure indicates the US grade level necessary to understand a given text and is based on the average number of letters and sentences per a hundred words [5], calculated as in Eq. 2:

$$Coleman = 0.0588 \times LP - 0.296 \times SP - 15.8 \qquad (2)$$

where LP means letters per a hundred words and SP means sentences per a hundred words.

Average word length and standard deviation (AWL): this measure is based on the assumption that more complex texts are likely to include longer words, due to the more frequent presence of prefixes and suffixes. These longer words are generally seen as more difficult, since they have combination of meanings (affix meaning plus base meaning), and they tend to be less frequent in simpler texts.

Average number of word senses and standard deviation (Senses): in this work this is implemented as the number of synsets in which each word appears according to the Portuguese data on BabelNet [19]. This measure is derived from the assumption that words which are more commonly used, and thus more easily understood, tend to have multiple meanings in the language.

Average word frequency in a general corpus and standard deviation (AFGC): based on the assumption that words with high frequencies are likely to be more familiar and well known to more readers, and consequently be included in more text levels, while rarer words are more likely to be restricted to more complex texts.

Incidence of unknown words (Unknown): the occurrence of words not present in a dictionary (in this work, a 3 million Portuguese unigram list) can be an indicative of more rare and complex, domain-specific words.

3.2   The Wikilivros Readability Corpus

A readability corpus composed of similar texts from at least three different reading levels was constructed by selecting the HTML book library from the Wikilivros website[5], the Portuguese version of the Wikibooks initiative. These books are separated in the following levels: 33 books used in the 1st to 9th grades in the Brazilian education system (from now on called *Level 1*), 65 books used in the 10th to 12th grades (*Level 2*) and 21 books used in college education (*Level 3*). Although they are divided into different levels, some overlaps between the levels were observed. Under the assumption that books present in more than one reading level would not be informative to determine text readability, these

---

[5] https://pt.wikibooks.org/.

Table 1. The Wikilivros Readability Corpus.

Metric	Level 1	Level 2	Level 3	All
Number of documents	15	45	17	77
Number of sentences	7061	17755	14049	38865
Average sentence size in words	15.70	15.72	17.20	16.20
Type	12622	26547	15293	54462
Token	111401	281436	243472	636309
TTR	0.11	0.09	0.06	0.08

overlaps were discarded. The resulting corpus, the Wikilivros Readability Corpus (WRC) is described in Table 1, while its readability features are shown in Table 2.

The corpus size per level was then normalized by randomly selecting 15 documents (the size of the smallest group) from each level as the training set.

Table 2. Readability features in the Wikilivros Readability Corpus. Standard Deviation is shown in brackets.

Feature	Level 1	Level 2	Level 3	All
Flesch	55.8	45.5	46.1	47.6
Coleman	10.3	11.7	11.6	11.4
AWL	4.82 (2.90)	4.99 (3.08)	4.97 (3.21)	4.95 (3.08)
AFGC	530181 (835457)	553806 (849364)	576357 (876828)	554183 (852718)
Senses	11.45 (10.18)	11.14 (9.67)	11.73 (10.08)	11.33 (9.86)
Unknown	0.2 %	0.6 %	0.5 %	0.5 %

4   Evaluation

The reference corpus presented in Sect. 3.2 and the features discussed in Sect. 3.1 were used to train a regression model (Sect. 4.1), which was evaluated in the construction of a web crawled corpus looking to lexical and syntactic features (Sect. 4.2).

4.1   Model

The WRC training set was used to build a classifier with SimpleLogistic [15] model from the Weka toolkit [14], with the readability levels (1 to 3) as classes. This linear model produces simple regression functions and applies automatic feature selection. A regression model is appropriate for the numeric nature of the classes, and the resulting equations where relevant features are weighted fit our requisite of low computational cost calculation for the classification of web corpora, as shown in Eqs. 3, 4 and 5. The formula with the higher value determines the appropriate class of a given document.

$$Level\ 1 = 18.43 + Unknown \times -89.44 + AWL_{STD} \times -6.94 + Senses_{STD} \times 0.32 \quad (3)$$

$$Level\ 2 = 17.49 + Flesch \times -0.03 + Senses \times -0.91 + Senses_{STD} \times -0.58 \quad (4)$$

$$Level\ 3 = -17.82 + AWL \times -1.43 + AWL_{STD} \times 7.94 \quad (5)$$

This model achieved an average F-measure of 0.691 (0.741 for level 1, 0.645 for level 2 and 0.688 for level 3), with precision of 0.702 and recall of 0.689. These results are compatible with those of Petersen and Ostendorf [21], where an SVM-based detector obtained an average F-measure of 0.609 for a 4 level classification. Both studies, however, are not directly comparable, given the different language and evaluation setup applied.

In a qualitative analysis of the Wikibooks corpus, we observed that the distinction between level 1 books against the other two levels can be seen in the lexical and syntactic level. It is possible to observe, for instance, the use of first person singular and the direct addressing to a second person (the reader) in the level 1 books. Sentences in this level tend also to be short and direct, presenting information in a very clear way. Texts from both level 2 and 3 have no clear difference in the way they were written, apart from the educational content they convey. For this reason, we selected level 2 as a negative class in the comparative study against Level 1 in Sect. 5.

4.2 Corpus

The pipeline for WaC crawling (Sect. 3) was used to collect more than 5000 web pages that compose our validation corpus, the readability-assessed WaC. This corpus was processed by the classifier (Sect. 4.1) and is described in Tables 3 and 4. The difference in proportions between the WRC and the readability-assessed WaC, the latter being almost a hundred times larger, illustrates the advantages of using automatically filtered web-crawled content to complement manually generated materials.

Table 3. Readability-assessed crawled WAC.

Metric	Level 1	Level 2	Level 3	All
Number of documents	1543	2881	1050	5474
Number of sentences	129323	236080	96498	461901
Average sentence size in words	13.59	15.27	17.40	15.42
Type	81018	151451	96322	328791
Token	1579323	3571962	1750491	6901776
TTR	0.051	0.042	0.055	0.049

Table 4. Readability features for the readability-assessed crawled WAC. Standard Deviation is shown in brackets.

Feature	Level 1	Level 2	Level 3	All
Flesch	59.1	47.5	40.4	48.9
Coleman	9.79	12.0	13.69	11.8
AWL	4.75 (2.75)	5.11 (3.06)	5.35 (3.47)	5.07 (3.06)
AFGC	485385 (810291)	510310 (840530)	569913 (880637)	516202 (841150)
Senses	10.67 (9.82)	10.11 (9.10)	11.12 (9.91)	10.45 (9.44)
Unknown	0.4 %	3.4 %	5.5 %	3.1 %

5   Results

Due to the lack of a gold standard for the evaluation of a focused Web as Corpus, for evaluating the generated WaC corpus we compared the linguistic properties between the readability-assessed WaC and the WRC, analyzing distributional, lexical and syntactic properties such as frequency, dependency tags and subcategorization frames. We compared level 1 class against level 2, as discussed in Sect. 4.1.

The first comparison uses the Jensen-Shannon divergence of the distributions, a symmetric and always finite variation of the Kullback-Leibler divergence. Given two probability distributions P and Q, the Kullback-Leibler divergence determines how much information is lost by using the latter to approximate the former. The results from this analysis are shown in Table 5. We also calculated the Spearman's rank correlation coefficient of the distributions[6]; results are shown in Table 6. These analyses where applied to both lexical (e.g. word surface forms, word lemmas) and syntactic (e.g. dependency tags, subcategorization frames) distributions.

Observing the lexical features in Table 5, there is a smaller divergence between the corpora in the same levels (WaC level 2 vs WRC level 2, and WaC level 1 vs WRC level 1) than between different levels (WaC level 1 and WRC level 2). Given that smaller Jensen-Shannon divergence values indicate more similar data, the inner level divergence is smaller than the inter level divergence. On the other hand, considering the syntactic features, WaC level 2 vs WRC level 2 presents a smaller divergence, but all the remaining scenarios are very close.

Table 5. Jensen-Shannon divergence analysis.

	Lexical	Syntactic
WaC Level 1 vs WRC Level 2	0.132	0.022
WaC Level 2 vs WRC Level 1	0.120	0.027
WaC Level 1 vs WRC Level 1	0.114	0.023
WaC Level 2 vs WRC Level 2	0.113	0.015

---

[6] All correlations presented a significance level higher than 99 %.

Table 6. Spearman's correlation analysis.

	Lexical	Syntactic
WaC Level 1 vs WRC Level 2	0.535	0.829
WaC Level 2 vs WRC Level 1	0.509	0.830
WaC Level 1 vs WRC Level 1	0.527	0.834
WaC Level 2 vs WRC Level 2	0.784	0.845

Table 7. Proportion of part-of-speech tags in different subcorpora of the WRC.

	Level 1	Level 2	Level 3
Nouns	19.8 %	24.9 %	26.3 %
Adjectives	6.3 %	7.7 %	7.8 %
Prepositions	16.1 %	16.3 %	16.4 %
Personal Pronouns	2.6 %	1.8 %	1.3 %
Relative	1.5 %	1.4 %	1.3 %
Verbs	15.7 %	13.7 %	15.1 %
Other	38 %	34.2 %	31.8 %

Table 8. Proportion of POS tags in different subcorpora of the readability-assessed WaC.

	Level 1	Level 2	Level 3
Nouns	21.3 %	23.5 %	25.2 %
Adjectives	5.4 %	6.5 %	8.1 %
Prepositions	14.8 %	16.4 %	17.2 %
Personal Pronouns	2.9 %	1.7 %	1.4 %
Relative	1.8 %	1.5 %	1.3 %
Verbs	16.7 %	14.4 %	12.4 %
Other	37.1 %	36 %	34.4 %

It is important to note, nonetheless, that, as our training features do not take into account this linguistic dimension, a lack of syntactic quality in the corpus is expected.

Finally, we also performed a comparative analysis of the part-of-speech distributions among the different corpora levels, identifying some interesting behavior patterns of the reference corpus that were also observed in the evaluation corpus. Nouns and adjectives are more frequent in the more advanced levels in both corpora levels, while personal pronouns are more frequent in the lower levels. Moreover, there is a more frequent use of prepositions in the more complex texts, and less frequent use of relative clauses. This is possibly explained by the more common supposition of previous knowledge in more advanced texts. These values are presented in more detail in Tables 7 and 8.

6 Conclusion

In this paper we presented a framework for the automatic generation of readability-assessed corpora, which equips the crawler with a classifier trained with various standard readability features to collect texts suitable for a given educational level. We evaluated the framework by collecting texts from the web, focusing on Brazilian Portuguese, and analyzing the correlation between the readability-assessed web crawled corpus and a reference corpus. These features are good predictors of level 1 texts. A qualitative analysis revealed that texts from other levels seem to have differences in content, but no clear lexical or syntactic differences. Furthermore, this work generated two corpora as results: the Wikilivros Readability Corpus, classified by grade level, and a readability-classified crawled WaC.

As future work we plan to incorporate additional measures, including those from the Coh-Metrix-Port system [23], to improve classification of text from levels 2 and 3, possibly as a post-processing step. Moreover, we intend to use the framework to collect corpora classified by readability on demand in real time.

Acknowledgments. This research was partially developed in the context of the project *Text Simplification of Complex Expressions*, sponsored by Samsung Eletrônica da Amazônia Ltda., in the terms of the Brazilian law n. 8.248/91. This work was also partly supported by CNPq (482520/2012- 4, 312114/2015-0) and FAPERGS AiMWEst.

References

1. Baroni, M., Bernardini, S., Ferraresi, A., Zanchetta, E.: The wacky wide web: a collection of very large linguistically processed web-crawled corpora. Lang. Resour. Eval. **43**(3), 209–226 (2009)
2. Boos, R., Prestes, K., Villavicencio, A., Padró, M.: *brWaC*: a wacky corpus for Brazilian Portuguese. In: Baptista, J., Mamede, N., Candeias, S., Paraboni, I., Pardo, T.A.S., Volpe Nunes, M.G. (eds.) PROPOR 2014. LNCS, vol. 8775, pp. 201–206. Springer, Heidelberg (2014)
3. Callan, J., Eskenazi, M.: Combining lexical and grammatical features to improve readability measures for first and second language texts. In: Proceedings of NAACL HLT, pp. 460–467 (2007)
4. Chall, J.S., Dale, E.: Readability Revisited: The new Dale-Chall readability formula. Brookline Books, Cambridge (1995)
5. Coleman, M., Liau, T.L.: A computer readability formula designed for machine scoring. J. Appl. Psychol. **60**(2), 283 (1975)
6. DuBay, W.H.: The principles of readability. Online Submission (2004)
7. Feng, L., Elhadad, N., Huenerfauth, M.: Cognitively motivated features for readability assessment. In: Proceedings of the 12th Conference of the European Chapter of the Association for Computational Linguistics, pp. 229–237. Association for Computational Linguistics (2009)
8. Feng, L., Jansche, M., Huenerfauth, M., Elhadad, N.: A comparison of features for automatic readability assessment. In: Proceedings of the 23rd International Conference on Computational Linguistics: Posters, COLING 2010, pp. 276–284. Association for Computational Linguistics, Stroudsburg (2010). http://dl.acm.org/citation.cfm?id=1944566.1944598

9. Ferraresi, A., Bernardini, S.: The academic web-as-corpus. In: Proceedings of the 8th Web as Corpus Workshop, pp. 53–62 (2013)

10. Flesch, R.F., et al.: Art of Plain Talk. Harper, New York (1946)

11. François, T., Miltsakaki, E.: Do nlp and machine learning improve traditional readability formulas? In: Proceedings of the First Workshop on Predicting and Improving Text Readability for target reader populations, pp. 49–57. Association for Computational Linguistics (2012)

12. Gasperin, C., Specia, L., Pereira, T., Aluísio, S.: Learning when to simplify sentences for natural text simplification. In: Proceedings of ENIA - Brazilian Meeting on Artificial Intelligence, pp. 809–818 (2009)

13. Graesser, A.C., McNamara, D.S., Louwerse, M.M., Cai, Z.: Coh-metrix: analysis of text on cohesion and language. Behav. Res. methods Instrum. comput. **36**(2), 193–202 (2004)

14. Hall, M., Frank, E., Holmes, G., Pfahringer, B., Reutemann, P., Witten, I.H.: The weka data mining software: an update. ACM SIGKDD Explor. Newsl. **11**(1), 10–18 (2009)

15. Landwehr, N., Hall, M., Frank, E.: Logistic model trees. Mach. Learn. **59**(1–2), 161–205 (2005)

16. Martins, T.B., Ghiraldelo, C.M., Nunes, M.d.G.V., de Oliveira Junior, O.N.: Readability formulas applied to textbooks in brazilian portuguese. Icmsc-Usp (1996)

17. McNamara, D.S., Louwerse, M.M., McCarthy, P.M., Graesser, A.C.: Coh-metrix: capturing linguistic features of cohesion. Discourse Processes **47**(4), 292–330 (2010)

18. McNamara, D., Louwerse, M., Cai, Z., Graesser, A.: Coh-metrix version 3.0 (2013). http://cohmetrix.com. Accessed 1 Apr 2015

19. Navigli, R., Ponzetto, S.P.: Babelnet: building a very large multilingual semantic network. In: Proceedings of the 48th Annual Meeting of the Association for Computational Linguistics, pp. 216–225. Association for Computational Linguistics (2010)

20. Neto, N., Rocha, W., Sousa, G.: An open-source rule-based syllabification tool for Brazilian Portuguese. J. Braz. Comput. Soc. **21**(1), 1–10 (2015)

21. Petersen, S.E., Ostendorf, M.: A machine learning approach to reading level assessment. Comput. Speech Lang. **23**(1), 89–106 (2009)

22. Pomikálek, J.: Removing boilerplate and duplicate content from web corpora. Ph.D. en informatique, Masarykova univerzita, Fakulta informatiky (2011)

23. Scarton, C., Aluisio, S.M.: Coh-metrix-port: a readability assessment tool for texts in Brazilian Portuguese. In: Proceedings of the 9th International Conference on Computational Processing of the Portuguese Language, Extended Activities Proceedings, PROPOR, vol. 10 (2010)

24. Scarton, C., Gasperin, C., Aluisio, S.: Revisiting the readability assessment of texts in Portuguese. In: Kuri-Morales, A., Simari, G.R. (eds.) IBERAMIA 2010. LNCS, vol. 6433, pp. 306–315. Springer, Heidelberg (2010)

25. Schwarm, S.E., Ostendorf, M.: Reading level assessment using support vector machines and statistical language models. In: Proceedings of the 43rd Annual Meeting on Association for Computational Linguistics, pp. 523–530. Association for Computational Linguistics (2005)

26. Stenner, A.J.: Measuring Reading Comprehension with the Lexile Framework. ERIC, Washington (1996)

27. Vajjala, S., Meurers, D.: On the applicability of readability models to web texts. In: Proceedings of the 2nd Workshop on Predicting and Improving Text Readability for Target Reader Populations, p. 59 (2013)

28. Vajjala, S., Meurers, D.: Exploring measures of readability for spoken language: analyzing linguistic features of subtitles to identify age-specific tv programs. In: Proceedings of the 3rd Workshop on Predicting and Improving Text Readability for Target Reader Populations (PITR)@ EACL, pp. 21–29 (2014)
29. Ziai, R., Ott, N.: Web as Corpus Toolkit: Users and Hackers Manual. Lexical Computing Ltd., Brighton (2005)

Towards a Statistical-Enriched Corpus Containing Portuguese Collocations in Use: Reviewing Possible Extraction Tools

Ângela Costa[1,3](✉) and Luísa Coheur[1,2]

[1] INESC-ID Lisboa, Lisbon, Portugal
{angela,luisa.coheur}@l2f.inesc-id.pt
[2] Instituto Superior Técnico, Lisbon, Portugal
[3] Centro de Linguística da, Universidade Nova de Lisboa, Lisbon, Portugal

Abstract. Collocations are a main problem for any natural language processing task, from machine translation to summarization. With the goal of building a corpus with collocations, enriched with statistical information about them, we survey, in this paper, four tools for extracting collocations. These tools allow us to collect sentences with collocations, and also to gather statistics on this particular type of co-ocurrences, like Mutual Information and Log likelihood values.

Keywords: Collocations · Wortschatz · DeepDict · CRPC · Sketch engine

1 Introduction

Collocations, here understood as "a privileged lexical co-ocurrence of two or more linguistic elements that establish between themselves a syntactic relation." [10], are a type of multiword expressions that have always caused difficulties to natural language processing tasks. Several approaches have been proposed to retrieve collocations from the analysis of large samples of textual data. These techniques automatically produce large numbers of collocations along with statistical figures intended to reflect the relevance of the associations. Some work has been done on the evaluation of these techniques. For instance, the work described in [1], working only with English, evaluated WordSmith Tools 4, Collocate, Xaira and Ngram Statistics Package, regarding the capacity to extract collocations without keywords, measures of association, capacity to handle xml files, extract multiword collocations, handle multiple files at the same time, and the presence of a Graphical User Interface. The work from [4] also focused on an evaluation of extraction tools, for Portuguese, using several technical parameters like, for instance, how the data are organized and stored. As a result of using extraction techniques, [8] extracted and manually evaluated possible candidates and created a large list of multiword expressions in Portuguese.

© Springer International Publishing Switzerland 2016
J. Silva et al. (Eds.): PROPOR 2016, LNAI 9727, pp. 319–329, 2016.
DOI: 10.1007/978-3-319-41552-9_32

In this paper, having in mind the construction of a statistical-enriched corpus containing Portuguese collocations in use, we survey four online tools that allow to automatically find collocation candidates: Wortschatz[1], DeepDict[2], Reference Corpus of Contemporary Portuguese[3] and Sketch Engine[4]. The evaluation of these tools allow us to understand which one better fits our purpose of creating a corpus of collocations in context that is enriched with statistical information. There are other tools to do queries on Portuguese corpora, like Cetempublico or Corpus do Português but we found the aforementioned ones more suitable for the extraction of collocations in context and with statistical information.

In Sect. 2, we present our methodology. The extraction tools are described in Sect. 3. In Sect. 4, we compare the tools. Finally, in Sect. 5, we highlight the main conclusions and point to future directions.

2   Methodology

In this section we present the methodology followed to collect collocations in context.

As frequency is not, by itself, a defining trait of the collocations, we decided to use nouns with high and low frequency in the language taken from a Portuguese corpus of frequency available in COMPARA[5]. We used a total of 10 nouns: 5 with high frequencies, and 5 with low frequencies. Then, we submitted these nouns to the surveyed tools. Each noun is being tested as the base of a collocation, and we target to find possible collocates.

Each one of the tools is then evaluated (description in Sect. 4.2), according to the parameters that we will now describe below, which we considered to be the most important regarding our research purposes.

Starting with the **processing time**, we want it to be fast, as we need to perform several searches. Having the option to establish a **frequency threshold** is also very important, so that we can find the most salient collocations, and not simply common combinations of words. The search should be done by **lemma**, because only searching one word form can produce limited results. Determine the **window span** is also relevant. For instance, we want to capture the collocation *strong tea* in a wider span, like in *the tea was very strong*. As using **syntactic patterns** has proven to be a successful way to extract collocations, we would like to have this option available. Moreover, we would like to use a tool that **distinguishes the varieties of Portuguese** and that **does not have repeated texts in its corpora**, in order to ensure the representativeness of the extracted collocations and statistics. It is also an advantage to be able to **analyse the collocations in their context**, rather than have access to a list of co-occurring words and a score. Also, we would like to be able to **download these sentences**,

---

[1] http://corpora.informatik.uni-leipzig.de.
[2] https://gramtrans.com/deepdict/.
[3] http://www.clul.ul.pt/pt/recursos/.
[4] https://www.sketchengine.co.uk.
[5] http://www.linguateca.pt/COMPARA/listas_freq.php.

and **upload different corpora** for analysis. Regarding metrics, we would like to have information about **Mutual Information, Log-likelihood and Dice coefficient**. Also, we would prefer to use a **freely available** tool. Finally, we will also compare the number of correct collocations that were selected by the systems (Sect. 4.1).

3   The Surveyed Tools

In this section, we present the tools that we have analyzed: Wortschatz, Deep-Dict, Reference Corpus of Contemporary Portuguese, and Sketch Engine. The word *momento* will be used for illustration proposes.

3.1   Wortschatz

Wortschatz is a system developed by the University of Leipzig that uses the Leipzig Corpora Collection. This corpus exists for more than 250 languages, all data is searchable. The Portuguese data consists of a newspaper corpus based on material crawled in 2011 and the Wikipedia also collected in 2011. For the European variety of Portuguese, there are 2.540.587 sentences and 53.879.750 tokens. The Brazilian variety has 25.008.883 sentences and 486.724.987 tokens. The variety of Macau has 392,371 sentences and 8.672.381 tokens. The search can be done in one of the varieties or all. Several tools are used for preprocessing the corpus (for instance, tokenization, word frequency calculation, word co-occurrence calculation). There are also post-processing tools, like POS tagging and lemmatization. All the tools are available for download from their website, as well as most of the corpora.

To do the search, typing the word is the only option, as Wortschatz is the only evaluated tool that does not allow to establish a frequency threshold, as well as perform the search by lemma. On Fig. 1, we can see the search result for the word *momento*. On the top, we can find the number of occurrences, the rank and frequency class and the sentences with the queried word. Bellow, the co-occurrences, the left neighbor co-occurrences and the right ones. If we click on one of the words that co-occurs, we can see the sentences where the candidate collocate occurs, but not the collocation. Finally, there is no download option.

3.2   DeepDict

DeepDict is a free tool that allows to build complex dictionary entries and context overviews for a given word on the fly. Word relations are based on Constraint Grammar[6] dependency analysis and grammatical functions, not just co-occurrence like Wortschatz. In the case of Portuguese, the multi-level Constraint Grammar parser used is PALAVRAS[7] and the corpus used is

---

[6] http://beta.visl.sdu.dk/constraint_grammar.html.
[7] http://linguateca.dei.uc.pt/Floresta/InicialFloresta.html.

Term: **momento**   Number of occurrences: 22,535   Rank: 203   Frequency class: 7

See also: Momento, MOMENTO

▾   Examples: A Deco pede, por isso, e para...

▾   Cooccurrences: neste | Neste | até | que | ,

▾   Left Neighbour Cooccurrences: neste | Neste | ao | num | um

▾   Right Neighbour Cooccurrences: , | em | histórico | certo | não

Fig. 1. Wortschatz results for the noun *momento.*

Floresta Sintá(c)tica[8], which is a collection of sentences that have been morfossyntactically analyzed, producing a treebank of 1.000.000 words collected from CETEM Publico [9].

To look up collocational candidates, we type the word, discriminate the word class and the language. There are also other advanced options available, like establishing a lexical frequency threshold (used to filter rare words), a minimum occurrence (rule out rare relations or include them) and a minimum relative frequency (set a threshold for co-occurrences). As for the window span, DeepDict as well as Wortschatz do not have that option.

Figure 2 shows how results are displayed, considering the word *momento.*

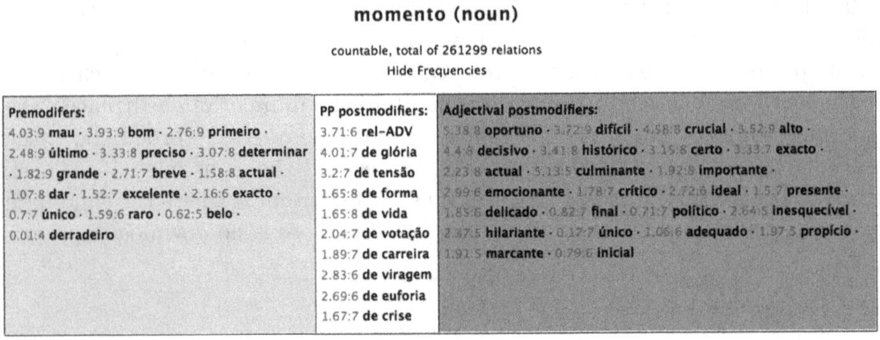

momento (noun)

countable, total of 261299 relations
Hide Frequencies

Premodifers:	PP postmodifiers:	Adjectival postmodifiers:
4.03:9 **mau** · 3.93:9 **bom** · 2.76:9 **primeiro** ·	3.71:6 **rel–ADV**	5.18:8 **oportuno** · 3.72:9 **difícil** · 4.58:8 **crucial** · 3.52:8 **alto** ·
2.48:9 **último** · 3.33:8 **preciso** · 3.07:8 **determinar**	4.01:7 **de glória**	4.4:8 **decisivo** · 3.41:8 **histórico** · 3.15:8 **certo** · 3.33:7 **exacto** ·
· 1.82:9 **grande** · 2.71:7 **breve** · 1.58:8 **actual** ·	3.2:7 **de tensão**	2.23:8 **actual** · 5.13:5 **culminante** · 1.92:5 **importante** ·
1.07:8 **dar** · 1.52:7 **excelente** · 2.16:6 **exacto** ·	1.65:8 **de forma**	2.99:6 **emocionante** · 1.78:7 **crítico** · 2.72:6 **ideal** · 1.5:7 **presente** ·
0.7:7 **único** · 1.59:6 **raro** · 0.62:5 **belo** ·	1.65:8 **de vida**	1.85:6 **delicado** · 0.82:7 **final** · 0.71:7 **político** · 2.64:5 **inesquecível** ·
0.01:4 **derradeiro**	2.04:7 **de votação**	2.87:5 **hilariante** · 0.17:7 **único** · 1.06:6 **adequado** · 1.97:5 **propício** ·
	1.89:7 **de carreira**	1.91:5 **marcante** · 0.79:6 **inicial**
	2.83:6 **de viragem**	
	2.69:6 **de euforia**	
	1.67:7 **de crise**	

Fig. 2. DeepDict results for the noun *momento.* (Color figure online)

For each search, relative and absolute frequency values are provided for each relation (premodifiers, PP postmodifiers and Adjectival postmodifiers). Frequency values (in red) can be clicked to see detailed statistics. Considering the

---

[8] http://www.linguateca.pt/floresta/principal.html.

co-occurring word mau as example, 4.03 is the co-occurrence strength between the lookup word and the relative frequency, and 9 is the dual logarithmic value of absolute frequency (scale is from 0 to 9). Red numbers can be clicked to show examples in concordance form, if available, and the statistics of these forms. Like Wortschatz, these data are not available for download.

3.3 Reference Corpus of Contemporary Portuguese

The Reference Corpus of Contemporary Portuguese (CRPC) is an electronic corpus of European Portuguese and varieties (Brazil, Angola, Cape Verde, Guinea-Bissau, Mozambique, S. Tome and Principe, Goa, Macao and East-Timor). It contains 311.4 million words and covers several types of written texts (literary, newspaper, technical, etc.) and spoken texts. Searches can be made online on the written subpart of the corpus (309 M). The texts were tokenized using the LX tokenizer [3]. The part-of-speech tagging was trained based on a memory-based tagger [5]. Finally, the lemmatization was done with a Portuguese version of the MBLEM lemmatizer [2].

First, we have to select a corpus, either the European Portuguese or the whole corpus, including varieties. Then, we type a word in the search box (or syntactic pattern). We select the number of hits per page and we can also restrict our search to a specific corpus (laws, newspapers, school books, etc.). This is the only tool that allows this specific type of search. We are then presented with the search word in context, after which we can create a collocation database with the list of words that co-occur with the retrieved word pattern.

Figure 3 shows the results given the word *momento*.

Collocation controls

Collocation based on:	Word form
Collocation window *from*:	3 to the Left
Freq(node, collocate) at least:	5
Filter results by:	specific collocate:

Statistic:	Log-likelihood
Collocation window *to*:	3 to the Right
Freq(collocate) at least:	5
and/or tag: (none)	Submit changed parameters Go!

There are 44,078 different words in your collocation database for "[lemma="momento"%c]". (Your query "{momento}" returned 131,381 matches in 80,302 different texts) [37.076 seconds – retrieved from cache]

No.	Word	Total no. in whole corpus	Expected collocate frequency	Observed collocate frequency	In no. of texts	Log-likelihood value
1	neste	174,686	444.47	47,567	34252	367279.952
2	em	3,031,659	7713.733	26,751	20915	29048.331
3	.	20,881,197	53129.979	86,393	44204	19061.758

Fig. 3. CRPC results for the noun *momento*.

We can decide the maximum window span to be used. On the collocation controls, unlike the previously described systems, we can change the statistical measure used and the frequency threshold. If we click on observed collocate frequency, we can see the collocation in context. Contrary to the other systems, we can download these sentences and all the statistics. Regarding the processing time for the word *momento*, Wortschatz and DeepDict took less than 6 s, while CRPC took approximately 46 s.

3.4 Sketch Engine

Sketch engine is a paid corpus software interface that works online. It allows to see concordances for any word, phrase or grammatical construction; it also shows each word grammatical and collocational behavior. It has 200 corpora in 82 languages, but we can also upload our own corpus, being the only tool that presents this feature. There are available corpora for two varieties of Portuguese: European and Brazilian. The parser used, like for Deep Dict, is PALAVRAS. There are several valences to this software, like creating a Thesaurus, word lists or comparing occurrences of two words, but we will focus on the options that allow extracting collocations: "word sketch" and "concordance". They will be detailed in the next paragraphs.

Word Sketch. Selecting "word sketch", will allow to see a word's grammatical and collocational behavior. We type the lemma and specify its part-of-speech. On the advanced options, we choose the European Portuguese corpus (ptTenTen11 with 3.245.834.337 tokens from web pages). We can select a minimum frequency and/or score (score is defined as logDice[9]) and the minimum value of co-occurrence. We can also choose to cluster[10], or not, the collocates and decide the maximum number of items in these grammatical relations. Collocations can also be sorted by salience[11] or by raw frequency. The word sketch, in addition to using a well-founded salience statistic and lemmatization, uses grammar patterns. Rather than looking at an arbitrary window of text around the headword, each grammatical relation that the word participates in is taken into consideration. For Portuguese, there are 11 grammatical relations. The word sketch then provides one list of collocates for each grammatical relation the word participates in, and the results will be presented in syntactic relation clusters (object_of, subject_of or n_modifier). CRPC also allows a search by a syntactic pattern, for instance, noun + adjective or a word, like *momento* + adjective (we have not used this option for the present evaluation). DeepDict does not allow a search by pattern, but the results are presented in relation clusters.

As usual, Fig. 4 shows the sketch for the word *momento*.

As mentioned before, the potential collocates are presented grouped according to the grammatical relation in which they occur. The first score in front of

---

[9] This measure is based only on a frequency of words w1 and w2 and bigram w1 w2, it is not affected by the size of the corpus.

[10] If the clustering option is selected, the collocates within a word sketch are clustered according to any such clusters from the distributional thesaurus that they appear in. The words from the thesaurus are clustered according to their distributional similarity scores.

[11] Salience is a statistical measure of how salient a word or lemma is in a given context, given the frequency of the word and the context. This is measured with logDice.

momento *(noun)*
ptTenTen [2011, Freeling v3] European freq = 372,624 (398.06 per million)

object_of			subject_of			n_modifier		
	56,968	12.00		10,228	10.40		92,362	12.30
proporcionar	2,798	7.77	suceder+se	9	4.36	difícil	2,946	8.61
atravessar	1,217	7.76	dar+se	34	4.24	convívio	1,370	8.32
partilhar	595	7.48	coincidir	12	3.82	marcante	1,452	8.29
viver	2,746	7.36	conservar+se	5	3.79	alto	3,251	8.16
recordar	604	7.33	ser+lhe	6	3.79	oportuno	942	8.15
viver+se	231	6.98	parecer+me	9	3.76	inesquecível	997	8.08
registar	417	6.75	viver+se	5	3.64	certo	2,191	8.06
proporcionar+lhe	196	6.63	intimar	6	3.62	actual	1,251	7.93
determinar	2,173	6.62	transformar+se	9	3.41	decisivo	1,136	7.84

Fig. 4. Word sketch results for the noun *momento*.

each collocate candidate is the word frequency. If we click on it, we can see the corpus contexts in which the node word and its collocate co-occur. The second number is the frequency within the cluster. The word sketches and the examples of the corpus can be downloaded in xml or txt.

Concordance. Another way to extract collocations, is to use the "concordance" option. We type the word, select the corpus, and attain all the occurrences of the searched word in the corpus. The next step is to build the collocation list. This is done by creating a list of words statistically associated with the words (node) in the query. In this search menu, we select the attribute (word, part of speech tag, lemma, etc.). We can specify the range (span of text) around our node word when considering candidates. The defaults are −3 (3 tokens before the node) and 3 (3 tokens after the node). And we can also specify thresholds on the frequency of the candidate in the whole corpus and the frequency within the range. We can stipulate which statistics are displayed, and the statistic to sort by. The statistics available are relative frequency, T-Score, MI-Score, MI3-Score, log-likelihood, minimum sensitivity, logDice and MI.log-f. More details on the metrics can be found in [6]. This is the extraction tool that provides more statistical information, although CRPC also gives a considerate amount of statistics.

Figure 5 shows the sketch for the word *momento*. To see the results ordered according to a metric, we just have to click on the chosen metric. We can see the word co-ocurrence on the corpus, if we click on "P". The metrics results and the corpora can be downloaded.

4   Comparison Between Systems

On the next sections, we will assess each tool according to the collocations they were able to find, the extracted corpora and statistics that resulted from that task.

Collocation candidates

Page 1   Go   Next >

	Frequency	T-score	MI	MI3	log likelihood	min. sensitivity	logDice	MI.log_f
P I N em+este	73,828	270.593	7.922	40.266	680,817.958	0.020	9.186	88.805
P I N partir	10,944	103.031	6.047	32.883	70,557.757	0.005	7.206	56.241
P I N em+um	15,784	123.562	5.922	33.814	99,265.330	0.005	7.164	57.247
P I N proporcionar	3,488	58.473	6.654	30.190	25,327.746	0.008	7.154	54.279
P I N em+aquele	3,545	58.890	6.517	30.100	25,074.195	0.007	7.089	53.269
P I N qualquer	11,239	104.171	5.846	32.758	69,377.319	0.005	7.039	54.526
P I N actual	2,366	48.276	7.058	29.474	18,492.902	0.006	7.031	54.835
P I N convívio	1,711	41.211	8.074	29.556	15,787.846	0.005	6.957	60.118
P I N atravessar	1,852	42.757	7.275	28.985	15,030.761	0.005	6.864	54.744
P I N marcante	1,587	39.652	7.751	29.015	13,927.506	0.004	6.809	57.127
P I N difícil	3,423	57.578	5.978	29.460	21,678.899	0.005	6.743	48.655
P I N inesquecível	1,310	36.073	8.225	28.936	12,362.293	0.004	6.653	59.045

Fig. 5. Concordance results for the noun *momento*.

4.1 Finding Collocations

As previously mentioned, we have done the searches, on each system, using 10
nouns as the base of the collocations. These nouns were selected from a list of
frequent nouns in Portuguese. These nouns are part of the current vocabulary of
Portuguese (opposed to specialized vocabulary) and, as tools corpora are mainly
extracted from newspapers and the web, we believe there were no problems of
low representativeness of those nouns.

For this paper, we have only analyzed the first 10 extractions from each search
tool. The setup used, when allowed by the system, was a minimum frequency
of 5 and a window span of 3 words to the right and left of the node. Table 1
shows the number of collocations extracted and validated by a linguist. The last
column shows examples of collocations that were extracted by more than one
system. As previously mentioned, the tools have different size internal corpora,
which obviously skews the comparison. Still as only one of the tools allows the
upload of a corpus, a comparison using the same texts would be impossible. That
said, we will interpret the size of the corpora as an idiosyncrasy of each tool and
the results should be interpreted in this light.

The system that was able to find more collocations was Word Sketch, showing that combining statistics and grammar patterns can be a successful way
to extract collocations, rather than simply count occurrences and frequencies,
like Wortschatz. Regarding the words that were selected but that were not collocates, we found among them articles (*o momento*), prepositions (*na verdade*) and
words that usually occur in the same semantic field (*autógrafo* and *fotografia*).
We should also point out that words that have a lower frequency score are the
ones that show fewer candidate options and more restriction in the choice of the
collocates (*diagnóstico precoce*, *escolaridade obrigatória*).

4.2 Collecting Corpora and Statistics

First of all, for the purpose of our research, we only wanted to use the European
variety of Portuguese. All engines distinguish between varieties and DeepDict

Table 1. Comparison between Wortschatz (Woc), DeepDict (DD), CRPC, Word Sketch (SE-1), Concordance (SE-2) extractions.

Word	WoC	DD	CRPC	SE-1	SE-2	Examples
momento	0	5	1	2	1	oportuno (2); decisivo (2)
fim	0	1	1	4	3	lucrativo (4); pôr (2)
verdade	1	1	0	5	2	absoluto (3)
certeza	1	2	3	4	2	absoluto (4); ter (4)
força	1	0	0	3	3	de vontade (3)
adversidade	1	2	3	6	5	superar (3); climatérico (3)
autógrafo	0	0	2	6	2	pedir (3); dar (2)
fumador	2	4	2	3	2	inveterado (5); passivo (5)
diagnóstico	1	1	1	2	1	precoce (5)
escolaridade	1	1	2	1	2	obrigatório (5)
TOTAL	**8**	**17**	**14**	**36**	**23**	

only has available an European Portuguese Corpus. Its is also important that the corpora used are not very repetitive in its constitution because this can bias the statistical results, but we only spotted repeated sentences in the CRPC corpus. We also wanted to be able to see the collocations in the context of the corpus, but this criteria was only not accomplished by DeepDict that only shows the collocate in context and not the entire collocation. Apart from visualizing the collocations in context, we also wanted to be able to download the sentences where they occurred. In this case, only Sketch Engine and CRPC allow download of the data. Sketch Engine is the only one that allows the user to upload its own corpus and do searches on it.

The purpose of this evaluation was to access which tool better served the purpose of collecting a corpus enriched with collocations but also gather all the statistical information available. DeepDict gives us the relative frequency of a given relation and the absolute frequency in a scale from 0 to 9. By using the "word sketch" in Sketch Engine, we can obtain the word frequency and the frequency within the cluster. With the "concordance" option, the collocation candidates can be put in order according to several metrics: relative frequency, T-Score, MI-Score, MI3 -Score, log-likelihood, minimum sensitivity, logDice and MI.log-f. CRPC provides information on the total number of occurrences in the whole corpus, the expected collocate frequency, the observed collocate frequency and number of texts of the occurrence. Then we can also change the metrics and see the ranking of results according to the Mutual information, MI13, Z-score, T-score, Log-likelihood, Dice coefficient or simple rank by frequency. The results will change according to the chosen metric. Wortschatz only features the number of occurrences. Finally, we also took into consideration if the tool was free. From the four we have used, only Sketch Engine needs to be paid. DeepDict is paid for some languages, but not for Portuguese. Table 2 sums up the evaluation of all systems. Although Sketch Engine has two search options, here was presented as one as the results are very similar.

Table 2. Comparison between systems features.

	Woc	DD	CRPC	SE
Processing time for the word *momento*	≈ 1 s	≈ 2 s	≈ 46 s	≈ 6 s
Frequency threshold	✗	✓	✓	✓
Search by lemma	✗	✓	✓	✓
Window span	✗	✗	✓	✓
Search by pattern	✗	✗	✓	✓
Distinguish varieties	✓	✓	✓	✓
No duplications in the corpus	✓	✓	✗	✓
Show co-ocurrences in the frpus	✗	✓	✓	✓
Download results	✗	✗	✓	✓
Upload corpus	✗	✗	✗	✓
Use several metrics	✗	✗	✓	✓
Free	✓	✓	✓	✗

5   Conclusions

In this paper we aimed at doing an assessment of four extracting tools: DeepDict, Sketch Engine, CRPC and Wortschatz. We started by describing each systems functionalities, then we used them to extract collocations, having as a base 10 words selected from the corpus of frequencies. We evaluated the resulting corpus and the static information provided by each tool. Based on the mentioned steps, we conclude that Sketch Engine, despite not being free, shows great potencial as collocation extractor, being able to find more correct collocations than the other systems. Of course this result is influenced by the size of the corpus that Sketch Engine uses. This tool outperforms the others in several aspects of our evaluation, for instance, allowing a fine-grained tuning of the search setup (frequency threshold, window span, search by lemma or pattern). The Portuguese corpus used is divided into BP and EP and, from our observations so far, does not seem to have repeated sentences. The results can be presented according to different metrics, allowing to understand which one better suits the research purpose[12]. Finally, the co-occurrences are shown in context, the results can be downloaded and different corpora can be uploaded to the tool. If using a paying tool is not an option, CRPC also shows good results, only having the disadvantage of the repetitions in the corpus. In the future, the aforementioned methodology used here for evaluation purposes, will be used to build a larger statistically-enriched corpus containing Portuguese collocations in use that will allow us to understand more about their contexts and will help us build rules to extract them.

---

[12] [7] suggests that MI is generally used for a lexicographical purpose, while MI3 is probably more useful for second language learning.

Acknowledgments. The work was partially supported by national funds through FCT - Fundação para a Ciência e a Tecnologia, reference UID/CEC/50021/2013. Ângela Costa is supported by PhD fellowship from FCT (SFRH/BD/85737/2012).

References

1. Anagnostou, N.K., Weir, G.R.S.: Review of software applications for derivingcollocations. In: ICT in the Analysis, Teaching and Learning of Languages, Preprints of the ICTATLL Workshop 2006, Glasgow, pp. 91–100 (2006)
2. van den Bosch, A., Daelemans, W.: Memory-based morphological analysis. In: Proceedings of the 37th Annual Meeting of the Association for Computational Linguistics on Computational Linguistics, ACL 1999, pp. 285–292. Association for Computational Linguistics, Stroudsburg (1999). http://dx.org/10.3115/1034678. 1034726
3. Branco, A., Silva, J.: Evaluating solutions for the rapid development of state-of-the-art pos taggers for Portuguese. In: Proceedings of the Fourth International Conference on Language Resources and Evaluation (LREC-2004). European Language Resources Association (ELRA), Lisbon. http://www.lrec-conf.org/proceedings/lrec2004/pdf/572.pdf, aCL Anthology Identifier: L04–1354
4. Correia, J.M.P.: Syntax Deep Explorer. Ph.D. thesis. Instituto Superior Técnico (2015)
5. Daelemans, W., Zavrel, J., Berck, P., Gillis, S.: Mbt: a memory-based part of speech tagger-generator. In: Proceedings of Fourth Workshop on Very Large Corpora, pp. 14–27. ACL SIGDAT (1996)
6. Kilgarriff, A., Rychly, P., Smrz, P., Tugwell, D.: The sketch engine. In: Proceedings of EURALEX (2004)
7. McEnery, T., Xiao, R., Tono, Y.: Corpus-Based Language Studies: An Advanced Resource Book. Taylor & Francis (2006)
8. Mendes, A., Antunes, S., do Nascimento, M.F.B., Miguel, J., Casteleiro, L.P., Sá, T.: Combina-pt: a large corpus-extracted and hand-checked lexical database of Portuguese multiword expressions. In: Proceedings of LREC, pp. 1900–1905 (2006)
9. Santos, D., Rocha, P.: Evaluating CETEMPúblico, a free resource for Portuguese. In: Proceedings of the 39th Annual Meeting on Association for Computational Linguistics, pp. 450–457. Association for Computational Linguistics (2001)
10. Tutin, A., Grossmann, F.: Collocations régulières et irrégulières: esquisse de typologie du phénomène collocatif. Revue française de linguistique appliquée **7**(1), 7–25 (2002)

Language Resources – Short Papers

The Portuguese B²SG:
A Semantic Test for Distributional Thesaurus

Rodrigo Wilkens[(✉)], Leonardo Zilio, Eduardo Ferreira, and Aline Villavicencio

Institute of Informatics, UFRGS, Porto Alegre, Brazil
{rodrigo.wilkens,lzilio,eduardo.ferreira,avillavicencio}@inf.ufrgs.br

Abstract. The lack of availability of gold standards for evaluation of distributional thesauri is a stumbling block that prevents a direct comparison of alternative approaches in a uniform way. Here we present B²SG, a TOEFL-like task for Portuguese that contains 2,875 tests with semantic relations (synonyms, antonyms and hypernyms) for nouns and verbs. The resource is validated by comparing it with lexical resources and by human judgment. The resource was used for evaluating two distributional thesauri: one built from lemmata and the other from surface forms. The evaluation of thesauri demonstrated that the use of lemmata is slightly more accurate than the use surface forms for building distributional thesauri. B²SG is readily available for download (http://www.inf.ufrgs.br/pln/resource/B2SG.zip).

Keywords: Gold standard · Semantic relations · Distributional thesauri

1 Introduction

The importance of resources such as WordNet [5] can be measured by the number of initiatives dedicated to (re)produce them in other languages, such as the EuroWordNet [15] and the Global WordNet Association [3]. For Portuguese, such initiatives include Onto.PT [7], OpenWN-PT [13], WordNet.PT [10], WordNet.Br [4]. Manual construction of such resources is costly and time-consuming, and normally ends up with low coverage. An alternative is the automatic construction of distributional thesauri, that present semantic similarity between words and are both language independent and applicable to any domain [9].

The automatic evaluation of such thesauri is a complex task, because of the lack of resources with information on the similarity of words. Moreover, due to the large scale of the resulting thesauri the evaluation cannot be done manually by judges as it would consume too much time. An alternative is the evaluation of thesauri based on their performance on a particular task. For instance, we can approximate the concept of similarity by presenting an explicit semantic relation between words, such as the TOEFL [8] and the WordNet-Based Synonymy Test (WBST) [6] for English. For Portuguese, there are no specific gold standards for the evaluation of distributional thesauri. Here we present the BabeNet-Based

© Springer International Publishing Switzerland 2016
J. Silva et al. (Eds.): PROPOR 2016, LNAI 9727, pp. 333–339, 2016.
DOI: 10.1007/978-3-319-41552-9_33

Semantic Gold Standard (B^2SG)[1], that builds upon the WBST, and we also use this resource as a means to evaluate two distributional thesauri.

We discuss the methodology used for developing and validating B^2SG in Sect. 2. The creation and evaluation of distributional thesauri is then described in Sects. 3 and 4. Conclusions and future work are presented in Sect. 5.

2   Constructing B²SG

The BabelNet-Based Semantic Gold Standard (B²SG) contains nouns and verbs involving antonym, hypernym and synonym relations. Similar to Toefl [8] and WBST [6], for each target word the B²SG lists 4 alternatives: one semantically related word, and 3 unrelated words. For instance, for the target noun *concorrente* and the synonym relation, there are four alternatives: *competidor, cortina, amurada*, and *carmesim*, among which the correct alternative is the first.

The data set was generated in 3 steps:

1. **Selection of target words:** we used a word frequency list from AC/DC project[2] to avoid low-frequency words. The number of senses of each word was annotated using BabelNet [12], and words not found on BabelNet were excluded from B²SG.
2. **Selection of semantically related words:** for each word in the frequency list from step one we selected a set of semantically related candidates from BabelNet. We then selected the candidate with closest frequency and number of senses regarding the target word. A total of 10,000 nouns and 5,000 verbs were chosen for synonymy and hypernymy, abiding to the restriction that they were the closest in frequency[3]. The antonym category for both verbs and nouns did not present the respective minimum of 10,000 and 5,000 words, so we used all candidates, without applying a frequency filter.
3. **Selection of unrelated words**: using the list of words from Step 2 we selected, for each target word, only words without explicit relation to it. These selected words were randomly divided in groups of 3 words, and we then selected the group with closest mean frequency and mean number of senses in regard to the target word.

Using this process, we ensured that the target, related and unrelated words were close in terms of frequency and polysemy. It is also important to clarify that the same group of words were used in multiple test items, either as target, related or unrelated word. After going through these three steps, we selected a list of test items containing 4,734 target words (1,200 verbs and 3,534 nouns), as shown in Table 1.

For a validation of B²SG, the data set was first evaluated against Onto.PT [7], a thesaurus for Portuguese. As a second step, any relation that was not found in

---

[1] A preliminary version of B^2SG, yet without validation, was presented in [16].

[2] Available at http://www.linguateca.pt/ACDC.

[3] This list of 10,000 words include both target and related words.

Table 1. B²SG per relation

	Synonyms	Hypernyms	Antonyms	*Total*
Verbs	500	500	200	*1,200*
Nouns	1,667	1,667	200	*3,534*
Total	*2167*	*2167*	*400*	*4,734*

Onto.PT was manually evaluated by two native speaker human judges. As a result 25.4 % of the resource was found in Onto.PT, and another 35.3 % were validated by human judges. From the initial 4,734 relations, 60.7 % in total were considered valid, resulting in a gold standard with 2,875 validated relations.

Table 2. Semi-automatic validation

	Antonym		Synonym		Hypernym		Total
	N	V	N	V	N	V	
Initial	200	200	1667	500	1667	500	4734
Onto.PT	40	51	676	244	191	0	1202
Human Judges	105	116	495	191	568	198	1673
Total Validated	145	167	1171	435	759	198	2875
% Correct	72,5 %	83,5 %	70,2 %	87,0 %	45,5 %	39,6 %	60,7 %

As can be seen in Table 2, the validation resulted in more true positive relations for verbs than for nouns, and more for synonyms and antonyms than hypernyms. In the case of nouns, many of the false positive candidates were proper nouns (e.g. *Martinho*), letters (e.g. *c*), abbreviations (e.g. *sr.*), and foreign words (e.g. *punch* and *eau*) present in BabelNet. When these words were among unrelated alternatives, they were replaced by other candidates following the same criteria for frequency and number of senses as before. However, when they were either among the target or related words, the whole relation was removed from the resource. For hypernyms, many of the false positives were candidates evaluated by the judges as synonyms, and, as they lacked the more general meaning of a hypernym, they were removed from the resource.

3   Generating Distributional Thesauri

Since one of the possible applications of resources such as B²SG is the evaluation of distributional thesauri, we decided to make an experiment that could be seen as a baseline for future distributional thesauri for Portuguese. So here we describe the corpus and methodology we used to generate two distributional thesauri that were later evaluated against B²SG.

To obtain a large representative corpus from which to generate the distributional thesauri we combined different Portuguese corpora and used their surface forms and lemmata (Table 3) for training[4]. The whole corpus was parsed with PALAVRAS [2], so that we had access to the lemmata.

Both distributional thesauri were then generated with *word2vec* [11], using Skip-Gram with the following parameters: a vector size of 300 dimensions, a context window of size 5, a downsampling threshold of 1e-5, a sampling of 5 for the negative training algorithm, and a minimum frequency of 10 in the corpus.

Table 3. Corpus information

Corpus	Surface		Lemma	
	Types	Tokens	Types	Tokens
brWaC	812 K	166.7 M	618 K	166.5 M
EuroPal	132 K	47.8 M	67 K	47.9 M
CETEN Folha	206 K	21.2 M	120 K	21.5 M
PLN-BR	582 K	34.1 M	479 K	34.1 M
CETEM Publico	611 K	166.6 M	330 K	138.9 M
Corpus Brasileiro	2.8 M	1 G	-	-
Total	3.7 M	1.5 G	1.5 M	409 M

4   Distributional Thesauri Evaluation

The evaluation of both distributional thesauri was made by obtaining the similarity values between the target word and alternatives for each test item in B²SG. If the related word (correct alternative) had the highest similarity score among the alternatives, the answer was considered correct.

We evaluated both thesauri using 2 criteria: a strict one, in which all 5 words in a test item (target and all alternatives) had to be in the thesauri; and a non-strict one, in which at least the target and related alternative had to be present.[5] The results of the evaluation are shown in Table 4, for the surface form thesaurus, and in Table 5, for the lemmatized thesaurus.

The results obtained with both thesauri were comparable, and the thesaurus built from the lemmatized corpus performed slightly better in general, even though the corpus was more than three times smaller[6].

---

[4] The parsed corpus does not include the *Corpus Brasileiro* [1] because the lemma information is different from the one in the other corpora, since it is not parsed with PALAVRAS.

[5] The results for the 2 criteria were very similar, because both thesauri had good coverage in relation to the test items.

[6] As Pennington, Socher and Manning [14] point out, larger corpora lead to better statistics, so we can assume that a larger corpus of lemmatized forms would present even better results.

Table 4. Evaluation of the surface form thesaurus: strict and non-strict criterium

Test	Type	Strict criterium			Non-strict criterium		
		Coverage	Correct	% Correct	Coverage	Correct	% Correct
Antonym	Noun	105	90	85.7 %	145	126	86.9 %
	Verb	143	100	69.9 %	167	118	70.7 %
Hypernym	Noun	545	432	79.3 %	756	606	80.2 %
	Verbs	167	115	68.9 %	198	138	69.7 %
Synonym	Noun	861	726	84.3 %	1167	997	85.4 %
	Verb	366	275	75.1 %	433	332	76.7 %

Table 5. Evaluation of the lemmatized thesaurus: strict and non-strict criterium

Test	Type	Strict criterium			Non-strict criterium		
		Coverage	Correct	% Correct	Coverage	Correct	% Correct
Antonym	Noun	98	82	83.7 %	143	123	86,0 %
	Verb	141	110	78.0 %	167	132	79,0 %
Hypernym	Noun	525	425	81.0 %	753	615	81,7 %
	Verb	166	118	71.1 %	198	141	71,2 %
Synonym	Noun	832	721	86.7 %	1162	1025	88,2 %
	Verb	366	267	73.0 %	433	320	73,9 %

5 Conclusions and Future Work

In this paper, we described the development of B²SG, a TOEFL-like task for Portuguese. We used a lexical resource for Portuguese to extract target words that were similar in frequency and polysemy, resulting in almost five thousand test items including nouns and verbs distributed among three relation types: synonymy, antonymy and hypernymy. After automatic and manual evaluation of the resource, B²SG presents 2,875 validated test items. Using an existing lexical resource for the validation of B²SG made the evaluation faster, since more than one fourth of the gold standard could be automatically validated before passing on to human judgments.

We also used B²SG in an experiment that serves as a baseline for Portuguese distributional thesauri. The test of B²SG was applied for the evaluation of two Portuguese distributional thesauri: one from surface forms and the other from lemmatized forms. On our results, the latter was in general slightly more accurate, even if smaller, than the former.

As future work we plan to apply this methodology to build the resource for other languages, and to extend the test items to include also adjectives and adverbs.

Acknowledgments. This research was partially developed in the context of the project *Text Simplification of Complex Expressions*, sponsored by Samsung Eletrônica da Amazônia Ltda., in the terms of the Brazilian law n. 8.248/91. This work was also partly supported by CNPq (482520/2012- 4, 312114/2015-0) and FAPERGS AiMWEst.

References

1. Berber Sardinha, T., Moreira Filho, J., Alambert, E.: O corpus brasileiro. Comunicaçao ao VII Encontro de Lingüística de Corpus (2008)
2. Bick, E.: The parsing system Palavras. Automatic Grammatical Analysis of Portuguese in a Constraint Grammar Framework (2000)
3. Bond, F., Paik, K.: A survey of wordnets and their licenses. In: Proceedings of the 6th Global WordNet Conference. pp. 64–71 (2012)
4. Dias-da-Silva, B.C., Felippo, A.D., das Graças Volpe Nunes, M.: The automatic mapping of Princeton WordNet lexical-conceptual relations onto the brazilian portuguese wordnet database. In: Proceedings of LREC 2008, European Language Resources Association, Marrakech, Morocco (2008)
5. Fellbaum, C.: WordNet. Wiley Online Library, New York (1998)
6. Freitag, D., Blume, M., Byrnes, J., Chow, E., Kapadia, S., Rohwer, R., Wang, Z.: New experiments in distributional representations of synonymy. In: Proceedings of the Ninth Conference on Computational Natural Language Learning. pp. 25–32. Association for Computational Linguistics (2005)
7. Gonçalo Oliveira, H., Gomes, P.: Towards the automatic creation of a wordnet from a term-based lexical network. In: Proceedings of the ACL Workshop TextGraphs-5: Graph-based Methods for Natural Language Processing. pp. 10–18. ACL Press (July 2010). http://eden.dei.uc.pt/~hroliv/pubs/GoncaloOliveira_Gomes2010_TextGraphs5_postconf.pdf
8. Landauer, T.K., Dumais, S.T.: A solution to plato's problem: the latent semantic analysis theory of acquisition, induction, and representation of knowledge. Psychol. Rev. **104**(2), 211 (1997)
9. Lin, D.: Automatic retrieval and clustering of similar words. In: Proceedings of the 36th Annual Meeting of the Association for Computational Linguistics and 17th International Conference on Computational Linguistics - vol. 2. pp. 768–774. ACL 1998, Association for Computational Linguistics (1998)
10. Marrafa, P.: WordNet do Português: uma base de dados de conhecimento linguístico. Instituto de Camões, Lisboa (2002)
11. Mikolov, T., Karafiát, M., Burget, L., Cernocký, J., Khudanpur, S.: Recurrent neural network based language model. In: 11th Annual Conference of the International Speech Communication Association, INTERSPEECH 2010, Makuhari, Chiba, Japan, pp. 1045–1048, 26–30 September 2010
12. Navigli, R., Ponzetto, S.P.: Babelnet: building a very large multilingual semantic network. In: Proceedings of the 48th Annual Meeting of the Association for Computational Linguistics. pp. 216–225. Association for Computational Linguistics (2010)
13. de Paiva, V., Rademaker, A., de Melo, G.: OpenWordNet-PT: an open Brazilian WordNet for reasoning. In: Proceedings of the 24th International Conference on Computational Linguistics (2012). http://www.coling2012-iitb.org (Demonstration Paper). Published also asTechreport http://hdl.handle.net/10438/10274
14. Pennington, J., Socher, R., Manning, C.D.: Glove: global vectors for word representation. EMNLP **14**, 1532–1543 (2014)

15. Vossen, P. (ed.): EuroWordNet: A Multilingual Database with Lexical Semantic Networks. Kluwer Academic Publishers, Norwell (1998)
16. Wilkens, R., Zilio, L., Gonçalves, G., Ferreira, E., Villavicencio, A.: Tesauros distribucionais para o português: avaliação de metodologias. In: Proceedings of STIL 2015. Sociedade Brasileira de Computação (2015)

Automatic Generation of Internet Memes from Portuguese News Headlines

Hugo Gonçalo Oliveira$^{(\boxtimes)}$, Diogo Costa, and Alexandre Miguel Pinto

CISUC, Department of Informatics Engineering,
University of Coimbra, Coimbra, Portugal
{hroliv,ampinto}@dei.uc.pt, dcosta@student.dei.uc.pt

Abstract. This paper presents MEMEGERA, a prototype tool that generates image-based memes from Portuguese news headlines. All is done automatically, with the help of computational linguistic resources, uncovered here with the rules for selecting images and adapting the text.

Keywords: Linguistic creativity · Computational humor · Internet memes

1 Introduction

Internet memes are a current trend in social media. They intend to spread an idea through the Web, in a mix of visual and verbal message. Classic memes are generally funny and combine an image macro with a piece of text (e.g. *"One does not simply X"*, *"What if I told you Y"*). The same macro is typically used to transmit messages that fall on a reusable pattern.

This paper presents exploratory work to mimic the creation of Internet memes automatically, with Portuguese text, and applied to the scope of news. Image macros are automatically assigned to news headlines, which are then adapted to the macro pattern and pasted in the image to produce a (proto)meme.

As memes are a product of human creativity, this work is in the domain of computational creativity [1]. It is a practical application of a set of linguistic resources for the computational processing of Portuguese.

Next, we enumerate related work on linguistic creativity and computational humor. Then, we describe the general lines of MEMEGERA, the developed prototype. Before concluding, a selection of examples is presented.

2 Related Work

The Web enables the fast spreading of all kinds of ideas, in many different ways. Internet memes are artefacts commonly shared in social networks, baptised after the original definition of meme – *an idea, behavior, or style that spreads from person to person within a culture* [2].

© Springer International Publishing Switzerland 2016
J. Silva et al. (Eds.): PROPOR 2016, LNAI 9727, pp. 340–346, 2016.
DOI: 10.1007/978-3-319-41552-9_34

We see the generation of meme phrases as a challenge for natural language generation and linguistic creativity. In this scope, researchers have tackled the generation of poetry [3], slogans [4], or verbal humor, closer to meme generation due to their funny aspect, but not covering the visual dimension. Work on computational humor covers the generation of punning-riddles [5,6], funny acronyms [7], or adult humor [8]. A chat system has also been developed to suggest humorous messages, often memes, to make conversations funnier [9]. Aforementioned works are in English and we are not aware of work of this kind for Portuguese.

We should add that humor has been studied from a variety of perspectives ranging from psychology and philosophy [10], sociological aspects in literature [11], or via the computational approach [12]. It occurs when there is a break of conventionality in language so, understanding humor is also a sign of fluency (see [13] on linguistic mechanisms for verbal humor in Portuguese).

3   Memes for News Headlines

MEMEGERA processes a news headline and, automatically: (i) selects a suitable image macro from a predefined set; (ii) adapts the text according to the selected macro; and (iii) pastes the text on the macro, with the help of the Imgflip API[1]. The result is ready to be consumed or posted in a social network. News titles can be obtained automatically from the Google News RSS feed[2].

3.1   Covered Macros

A broad range of image macros is typically used as memes, each with their own semantics for transmitting a singular kind of message. We have looked both at popular memes and at a sample of news to manually identify text patterns that would suit certain macros. Currently, MEMEGERA covers the following, for which we describe the meaning, according to the *KnowYourMeme* website[3]:

- *Brace Yourselves* works as an announcement of something.
- *One Does Not Simply* points out a difficult task.
- *Not Sure If* represents an internal monologue with underlying uncertainty.
- *Success Kid* transmits a successful achievement.
- *Sad Keanu* transmits a sad event.
- *Bad Luck Brian* transmits an embarrassing event.
- *Condescending Wonka* transmits a sarcastic message.
- *Ancient Aliens* explains inexplicable phenomena as the direct result of aliens.
- *Money Money* is related to money.
- *Matrix Morpheus* reveals something unexpected.
- *Wise Confucius* gives an advice that turns out to be a pun.
- *Am I The Only One* expresses the feeling of not following a trend.
- *X, X Everywhere* points out an emerging trend.

---

[1] https://api.imgflip.com/.

[2] https://news.google.com/news?cf=all&hl=pt-PT&pz=1&ned=pt-PT_pt&output=rss.

[3] http://knowyourmeme.com/.

3.2   Linguistic Resources

In order to select a suitable meme for a news story, the headline is first part-of-speech (POS) tagged and lemmatised. A tagger based on the OpenNLP[4] toolkit is used with the Portuguese models[5], and with the LemPORT [14] lemmatiser.

To identify the sentiment of the words in the headline, we use SentiLex [15], where Portuguese words have their polarity annotated. When inflections are required to produce the resulting text, we resort to LABEL-LEX[6], a morphological lexicon for Portuguese with a large coverage of inflected word forms. When verbs need to be nominalised, this task is performed with the help of Nomlex-PT [16], a nominalisation lexicon for Portuguese.

The identification of the most relevant word in the headline is simplified by the selection of the less frequent noun, verb or adjective, according to the frequency lists of the AC/DC project [17]. The selected word has still to be in those lists. We also use the proverbs available in the scope of project Natura[7]. To measure the semantic similarity between the headline and a proverb, we compute the average similarity between the nouns, verbs and adjectives they contain, using the PMI-IR [18] method on the Portuguese Wikipedia.

3.3   Rule-Based Classifier

In order to assign one of the covered macros to a news headline and produce a meme, a classifier is run on the headline. It is currently based on a set of trigger rules over features extracted by the aforementioned linguistic resources. Table 1 displays the rules applied for each macro and the text resulting after the adaptation to the macro. Some rules are very simple, such as those for *Am I The Only One* and *X, X Everywhere*, which are only based on Portuguese trends in the Twitter network and do not even use headlines as input. All the other rules require the POS-tagging and lemmatisation of the news headline. They may rely on the occurrence of specific tokens (eg. *One Does Not Simply, Not Sure If*), linguistic constructions (eg. *Brace Yourselves, Condescending Wonka*), or sentiment-related features (eg. *Success Kid, Bad Luck Brian*).

Besides those of the table, two additional macros are used as a fallback. They are applicable to all pieces of text, although they may result in non-sense:

- *Matrix Morpheus* is used with proverbs that mention the most relevant word of the text. If there is more than one such proverb, the most semantically-similar with the headline (higher PMI) is selected.
- *Wise Confucius* is applied to headlines without matching proverbs and can be seen as an application of lexical replacement humor [8]. It selects a proverb with a termination that rhymes with the most relevant headline word r and replaces its last word with r. Ties are also solved with the PMI.

---

[4] https://opennlp.apache.org/.
[5] https://github.com/rikarudo/TagPORT.
[6] http://label.ist.utl.pt/pt/labellex_pt.php.
[7] http://natura.di.uminho.pt/~jj/pln/proverbio.dic.

Table 1. Used image macros, triggers and resulting text.

Macro	Trigger (in news headline h)	Resulting text
Brace Yourselves	h mentions an announcement, expressed by verbs in the present or future, eg.: X *preparar/pleanear/projectar/anunciar* Y	*Preparem-se/Acautelem-se/Atenção ... Y (está a chegar)*
One Does Not Simply	h refers to an unfinished action, expressed by the adverb *não* followed by a verb v, eg.: X *não* v Y	*Simplesmente não se ... v Y / Y*
Not Sure If	h contains the alternative conjunction *ou* opposing two ideas, eg.: ... X *ou* Y ...	*Não sei se X ... ou Y*
Success Kid	h either: expresses a highly positive sentiment (at least three positive words); has a negative phrase $(P-)$ followed by an adversative conjunction *conj* (*mas*) and a positive phrase $(P+)$	$P-$ *... conj* $P+$
Sad Keanu	h is highly negative because it has at least three negative words	h
Bad Luck Brian	h has a positive phrase $(P+)$ followed by an adversative conjunction *conj* (*mas*) and a negative phrase $(P-)$	$P+$ *... conj* $P-$
Condescending Wonka	h mentions someone's opinion by the linguistic constructions: X *dizer/achar/acreditar que** Y	*Então achas que Y? ... Por favor, fala-me mais sobre isso*
Ancient Aliens	h contains words in the space domain (eg. *NASA*, planet names, *extraterrestre, ovni, astronauta, espacial, ...*)	*h ... Aliens*
Money Money	h mentions a large amount of money through the expressions: *milhão de euros/dálares*	h
Am I The Only One?	Twitter trend T	*Mas serei o único ... que não está a falar sobre T?*
X, X Everywhere	Twitter trend T	*T ... fala-se sobre T em todo lado*

4 Examples

Figure 1 illustrates the results of MEMEGERA with a selection of good examples, originally posted on the Twitter social network. More examples are posted every hour by our Twitterbot, and can be found in the Twitter account *@memegera*.

Fig. 1. Examples of produced and published memes of different types.

5   Concluding Remarks

In addition to the research challenges involved, in the scope of natural language generation and computational creativity, MEMEGERA provides an alternative funny way of reading current news, through the @memegera Twitter feed. Although most of the humor value is provided by the image macros, the memes produced so far show that we are heading towards the right direction.

In the future, besides increasing the number of covered macros and improving the classifier with more fine-grained rules, we are planning to perform an evaluation of the results. It should cover the dimensions of syntactic and semantic coherence, suitability for the news, humor value and surprise level. Given the subjectivity of this task, we will rely on several human judgements.

Acknowledgements. This work was supported by the project ConCreTe. The project ConCreTe acknowledges the financial support of the Future and Emerging Technologies (FET) programme within the Seventh Framework Programme for Research of the European Commission, under FET grant number 611733.

References

1. Colton, S., Wiggins, G.A.: Computational creativity: the final frontier? In: Proceedings of the 20th European Conference on Artificial Intelligence (ECAI 2012), Montpellier, France, pp. 21–26. IOS Press (2012)
2. Dawkins, R.: The Selfish Gene. Oxford University Press, Oxford (1976)
3. Gervás, P.: An expert system for the composition of formal Spanish poetry. J. Knowl. Based Syst. **14**, 181–188 (2001)
4. Tomašič, P., Žnidaršič, M., Papa, G.: Implementation of a slogan generator. In: Proceedings of the 5th International Conference on Computational Creativity, ICCC 2014, Ljubljana, Slovenia, pp. 340–343, June 2014
5. Binsted, K., Ritchie, G.: An implemented model of punning riddles. In: Proceedings of the 12th National Conference on Artificial Intelligence, AAAI 1994, vol. 1, Menlo Park, CA, USA, pp. 633–638. AAAI Press (1994)
6. Manurung, R., Ritchie, G., Pain, H., Waller, A., O'Mara, D., Black, R.: The construction of a pun generator for language skills development. Appl. Artif. Intell. **22**(9), 841–869 (2008)
7. Stock, O., Strapparava, C.: The act of creating humorous acronyms. Appl. Artif. Intell. **19**(2), 137–151 (2005)
8. Valitutti, A., Toivonen, H., Doucet, A., Toivanen, J.M.: Let everything turn well in your wife: generation of adult humor using lexical constraints. In: Proceedings of the 51st Annual Meeting of the Association for Computational Linguistics, vol. 2, Sofia, Bulgaria, pp. 243–248. ACL Press, August 2013
9. Wen, M., Baym, N., Tamuz, O., Teevan, J., Dumais, S., Kalai, A.: OMG UR funny! computer-aided humor with an application to chat. In: Proceedings of the 6th International Conference on Computational Creativity, ICCC 2015, Park City, Utah, pp. 86–93. Brigham Young University, June–July 2015
10. Morreall, J.: Philosophy of humor. In: Zalta, E.N. (ed.) The Stanford Encyclopedia of Philosophy, Spring 2013 edn. (2013)

11. Kuipers, G.: Humor styles and symbolic boundaries. J. Literary Theor. **3**, 219–239 (2010)
12. Suslov, I.M.: Computer model of a "sense of humour". I. general algorithm. Biophysics **37**(2), 242–248 (1992)
13. Tagnin, S.E.O.: O humor como quebra da convencionalidade. Rev. Bras. de Lingüística Apl. **5**(1), 247–257 (2005)
14. Rodrigues, R., Gonçalo Oliveira, H., Gomes, P.: LemPORT: a high-accuracy cross-platform lemmatizer for Portuguese. In: Proceedings of the 3rd Symposium on Languages, Applications and Technologies (SLATE 2014), Bragança, Portugal, pp. 267–274. OASICS, Schloss Dagstuhl (2014)
15. Silva, M.J., Carvalho, P., Sarmento, L.: Building a sentiment lexicon for social judgement mining. In: Caseli, H., Villavicencio, A., Teixeira, A., Perdigão, F. (eds.) PROPOR 2012. LNCS, vol. 7243, pp. 218–228. Springer, Heidelberg (2012)
16. de Paiva, V., Real, L., Rademaker, A., de Melo, G.: Nomlex-pt: a lexicon of Portuguese nominalizations. In: Proceedings of the 9th International Conference on Language Resources and Evaluation (LREC 2014), Reykjavik, Iceland. ELRA, May 2014
17. Santos, D., Bick, E.: Providing Internet access to Portuguese corpora: the AC/DC project. In: Proceedings of the 2nd International Conference on Language Resources and Evaluation, LREC 2000, pp. 205–210 (2000)
18. Turney, P.D.: Mining the web for synonyms: PMI-IR versus LSA on TOEFL. In: De Raedt, L., Flach, P. (eds.) ECML 2001. LNCS, vol. 2167, pp. 491–502. Springer, Heidelberg (2001)

FrameNet-Based Automatic Suggestion of Translation Equivalents

Simone Peron-Corrêa, Alexandre Diniz, Meire Lara, Ely Matos, and Tiago Torrent[✉]

Universidade Federal de Juiz de Fora, FrameNet Brasil, Rua José Lourenço Kelmer, s/n°,
Faculdade de Letras, sl.1411, Juiz de Fora, Brazil
speronjf@yahoo.com.br, alexdiniz5@gmail.com,
meire.s.lara@gmail.com, {ely.matos,tiago.torrent}@ufjf.edu.br
http://www.framenetbr.ufjf.br

Abstract. This paper presents an application developed for automatically suggesting translation equivalents in a frame-based domain specific trilingual electronic dictionary covering the domains of the World Cup and Tourism. By comparing the syntactic and semantic affordances of a lexical unit in the source language with those shown by all lexical units evoking the same frame in a target language, the application suggests which of them is the best-fit translation equivalent. The application contributes to the purpose of bringing to scale the development of frame-based multilingual lexical databases. We discuss the current limitations of the application, especially those regarding entity nouns, and propose the use of ontologies and qualia structure as a means of enhancing machine translation for entity nouns.

Keywords: FrameNet · Machine translation · Ontology · Qualia structure

1 Introduction

FrameNet Brasil has been developing multilingual computational applications for human users based on Frame Semantics [1, 2]. In the FrameNet Brasil World Cup Dictionary (WCD) [3], frames were used as interlingual representations connecting the lexica of tourism and the World Cup in Brazilian Portuguese (PTB), English (EN) and Spanish (ES) [4]. FrameNet (FN) valences can provide relevant information for automatically suggesting terminologically accurate translation equivalents for verbs and nouns denoting events [5]. However, for nouns denoting entities (i.e. objects, people, places etc.), valences are unable to differentiate lexical units. This paper aims (i) to present such a computational solution and (ii) to discuss how to enrich FN with

---

Authors are thankful to FAPEMIG (Grants # CHE-APQ-00567-12 and CHE-APQ-00471-15) and CNPq (Grant # 448990/2014-8) for funding the development of the applications described in this paper. The presentation of this paper at PROPOR 2016 was made possible by FAPEMIG (Grant # PEE-00499-16).

J. Silva et al. (Eds.): PROPOR 2016, LNAI 9727, pp. 347–352, 2016.
DOI: 10.1007/978-3-319-41552-9_35

ontologies [6] and qualia structure [7], so that it can be used as a semantically grounded tool for enhancing the results of machine translation algorithms [8, 9].

2   Frame-Based Suggestion of Translation Equivalents

In WCD, 128 frames created for representing the Football World Cup and Tourism served as an interlingua connecting 1.225 lexical units (LUs) in three languages. Since none of the domains covered by the application vary across cultures in a way that calls for the proposition of language-specific frames [4, 5], the adoption of interlingual frames was possible.

LUs, the pairing between one lemma and one frame, are associated to the frames they evoke and to annotation sets containing sentences extracted from corpora. Sentences are annotated for the Frame Elements (FEs) in the evoked frame and also for the Grammatical Functions (GFs) and Phrase Types (PTs) of the linguistic material instantiating the FEs. The set of semantic and syntactic affordances of each LU forms its valence description.

Unlike FEs, GFs and PTs used in the dictionary are not the same across languages. GF and PT labels for each language were defined by their respective FN projects [10–12]. The result of the language-specific annotation is summarized in patterns, as shown in Fig. 1.

Número Anotado [Number Annotated]	Patterns				
1 TOTAL	Attraction	Attraction	Place	Time	Tourist
(1)	NP Obj	NP Obj	PP Dep	AVP Dep	NP Ext
1 TOTAL	Attraction	Depictive	Tourist		
(1)	NP Obj	Srel Dep	NP Ext		
1 TOTAL	Attraction	Place	Purpose	Tourist	
(1)	NP Obj	PP Dep	VPto Dep	CNI --	
5 TOTAL	Attraction	Place	Tourist		
(2)	NP Obj	AVP Dep	NP Ext		
(1)	NP Obj	PP Dep	NP Ext		
(2)	NP Obj	VPed Dep	NP Ext		

Fig. 1. Valence Patterns for *visit.v* in English.

2.1   Automatic Suggestion of Translation Equivalents in WCD

For automatically suggesting translation equivalents for an LU, WCD uses as input: (i) the frame evoked by the LU, (ii) its valence patterns and (iii) other LUs evoking the same interlingual frame. Annotated sentences provide a set of data regarding how each LU organizes the FEs in its local syntax. The system uses this information to assess the similarities and differences in the valence patterns. The main claim is: the more similar the valence descriptions of two words are, the greater the possibility of them being good translation equivalents for each other in that specific domain.

WCD compares the core FEs instantiated in the sentences containing the LU, as well as the GFs and PTs of each FE. Since GF and PT labels are language specific, correspondence tables provide the bases for comparison. Each label from a language is associated with a label of the other language. Different weights are applied for GFs pairs to account for their prominence in the syntactic valence and for structural differences between the languages. For instance, FN Brasil differentiates Direct Objects, Indirect Objects and Dependents [12] while Berkeley FN only makes a difference between (Direct) Objects and Dependents [10].

Given a Source LU (the LU in the source language), the evaluation algorithm compares its valence descriptions to valence descriptions of Candidate LUs (LUs evoking the same frame in the target language). The algorithm chooses the Candidate LU with the valence description most similar to that of the Source LU as the best-fit translation equivalent (the Target LU). To illustrate, consider a Source LU and two Candidate LUs, with two different valence patterns each one. The algorithm would carry out eight valence comparisons (four for each Candidate LU). Each comparison receives a partial score. The partial score is added in one point for each exact equivalence found between the valence patterns. The final score is the division of the partial score by the total of possible points that would be added if valence patterns were 100 % correspondent to each other. After the comparison, the algorithm ascribes a final average score for each Candidate LU. The Candidate LU presenting the highest final score is then suggested as the Target LU.

The Touring frame models the experience of a TOURIST at an ATTRACTION or PLACE. In this frame, one would find the following LUs for EN and PTB:

– EN: enjoy.v, see.v, tour.v, tour.n, tourist.n, visit.v, visitor.n
– PTB: apreciar.v, conhecer.v, desfrutar.v, tour.n, turista.n, visitante.n, visitar.v

Table 1. Average scores for each PTB Candidate LU in regard to EN Source LUs evoking the Touring frame. Variance is presented in brackets below the average score.

	apreciar.v	conhecer.v	desfrutar.v	visitar.v	tour.n	turista.n	visitante.n
enjoy.v	0.523	0.435	0.453	0.384	0.342	0.250	0.253
	(0.041)	(0.029)	(0.036)	(0.023)	(0.013)	(0.000)	(0.001)
see.v	0.500	0.437	0.405	0.371	0.328	0.250	0.252
	(0.039)	(0.025)	(0.034)	(0.023)	(0.015)	(0.000)	(0.000)
tour.v	0.432	0.350	0.397	0.390	0.296	0.250	0.249
	(0.048)	(0.042)	(0.043)	(0.040)	(0.017)	(0.000)	(0.000)
visit.v	0.498	0.423	0.419	0.412	0.294	0.250	0.251
	(0.043)	(0.037)	(0.032)	(0.036)	(0.014)	(0.000)	(0.000)
tour.n	0.249	0.207	0.389	0.215	0.541	0.250	0.255
	(0.011)	(0.006)	(0.025)	(0.005)	(0.054)	(0.000)	(0.002)
tourist.n	0.303	0.325	0.267	0.298	0.258	0.738	0.712
	(0.001)	(0.003)	(0.001)	(0.001)	(0.000)	(0.005)	(0.016)
visitor.n	0.298	0.318	0.263	0.297	0.255	0.738	0.712
	(0.002)	(0.004)	(0.001)	(0.002)	(0.001)	(0.005)	(0.016)

Table 1 presents the final scores (and the corresponding variance) for each LU in this frame in the EN-PTB direction. The initial idea is that the higher the score, the better the translation equivalent suggested. However, variance is important. Because the score for each Candidate LU is an average of all scores ascribed individually to each sentence instantiating the LU, a lower variance indicates less discrepancy between the scores. If two Candidate LUs, e.g. A and B, present very close average scores and B presents lower variance, B may be a better translation equivalent to the Source LU than A, because it presents similar valence patterns in the majority of its annotated sentences, as opposed to presenting some sentences with a very high score and some others with a very low score. The results displayed in Table 1 support the following claims:

1. There is a clear-cut difference between entity nouns (*tourist.n, visitor.n, turista.n* and *visitante.n*), eventive nouns (*tour.n* – in both languages), and verbs.
2. Scores for the word *tour.n* both in EN and PTB indicate a very clear equivalence between eventive nouns in the two languages. The higher variance points to discrepancies in valences, because in EN a FE such as ATTRACTION can be instantiated before the noun while in PTB this FE must be instantiated after the noun.
3. Because scores between verb sources and verb candidates can be very similar to each other, a rank is built by dividing scores by variances. All high-ranked candidates are shown to the user, from the most general to the least general equivalent.

2.2 Advantages and Limitations

The solution proposed has the advantage of being automatic and easily extendable to other languages or other frame-based multilingual applications. The only adaptation needed would be the inclusion of equivalences between the GFs and PTs defined for those languages in the database. In this sense, it contributes for the purpose of bringing to scale the development of frame-based multilingual lexical databases.

On the other hand, four important limitations must be considered: (i) the dictionary is domain-specific and it may not be fully extendable to general vocabularies; (ii) the influence of the model over the application, since if different frames were defined, different translation equivalences would be suggested; (iii) the partial correspondence between some GFs and PTs must be checked against more data and (iv) the low relevance of valence descriptions for noun targets evoking entity frames.

Solutions for (i) and (iii) can be investigated in the future, when the application is extended to other domains and languages. Gamonal [4] discusses a methodology for reducing the problem in (ii). A solution to (iv) must include alternative modeling approaches for entity nouns, which will be outlined next.

3   Towards Semantically-Enhanced Machine Translation

Qualia structure can add semantic information to a computational lexicon. Pustejovsky [7] proposes four qualia roles: (a) formal, distinguishing the entity in a larger or more general domain; (b) constitutive, expressing the relationship between an entity and its constituents; (c) telic, indicating the function or purpose of the entity, and (d) agentive, indicating the factors involved in the origin of an entity. The use of qualia structures allows for the ontological organization of the lexicon and provides an applicable semantic and computational representation model, capable of differentiating entity nouns evoking one same frame. Bilingual examples (1–3) demonstrate how telic roles can be used to disambiguate entity nouns.

(1) The [striker$_{\text{FE:POSITION/TELIC:SCORE\_GOAL}}$] was suspended for two matches.
O [atacante$_{\text{FE:POSITION/TELIC:SCORE\_GOAL}}$] foi suspenso por dois jogos.
(2) The [defender$_{\text{FE:POSITION/TELIC:MARK\_PLAYER}}$] was suspended for two matches.
O [zagueiro$_{\text{FE:POSITION/TELIC: MARK\_PLAYER}}$] foi suspenso por dois jogos.
(3) The [goalkeeper$_{\text{FE:POSITION/TELIC:AVOID\_GOAL}}$] was suspended for two matches.
O [goleiro$_{\text{FE:POSITION/TELIC:AVOID\_GOAL}}$] foi suspenso por dois jogos.

The FE POSITION is the same for the three different player positions, but their telic qualia roles are different. In fact, these roles correspond to frames in the WCD database. If one models qualia roles as relations between LUs and frames, one can disambiguate entity nouns evoking one same frame. Hence, a system consulting such an enriched database could suggest translation equivalents even when the valence descriptions of the LUs are not informative.

A lexicographic database focused on domain-specific terminology would not include proper nouns. However, ontologies [6] can be used to relate entity and proper nouns to categories and instances, i.e. it is possible to associate the proper noun *Neymar* with a category *player* and to store the property of *Neymar* being a *striker.n*. This information can be associated to both frames and qualia structure, which could positively impact hybrid machine translator algorithms [9]: after generating the statistically based translation alternatives for a given sentence, the algorithm could consult the best-fit equivalents suggested by the ontology-qualia-structure-enriched frame database. This is especially relevant for sentences like (5–6), which feature the same verbal lexeme in PTB, but two different translations in EN.

(4) Neymar **marcou** dois. | *Neymar **scored** two [goals]*.
(5) Cada zagueiro alemão **marca** dois. | *Each German defender **marks** two*.

Although (4–5) refer to very different frames (Score_goal and Mark_player, respectively) the statistically grounded algorithm in Google Translate translates (5) as *Each German defender scores two*. By querying the enriched frame database for the automatically generated translation equivalents, the algorithm could "learn", using an ontology [13], that Neymar is a striker (the PLAYER FE who scores the GOAL FE in the Score_goal frame). On the other hand, it could also infer that the Telic Role of *zagueiro.n* is precisely Mark_player. Thus, an EN verb evoking this frame and featuring the DEFENDER FE in the Subject position of its valence, such as *mark.v* would be a best-fit translation equivalent for *marcar.v*.

4   Conclusion

In this paper we showed how valence patterns can be used for automatically suggesting translation equivalents for verbs and eventive nouns. After discussing the limitations of such an application, we argued that enriching FrameNet with qualia structure and ontologies could improve machine translation for entity nouns as well.

References

1. Fillmore, C.J.: Frame semantics. In: The Linguistics Society of Korea (ed.): Linguistics in the Morning Calm, pp. 111–137. Hanshin Publishing Co., Seoul (1982)
2. Fillmore, C.J.: Frames and the semantics of understanding. Quaderni di Semantica **6**(2), 222–254 (1985)
3. Torrent, T.T., Salomão, M.M., Campos, F.C.A., Braga, R.M.M., Matos, E.E.S., Gamonal, M.A., Gonçalves, J.A., Souza, B.C.P., Gomes, D.S., Peron, S.R.: Copa 2014 FrameNet Brasil: a frame-based trilingual electronic dictionary for the Football World Cup. In: Proceedings of COLING 2014, The 25th International Conference on Computational Linguistics: System Demonstrations, Dublin, Ireland, pp. 10–14. ACL (2014)
4. Gamonal, M.A.: Copa 2014 FrameNet Brasil: diretrizes para a constituição de um dicionário eletrônico trilíngue a partir da análise de frames da experiência turística. M.A. Thesis in Linguistics. Universidade Federal de Juiz de Fora, Juiz de Fora (2013)
5. Peron-Corrêa, S.R.: Copa 2014 FrameNet Brasil: *frames* secundários em unidades lexicais evocadoras da experiência turística em português e em espanhol. M.A. Thesis in Linguistics. Universidade Federal de Juiz de Fora, Juiz de Fora (2014)
6. Gruber, T.: Ontology. In: Liu, L., Tamer Özsu, M. (eds.): Encyclopedia of Database Systems, pp. 1963–1965. Springer, New York (2009)
7. Pustejovsky, J.: The Generative Lexicon. MIT Press, Cambridge (1995)
8. Koehn, P.: Statistical Machine Translation. Cambridge University Press, Cambridge (2009)
9. Koehn, P., Hoang, H. Birch, A., Callison-Burch, C., Frederico, M., Bertoldi, N., Cowan, B., Shen, W., Moran, C., Zens, R., Dyer, C., Bojar, O., Constantin, A., Herbst, E.: Moses: open source toolkit for statistical machine translation. In: Proceedings of the Annual Meeting of the Association for Computational Linguistics, Demonstration Session. ACL, Prague (2007)
10. Ruppenhofer, J., Ellsworth, M., Petruck, M.R.L., Johnson, C.R., Scheffczyk, J.: FrameNet II: Extended Theory and Practice. ICSI, Berkeley (2010)
11. Spanish FrameNet. http://sfn.uab.es:8080/SFN/
12. Torrent, T.T., Ellsworth, M.: Behind the Labels: criteria for defining analytical categories in FrameNet Brasil. Veredas **17**(1), 44–65 (2013)
13. Bouayad-Agha, N., Casamayor, G., Wanner, L., Díez, F., López Hernández, S.: FootbOWL: using a generic ontology of football competition for planning match summaries. In: Antoniou, G., Grobelnik, M., Simperl, E., Parsia, B., Plexousakis, D., De Leenheer, P., Pan, J. (eds.) ESWC 2011, Part I. LNCS, vol. 6643, pp. 230–244. Springer, Heidelberg (2011)

Building a Question-Answering Corpus Using Social Media and News Articles

Paulo Cavalin$^{(\boxtimes)}$, Flavio Figueiredo, Maíra de Bayser, Luis Moyano,
Heloisa Candello, Ana Appel, and Renan Souza

IBM Research, São Paulo, Brazil
pcavalin@br.ibm.com

Abstract. Is it possible to develop a reliable QA-CORPUS using social media data? What are the challenges faced when attempting such a task? In this paper, we discuss these questions and present our findings when developing a QA-CORPUS on the topic of Brazilian finance. In order to populate our corpus, we relied on opinions from experts on Brazilian finance that are active on the Twitter application. From these experts, we extracted information from news websites that are used as answers in the corpus. Moreover, to effectively provide rankings of answers to questions, we employ novel word vector based similarity measures between short sentences (that accounts for both questions and Tweets). We validated our methods on a recently released dataset of similarity between short Portuguese sentences. Finally, we also discuss the effectiveness of our approach when used to rank answers to questions from real users.

Keywords: Question and Answer · Social media · Finance

1 Introduction

The availability of corpora to drive and sustain Question-Answering (QA) systems [8] is of fundamental importance. Such corpora are generally obtained from various sources, normally large collections of text, such as online news [1], or Wikipedia [6]. Some authors have put forward the advantages of using social media in the construction of certain types of corpora, e.g., in [3], the authors propose building comparable corpora from social networks, in particular, Twitter. In the same line, the authors of [5] propose using Twitter as an alternative in short sentences settings. Finally, authors in [7] address QA in social media mainly as a characterization effort on the type and frequency of questions and answers that may be found in Twitter, even though they don't address the construction of a corpus nor the targeting of any specific domain.

Motivated by the above setting, we study the potential of using social media data to create a QA corpus for specific fields. We refer to the corpus simply as QA-CORPUS. In details, we study the viability of how can social media information be explored to create a QA-CORPUS on the topic of Brazilian finance.

© Springer International Publishing Switzerland 2016
J. Silva et al. (Eds.): PROPOR 2016, LNAI 9727, pp. 353–358, 2016.
DOI: 10.1007/978-3-319-41552-9_36

To our knowledge, this is the first work to study the possibility of building a domain-specific corpus from social media. We believe that social media data can provide a proxy to field experts who can provide timely, possibly spam free (using the right techniques), easy to understand, reliable information [3,5,11]. As previous work have showed, this information can be acquired by the careful choice of domain experts to follow in magazines, newspapers, or in a social media application such as Twitter [11]. In this sense, using social media we can bypass or tackle a major bottleneck on the creation of a QA-CORPUS, of the gathering of candidate and reliable answers to possible user questions.

We combine the use of social media data together with novel, word vector based [4], short-sentence similarity measures to create a QA-CORPUS that can answer user question in free text form. Using the method of [4], we match questions on Twitter, containing a URLs pointing out to a news article with the supposed answer, to real user questions.

We can summarize the major contribution of this work as the creation and evaluation of a method that automatically creates a QA-CORPUS using social media data, based only on a seed set of Twitter ids. Using a novel similarity measure, this corpus can be used to provide answer to questions from real world users. We evaluate our steps on a recently released dataset of similarity between short Portuguese sentences. More importantly, we also show how QA-CORPUS provides significant answers to questions provided from a user study performed by us. Even though we deal with texts in Portuguese and for the financial domain as a case study, the system here described can be easily applied to other languages and domains.

2   QA-Corpus Creation

In this section we describe the steps we took in building our QA-CORPUS using social media data. Starting with Fig. 1a, we present an overview of the main workflow of our QA-CORPUS creation method. The main input is a list of social media accounts of experts of a given subject.

Given these users, the system collects all posts they submit on the social media service, for instance Twitter, and save them into a database, namely *Tweets DB*. Then, the module *Find Questions*, with the aid of the *Question Classifier*, finds all tweets that contain a question, and saves them into *Question tweets*. Finally, the *Get answers* module extracts the URL[1] of the news article linked in the text, and saves both the question tweet and its corresponding news article that answer the question into the QA-CORPUS.

Our first approach to develop the question classifier was to train a supervised classifier to identify questions from tweets. However, as it has been discussed by previous work, simple heuristics can achieve over 90 % accuracy [7] given the short nature of microblogging texts. Therefore, we considered a text as a question if it contains the question mark symbol ('?'). In addition, in order to populate our corpus with answers, we consider that any URL that follows a question on the tweet text is a candidate answer to that question.

---

[1] Uniform Resource Locator.

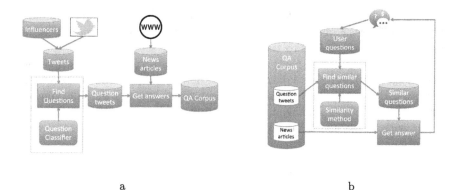

a   b

Fig. 1. In (a), the main workflow of the QA-CORPUS creation method making use of social media posts and news articles from the web. And in (b), an illustration of the use of the QA-CORPUS in a question answering system.

2.1 Social Media Data Extraction

To generate a corpus of social posts that is used as input for the QA-CORPUS creation method, we have manually-selected a list with 104 Twitter users that are considered experts in Brazilian finance. This list includes journalists, bloggers, and professors. By means of the Twitter REST API, up to 30 November 2015, 184,001 posts could be collected. After applying the question classifier on these posts, the resulting QA-CORPUS contained a total of 18,491 pairs of questions and answers, which corresponds to 10 % of the total of posts.

3 Using QA-Corpus for a QA System

In this section, we present the results of using the QA-CORPUS. described in Sect. 2.

For doing so, we rely on real user questions extracted from our alpha version QA financial application. In details, we performed a user study with 7 users, following the Wizard of Oz protocol [2], and extracted 124 questions on the topics of two savings investments available in most Brazilian banks: Savings Account and CDB (Bank Deposit Certificate) investments.

Moreover, we employ a word vector based similarity measure to find answers to questions on QA-CORPUS. We validate the similarity measure on a dataset of similarity scores between pairs of Portuguese sentences. Finally, we also show how this method provides accurate answers to the real user questions.

3.1 QA System

Figure 1b depicts our proposed QA system based on the use of the QA-CORPUS created in Sect. 2. The method works as follows. Given a question from the *User questions* database, the module *Find similar questions*, with the aid of *Similarity method*, looks for the question tweets which are the most similar ones to the user question, and saves them into the *Similar questions* set. Next, the *Get answer* module gets corresponding news article that are related to the similar questions found in the *Question tweets* dataset. These news articles are then retrieved.

It is worth mentioning that all texts are pre-processed in the following way. First, the text is case-normalized, then it is tokenized. Next, all hashtags (tokens starting with an #) and URLs are removed.

3.2 Similarity Method

Our similarity method is based on the approach described in [4], making use of word vector representations, which is currently a well-known deep learning approach to extract the semantic meaning of the words [9,10]. In this work, though, a regression model has been trained instead of a binary classifier, so that continuous values of similarity can be used to rank the most similar sentences. In this case, texts with higher values are considered as more similar.

The regression model has been trained in the ASSIN similarity dataset, which has been released as part of the PROPOR Semantic Similarity and Textual Inference[2].

We made use of the set of 3,000 pairs of sentences in Brazilian portuguese to train a Support Vector Regression (SVR) model by considering a 30-dimensional feature set defined with both domain-independent data, from word vectors created with most recent dump of the Wikipedia in Portuguese[3], plus domain-specific data coming from all of the 184,001 originally crawled tweets and 64,646 news articles.

Based on a 70/30 division of the set, the SVR achieved Person correlation of 0.61, and mean squared error of 0.47. By considering a 3.5 threshold to convert the set into a binary classification problem, an SVM trained with the same configuration reached an F-score of 0.65.

3.3 Results

In Fig. 2, we present the histogram according to the number of similar questions found for the 124 user questions. We can observe that, for the largest portion of questions, i.e. 47 questions or 38 %, 0 to 4 similar questions are found in the dataset. From these, the system has not been able to find any similar question for 19 questions (40 %). On the other tail, we observe that 39 questions had more than 20 similar questions, being in some cases very large sets ranging from 105 to 214 similar questions.

In order to better understand the results, we have also conducted a qualitative analysis regarding the most similar questions that were found for each user

---

[2] ASSIN: Avaliação de Similaridade Semântica e Inferência Textual - http://propor2016.di.fc.ul.pt/?page_id=381.

[3] Dump of 12 December 2015.

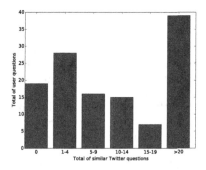

Fig. 2. Histogram of the number of similar questions found for the user questions.

question. We observe that the QA system can find pretty good matches when the question is very clear and direct, such as *What is CDB?*. Even when the question is not very clear, for instance *o rendimento em poupança é melhor*, the system has been able to point out some similar questions that might link to an answer. In some cases such as *qual a diferença entre poupança e cdb?* and *Então em curto prazo a poupança é mais rentável?* show that, even though the similarity method has captured the main meaning behind the question, lack of data probably contributed to not bringing any accurate similar question. Finally, no similar question was found for *investimento na poupana é seguro, mas existem outras opções que também são de baixo risco, mas com rentabilidade melhor.*, and the reason is not straightforward. This may have happend because: (a) the question is not very clear; (b) the text is too long; (c) lack of data; or (d) it is not a question but only a opinion.

4 Conclusions and Future Work

In this paper we presented a methodology to automatically create a QA-CORPUS on the topic of Brazilian finance. The QA-CORPUS is built through the use of social media data. We employ novel deep-learning based similarity measures to match questions from users and rank candidate answers. We validate our method on a novel dataset, as well as present a qualitative discussion of how our QA-CORPUS can benefit real users for a financial advisor application.

As we have shown, with simple heuristics and state-of-the-art similarity measures, we can create the QA-CORPUS. Nevertheless, one direction for future work is to improve the selection of social media posts to be included in QA-CORPUS. Currently, we make use of every post in the form of a question. The aim is to increase the possibility of finding better answers to more specific user questions. Also, investigating other similarity measures for comparisons is a promising task for further study.

References

1. Dolan, B., Quirk, C., Brockett, C.: Unsupervised construction of large paraphrase corpora: exploiting massively parallel news sources. In: Proceedings of the 20th International Conference on Computational Linguistics, p. 350. Association for Computational Linguistics (2004)
2. Dow, S.P., Mehta, M., MacIntyre, B., Mateas, M.: Eliza meets the wizard-of-oz: blending machine and human control of embodied characters. In: Proceedings of the SIGCHI Conference on Human Factors in Computing Systems, pp. 547–556. ACM (2010)
3. Hajjem, M., Trabelsi, M., Latiri, C.: Building comparable corpora from social networks. In: BUCC, 7th Workshop on Building and Using Comparable Corpora, LREC, Reykjavik, Iceland (2013)
4. Kenter, T., de Rijke, M.: Short text similarity with word embeddings. In: CIKM 2015: 24th ACM Conference on Information and Knowledge Management. ACM, October 2015
5. Ljubešic, N., Fišer, D., Erjavec, T.: Tweet-cat: a tool for building twitter corpora of smaller languages. In: Proceedings of the Ninth International Conference on Language Resources and Evaluation (LREC 2014), Reykjavik, Iceland. European Language Resources Association (ELRA) (2014)
6. Nothman, J., Murphy, T., Curran, J.R.: Analysing wikipedia and gold-standard corpora for ner training. In: Proceedings of the 12th Conference of the European Chapter of the Association for Computational Linguistics, pp. 612–620. Association for Computational Linguistics (2009)
7. Paul, S., Hong, L., Chi, E.: Is twitter a good place for asking questions? a characterization study. In: International AAAI Conference on Web and Social Media (2011)
8. Singh, V., Dwivedi, S.K.: Question answering: a survey of research, techniques and issues. Int. J. Inf. Retrieval Res. (IJIRR) 4(3), 14–33 (2014)
9. Socher, R., Chen, D., Manning, C.D., Ng, A.: Reasoning with neural tensor networks for knowledge base completion. In: Advances in Neural Information Processing Systems, pp. 926–934 (2013)
10. Socher, R., Huang, E.H., Pennin, J., Manning, C.D., Ng, A.Y.: Dynamic pooling and unfolding recursive autoencoders for paraphrase detection. In: Advances in Neural Information Processing Systems, pp. 801–809 (2011)
11. Zafar, M.B., Bhattacharya, P., Ganguly, N., Gummadi, K.P., Ghosh, S.: Sampling content from online social networks: comparing random vs. expert sampling of the twitter stream. ACM Trans. Web (TWEB) 9(3), 12 (2015)

Speech Processing – Full Papers

Evaluating Phonetic Spellers for User-Generated Content in Brazilian Portuguese

Gustavo Augusto de Mendonça Almeida, Lucas Avanço,
Magali Sanches Duran, Erick Rocha Fonseca[(✉)],
Maria das Graças Volpe Nunes, and Sandra Maria Aluísio

Interinstitutional Center for Computational Linguistics (NILC),
Institute of Mathematical and Computer Sciences
University of São Paulo, São Paulo, Brazil
gustavoauma@gmail.com, avanco89@gmail.com,
magali.duran@uol.com.br, erickrfonseca@gmail.com,
{gracan,sandra}@icmc.usp.br

Abstract. Recently, spell checking (or spelling correction systems) has regained attention due to the need of normalizing user-generated content (UGC) on the web. UGC presents new challenges to spellers, as its register is much more informal and contains much more variability than traditional spelling correction systems can handle. This paper proposes two new approaches to deal with spelling correction of UGC in Brazilian Portuguese (BP), both of which take into account phonetic errors. The first approach is based on three phonetic modules running in a pipeline. The second one is based on machine learning, with soft decision making, and considers context-sensitive misspellings. We compared our methods with others on a human annotated UGC corpus of reviews of products. The machine learning approach surpassed all other methods, with 78.0 % correction rate, very low false positive (0.7 %) and false negative rate (21.9 %).

1 Introduction

Spell checking is a very well-known and studied task of natural language processing (NLP), being present in applications used by the general public, including word processors and search engines. Most of spell checking methods are based on large dictionaries to detect non-words, mainly related to typographic errors caused by key adjacency or fast key stroking. Currently, with the recent boom of mobile devices with small touchscreens and tiny keyboards, one can miss the keystrokes, and thus spell checking has regained attention [1].

Dictionary-based approaches can be ineffective when the task is to detect and correct spelling mistakes which coincidentally correspond to existing words (real-word errors). Different from non-word errors, real-word errors generate variant forms that are ambiguous with other words in the language and must be addressed considering context. For instance, in "eu vou comprar" (*I will buy-*INF), if the last character from "comprar" (*to buy-*INF) is deleted, it will produce "compra" (*buys-*PRES.3SG) which is also a valid verb form in BP. Therefore,

© Springer International Publishing Switzerland 2016
J. Silva et al. (Eds.): PROPOR 2016, LNAI 9727, pp. 361–373, 2016.
DOI: 10.1007/978-3-319-41552-9_37

disambiguation must be performed taking into account the surrounding tokens: it is rare to have an inflected verb form after the auxiliary "vou", much more likely is to observe an infinitive in such context to make the compound future tense "vou comprar" (will buy).

Several approaches have been proposed to deal with these errors: mixed trigram models [2], confusion sets [3], improvements on the trigram-based noisy-channel model [4,5], use of GoogleWeb 1T 3-gram data set and a normalized and modified version of the Longest Common Subsequence string matching algorithm [6], a graph-based method using contextual and PoS features and the double metaphone algorithm to represent phonetic similarity [7]. As an example, although MS Word (from 2007 version onwards) claims to include a contextual spelling checker, an independent evaluation of it found high precision but low recall in a sample of 1400 errors [8]. Hunspell [9], the open-source speller for LibreOffice, uses n-gram similarity, rules and dictionary based pronunciation data in order to provide suggestions for spelling errors.

Errors due to phonetic similarity also impose difficulties to spell checkers. They occur when a writer knows well the pronunciation of a word but not how to spell it. This kind of error requires new approaches to combine phonetic models and models for correcting typographic and/or real-word errors. In [10], for example, the authors use a linear combination of two measures – the Levenshtein distance between two strings and the Levenshtein distance between their Soundex [11] codes.

In the last decade, some researchers have revisited spell checking issues motivated by web applications, such as search query engines and sentiment analysis tools based on UGC, e.g. Twitter data or product reviews. Normalization of UGC has received great attention also because the performance of NLP tools (e.g. taggers, parsers and named entity recognizers) is greatly decreased when applied to it. Besides misspelled words, this kind of text presents a long list of problems, such as acronyms and proper names with inconsistent capitalization, abbreviations introduced by chat-speak style, slang terms mimicking the spoken language, loanwords from English as technical jargon, as well as problems related to ungrammatical language and lack of punctuation [12–15].

In [15] the authors propose a spell checker for Brazilian Portuguese (BP) to work on the top of Web text collectors, and tested their method on news portals and on informal texts collected from Twitter in BP. However, they do not inform the error correction rate of the system. Furthermore, while their focus is on the response time of the application, they do not address real-word errors.

Spell checking is a well-developed area of research within NLP, which is now crossing the boundaries between science and engineering, by introducing new sources of knowledge and methods which were successfully applied to different scenarios. This paper presents two new spell checking methods for UGC in BP. The first one deals with phonetically motivated errors, a recurrent problem in UCG not addressed by traditional spell checkers, and the second one deals additionally with real-word errors. We present a comparison of these methods with a baseline system and JaSpell over a new and large benchmark corpus for this task. The corpus is also a contribution of our study[1], containing product reviews with 38,128 tokens and 4,083 annotated errors.

This paper is structured as follows. In Sect. 2 we describe our methods, the setup of the experiments and the corpus we compiled. In Sect. 3 we present the results. In Sect. 4 we discuss related work on spelling correction of phonetic and real-word errors. To conclude, the final remarks are outlined in Sect. 5.

2   Experimental Settings and Methods

In this Section we present the four methods compared in our evaluation. Two of them are used by existing spellers; one is taken as baseline and the other is taken as benchmark. The remaining two are novel methods developed within the project reported herein. After describing in detail the novel methods, we present the corpus specifically developed to evaluate BP spellers, as well as the evaluation metrics.

2.1   Method I - Baseline

We use as a baseline the open source Java Spelling Checking Package, JaSpell[2]. JaSpell can be considered a strong baseline and is employed at the tumba! Portuguese Web search engine to support interactive spelling checking of user queries. It classifies the candidates for correcting a misspelled word according to their frequency in a large corpus together with other heuristics, such as keyboard proximity or phonetic keys, provided by the Double Metaphone algorithm [18] for English. At the time this speller was developed there was no version of these rules for the Portuguese language[3].

2.2   Method II - Benchmark

The method presented in [19] is taken as benchmark. It combines phonetic knowledge in the form of a set of rules and the algorithm Soundex, and was inspired by the analysis of errors of the same corpus of products' reviews [20] that inspired our proposals.

Furthermore, as this method aims to normalize web texts, it performs automatic spelling correction. To increase the accuracy of the first hit, it relies on some ranking heuristics, which consider the phonetic proximity between the wrong input word and the candidates to replace it. If the typed word does not belong to the lexicon, a set of candidates if generated with words from the lexicon having an edit distance to the original word of one or two.

---

[1] The small benchmark of 120 tokens used in [16,17] is not representative of our scenario.

[2] http://jaspell.sourceforge.net/.

[3] Currently, a BP version of the phonetic rules can be found at http://sourceforge.net/projects/metaphoneptbr/.

Then, a set of phonetic rules for Brazilian Portuguese codifies letters and digraphs which have similar sounds in a specific code. If necessary, the next step performs the algorithm Soundex, slightly modified for BP. Finally, if none of these algorithms is able to suggest a correction, the candidate with the highest frequency in a reference corpus among the ones with the least edition-distance is suggested. The lexicon used is the Unitex-PB[4] and the frequency list was taken from Corpus Brasileiro[5].

2.3 Method III - Grapheme-to-Phoneme Based Method (GPM)

By testing the benchmark method, we noticed that many of the wrong corrections were related to a gap between the application of phonetic rules and the Soundex module. The letter-to-sound rules were developed specially for the spelling correction, therefore, they are very accurate for the task but have a low recall, since many words do not possess the misspelling patterns which they try to model. In contrast, the transcriptions generated by the adapted Soundex algorithm are too broad and many phonetically different words are given the same code. For instance, the words "perto" (*near*) and "forte" (*strong*) are both transcribed with the Soundex code "1630", in spite of being very distinct phonetically: "perto" corresponds to ['pɛx.tʊ], and "forte" to ['fɔx.tʃɪ].

To fill this gap, we propose the use of a general-purpose grapheme-to-phoneme converter to be executed prior to the Soundex module. We employed Aeiouado's grapheme-to-phoneme converter [21], since it is the state of the art in grapheme-to-phoneme transcription for Brazilian Portuguese.

The usage of the grapheme-to-phoneme converter is a bit different from a simple pipeline. According to Toutanova [22], phonetic-based errors usually need larger edit distances to be detected. For instance, the word "durex" (*sellotape*) and one of its misspelled forms "duréquis" have an edit distance of 5 units, despite having very similar phonetic forms: [du'rɛks] ~ [du'rɛkɪs].

Therefore, instead of simply increasing the edit distance, which would imply in having a larger number of candidates to filter, we decided to do the reverse process. We transcribed the Unitex-PB dictionary and stored it into a database, with the transcriptions as keys. Thus, in order to obtain words which are phonetically, we transcribe the input and look it up in the database. Considering the "duréquis" example, we would first transcribe it as [du'rɛ.kɪs], and then check if there are any words in the database with that transcription. In this case, it would return "durex", the expected form.

The only difference of GPM in comparison with Method II lies in the G2P transcription match, which takes place prior to Soundex. In spite of being better than the baseline because they tackle phonetic-motivated errors, Method II and GPM have a limitation: they do not correct real word errors. The following method is intended to overcome this shortcoming by using context information.

---

[4] http://www.nilc.icmc.usp.br/nilc/projects/unitex-pb/web/.

[5] http://corpusbrasileiro.pucsp.br/cb/.

2.4 Method IV – GPM in a Machine Learning Framework (GPM-ML)

Method IV has the advantage of bringing together many approaches to spelling correction into a machine learning framework. Its architecture is described in Fig. 1.

The method is based on three main steps: (i) candidate word generation, (ii) feature extraction and (iii) candidate selection. The first encompasses three modules which produce a large number of suggestions, considering orthographic, phonetic and diacritic similarities. For producing suggestions which are typographically similar, the Levenshtein distance is used: for each input word, we select all words in a dictionary which diverge by at most 2 units from the input. For instance, suppose the user intended to write "mesa" (*table*), but missed a keystroke and typed "meda" instead. The Levenshtein module would generate a number of suggestions including an edit distance of 1 or 2, such as "medo" (*fear*), "meta" (*goal*), "moda" (*fashion*), "nada" (*nothing*), "mexe" (*he/she moves*) etc.

For computational efficiency, we stored the dictionary in a trie structure. A revised version of the Unitex-PB was employed as our reference dictionary (*circa* 550,000 words)[6].

As for phonetic similarity, we used the same procedure as with GPM. Thus, in order to generate suggestions which are phonetically similar to the word typed by the user, we look up in the database.

The diacritic module generates words which differ from the input in terms of diacritic symbols. This module was proposed because we observed that most of the misspellings in the corpus were caused by a lack or misuse of diacritics. BP has five types of diacritics: accute (´), cedilla (ç), circumflex (ˆ), grave (`) and tilde (˜), and they often indicate different vowel quality, timbre or stress.

However, they are rarely used in UGC, and the reader uses the context to disambiguate the intended word. In order to deal with this problem, the diacritic model generates, given a input word, all possible word combinations of diacritics. Once more, the Unitex-PB is used as reference.

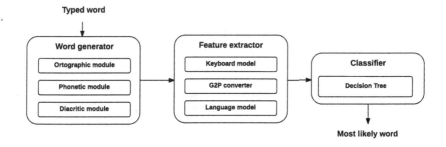

Fig. 1. Architecture of the GPM-ML

---

[6] The dictionary is available upon request.

Table 1. List of features

Feature	Description
1. TYPEDORGEN	Whether the word was typed by the user or was produced in the word generation phase
2. ISTYPO	1 if the word was generated by the typographical module; 0 otherwise
3. ISPHONE	1 if the word was generated by the phonetic module; 0 otherwise
4. ISDIAC	1 if the word was generated by the diacritic module; 0 otherwise
5. TYPEDPROB	The unigram probability of the word typed
6. GENUNIPROB	The unigram probability of the word suggestion
7. TYPEDTRIPROB	The trigram probability of the word typed
8. GENTRIPROB	The trigram probability of the word suggestion
9. TYPOLEVDIST	The levenshtein distance between the typed word and the suggestion
10. INSKEYDIST	The sum of the key insertion distances
11. DELKEYDIST	The sum of the key deletion distances
12. REPLKEYDIST	The sum of the key replacement distances
13. KEYDISTS	The sum of all previous three types of key distances
14. PHONELEVDIST	The levenshtein distance between of the phonetic transcription of the typed word and of the suggestion

After candidate generation, the feature extraction phase takes place. The aim is to allow the classifier to compare these words with the one typed by the user, in such a way that the classifier can choose to keep the typed word or to replace it with one of generated suggestions.

As misspelling errors may be of different nature (such as typographical, phonological or related to diacritics), we try to select features that encompass all these phenomena. For each word suggestion produced in the word generation phase, we extract 14 features, as described in Table 1.

The probabilities come from a language model trained over a subset of the Corpus Brasileiro (*circa* 10 million tokens). Good-Turing smoothing is used to estimate the probability of unseen trigrams.

After feature extraction, the word selection phase comes into play. The classifier was implemented through scikit-learn [23] and comprises an optimized version of the CART algorithm, trained over the dataset presented in Sect. 2.5, with the features we discussed, and evaluated through 5-fold cross-validation. Several other classification algorithms were tested, but since our features contain both nominal and numerical data, and some of them are dependent, the decision tree classifier achieved the best performance.

2.5   Dataset

The evaluation corpus was compiled specially for this research and is composed
of a set of annotated product reviews, written by users on Buscapé[7], a Brazilian
price comparison search engine. All misspelled words were marked, the correct
expected form was suggested and the misspelling category was indicated.

Considering that the amount of errors is not very large, we used a non-
probabilistic sampling technique similar to snowball sampling [24], in order to
obtain a reasonable amount of data with incorrect orthography. A list of orthographic errors with frequency greater than 3 in the the corpus of product reviews
compiled by [20] was used to pre-select from the same corpus sentences with at
least one incorrect word. Among those, 1,699 sentences were randomly selected
to compose the corpus (38,128 tokens). All these sentences were annotated by
two linguists with prior experience in corpus annotation. The inter-rater agreement for the error detection task is described in Table 2.

Table 2. Inter-rater agreement for the error detection task

		Annot. B		Total
		Correct	Wrong	
Annot. A	Correct	33, 988	512	34, 500
	Wrong	76	3, 559	3, 635
	Total	34, 064	4, 071	38, 135

The agreement was evaluated by means of the kappa test [25]. The κ value
for the error detection task was 0.915, which stands for good reliability or almost
perfect agreement [26]. The final version of the corpus used to evaluate all methods was achieved by submitting both annotations to an adjudication phase, in
which all discrepancies were resolved. We noticed that most annotation problems
consisted of whether or not to correct abbreviations, loanwords, proper nouns,
internet slang, and technical jargon. In order to enrich the annotation and the
evaluation procedure, we classified the misspellings into five categories:

1. TYPO: misspellings which encompass a typographical problem (character
   insertion, deletion, replacement or transposition), usually related to key adjacency or fast typing; e.g. "obrsevei" instead of "observei" (*I noticed*) and
   real-word errors such as "clama" (*cries out*-PRES3.SG) instead of "calma"
   (*calm*).
2. PHONO: cognitive misspellings produced by lack of understanding of letter-to-
   sound correspondences, e.g. "esselente" for "excelente" (*excelent*), since both
   "ss" and "xc", in this context, sound like [s]. Real-word errors also occur,
   such as "compra" (*buys*-PRES3.SG) instead of "comprar" (*to buy*).

---

[7] http://www.buscape.com.br/.

3. DIAC: this class identifies misspellings which are related to the inserting, deleting or replacing diacritics in a given word, e.g. "organizacao" instead of "organização" (*organization*) and real-word errors, such as "sabia" (*knew*-IMPF.3SG) instead of "sábia" (*wise*-FEM.SG).
4. INT_SLANG: use of internet slang or emoticons, such as "vc" instead of "você" (*you*), "kkkkkk" (to indicate laughter) or ":-)".
5. OTHER: other types of errors that do not belong to any of the above classes, such as abbreviations, loanwords, proper nouns, technical jargon; e.g. "aprox" for "aproximadamente" (*approximately*).

The distribution of each of these categories of errors can be found in Table 3. Interestingly, Internet slang, which is usually assumed to be a core problem in UGC, was not a frequent error pattern, accounting for only 4.9 % of the occurrences.

The difference between the total number of counts in Tables 2 and 3 is caused by spurious forms which were reconsidered or removed in the adjudication phase. In addition to the five categories previously listed, we also classified the misspellings into either contextual or non-contextual, i.e. if the misspelled word corresponds to another existing word in the dictionary, it is considered a contextual error (or real-word error). For instance, if the intended word was "está" (*he/she/it is*), but the user typed "esta", without the acute accent, it is classified as a contextual error, since "esta" is also a word in BP which means *this* FEM.

The corpus has been made publicly available[8] and intends to be a benchmark for future research in spelling correction for user generated content in BP.

Table 3. Error distribution in corpus by category

Misspelling type		Counts	% Total
TYPO	-	1,027	25.2
PHONO	Contextual	49	1.2
	Non-contextual	683	16.7
DIAC	Contextual	411	10.1
	Non-contextual	1,626	39.8
INT_SLANG	-	201	4.9
OTHER	-	86	2.1
Total/Avg		4,083	100.0

2.6 Evaluation Metrics

Four performance measures are used to evaluate the spellers: *Detection rate* is the ratio between the number of detected errors and the total number of existing errors. The *Correction rate* stands for the ratio between the number of corrected

---

[8] https://github.com/gustavoauma/propor_2016_speller.

errors and the total number of errors. *False positive rate* is the ratio between the number of false positives (correct words that are wrongly treated as errors) and the total number of correct words. The *False negative rate* consists of the ratio between the number of false negatives (wrong words that are considered correct) and the total number of errors.

In addition, the correction hit rates are evaluated by misspelling categories. In the analysis, we do not take into account the "int_slang" and "other" categories, since both show a very irregular behavior and constitute specific types of spelling correction.

3   Discussion

In Table 4, we summarize all methods' results. As one can observe, the GPM-ML achieved the best overall performance, with the best results in at least three rates: detection, correction and false positive rate. Both methods we proposed in this paper, GPM and GPM-ML, performed better than the baseline in all metrics; however, GPM did not show any improvement in comparison with the benchmark.

In fact, the addition of the grapheme-to-phoneme converter decreased the correction rate. By analyzing the output of GPM, we noticed that there seems to be some overlapping information between the phonetic rules and the grapheme-to-phoneme module. Apparently, the phonetic rules were able to cover all cases which could be solved by adding the grapheme-to-phoneme converter, and therefore our hypothesis was not supported.

Table 4. Comparison of the Methods

Method	Rate			
	Detection	Correction	FP	FN
Baseline JaSpell	74.0 %	44.7 %	5.9 %	26.0 %
Benchmark Rules&Soundex	83.4 %	68.6 %	1.7 %	**16.6 %**
GPM	83.4 %	68.2 %	1.7 %	**16.6 %**
GPM-ML	**84.9 %**	**78.1 %**	**0.7 %**	21.9 %

All methods showed a low rate of false positives, and the best value was found in GPM-ML (0.7 %). The false positive rate is very important for spelling correction purposes and is related to the reliability of the speller. In the following we discuss the correction hit rates by misspelling categories, presented in Table 5.

The baseline JaSpell (Method I) presented an average correction rate of 44.7 %. Its best results comprise non-contextual diacritic misspellings with a rate of 64.0 %; worst results are found in contextual phonological errors, in which not a single case was corrected by the speller. Typographical misspellings were also very troublesome, with a correction hit rate of 28.3 %. These results indicate

Table 5. Comparison of Correction Rates

Misspelling type	Errors	Correction rate by method			
		I	II	III	IV
Typo	1,027	28.3%	**56.3%**	53.0%	55.4%
Phono Contextual	49	0.0%	0.0%	0.0%	**8.1%**
Phono Non-contextual	683	48.2%	85.1%	**87.1%**	81.1%
Diac Contextual	411	9.2%	26.5%	26.5%	**64.5%**
Diac Non-contextual	1626	64.0%	82.2%	82.4%	**96.6%**
Total/Weighted Avg	3,796	44.7%	68.6%	68.1%	**78.0%**

that this method is not suitable for real world applications which deal with user generated content (it is important to notice that the JaSpell was not developed for this text domain).

The benchmark Rules&Soundex (Method II) achieved a correction rate of 68.6%, a relative gain of 53.4% in comparison to the baseline. The best results are, once more, related to the non-contextual diacritic misspellings (82.2%), the major error class. The best improvements compared to the baseline appear in phonological errors influenced by context (85.1%), with a relative increase of 76.6%. These results are coherent with the results reported by [19], since they claim that the method focuses on phonetically motivated misspellings.

As already mentioned, GPM (Method III) did not show any gain in comparison with the benchmark. As can be noticed, it had a small positive impact in what regards to the phonological errors, raising the correction rate of non-contextual phonological misspellings from 85.1% to 87.1% (2.3% gain).

GPM-ML (Method IV) achieved the best performance among all methods in what regards correction hit rate (78.0%). A very high correction rate was achieved in non-contextual diacritic errors (96.6%) and non-contextual phonological errors (81.1%). The trigram Language Model proved to be effective for capturing some contextual misspellings, as can be seen by the contextual diacritic correction rate (64.5%). However, the method was not able to properly infer contextual phonological misspellings (8.1%). We hypothesize this result might be caused by the small number of contextual phonological errors in the training corpus (there were only 49 cases of contextual phonological misspellings). No significant improvement was found with respect to typographical errors (55.4%) in comparison to the other methods.

4   Related Work

The first approaches to spelling correction date back to Damerau [27] and address the problem by analyzing the edit distances. He proposes a speller based on a reference dictionary and on an algorithm to check for out-of-vocabulary (OOV) words. The method assumes that words not found in the dictionary have at most one error (a threshold established to avoid high computational cost), which was caused by a letter insertion, deletion, substitution or transposition.

An improved error model for spelling correction, which works for letter sequences of lengths up to 5 and is also able to deal with phonetic errors was proposed by [28]. It embeds a noisy channel model for spell checking based on string to string edits. This model depends on the probabilistic modeling of sub-string transformations.

As texts present several kinds of misspellings, no single method will cover all of them, and therefore it is natural to combine methods which supplement each other. This approach was pursued by [22], who included pronunciation information to the model of typographical error correction. [22] and also [29] took pronunciation into account by using grapheme-to-phoneme converters. The latter proposed the use of triphone analysis to combine phonemic transcription with trigram analysis, since they performed better than either grapheme-to-phoneme conversion or trigram analysis alone.

Our GPM method also combines models to correct typographical errors by using information on edition distance, pronunciation, grapheme-to-phoneme conversion and the output of the Soundex method. In this two-layer method, these modules are put in sequence, as we take advantage of the high precision of the phonetic rules before trying the converter; typographical errors are corrected in the last pass of the process.

5   Final Remarks

We compared four spelling correction methods for UGC in BP, two of which consisting of novel approaches proposed in this paper. The Method III (GPM) consisted of an upscale version of the benchmark method, containing an additional module with a grapheme-to-phoneme converter. This converter was intended to provide the speller with transcriptions that were not so fine-grained or specific as those generated by the phonetic rules and also not so coarse-grained as those created by Soundex; however, it didn't work as well as expected.

The Machine Learning version of GPM, the GPM-ML, instead, presented a good overall performance, as it is the only one that addresses the problem of real word errors. It surpasses all other methods in most situations, reaching a correction rate of 78.0 %, with very low false positive (0.7 %) and false negative (21.9 %), thus establishing the new state of the art in spelling correction for UGC in BP. As for future work, we intend to improve GPM-ML by expanding the training database, testing other language models as well as new phone conventions. In addition, we plan to more thoroughly evaluate it on different testing corpora. We also envisage, in due course, the development of an Internet slang module.

Acknowledgments. Part of the results presented in this paper were obtained through research activity in the project titled "Semantic Processing of Brazilian Portuguese Texts", sponsored by *Samsung Eletrônica da Amazônia Ltda.* under the terms of Brazilian federal law number 8.248/91.

References

1. Duan, H., Hsu, B.P.: Online spelling correction for query completion. In: Proceedings of the 20th International Conference on World Wide Web, WWW 2011, NY, USA, pp. 117–126. ACM (2011)
2. Fossati, D., Di Eugenio, B.: A mixed trigrams approach for context sensitive spell checking. In: Gelbukh, A. (ed.) CICLing 2007. LNCS, vol. 4394, pp. 623–633. Springer, Heidelberg (2007)
3. Fossati, D., Di Eugenio, B.: I saw TREE trees in the park: how to correct real-word spelling mistakes. In: Proceedings of the Sixth International Conference on Language Resources and Evaluation LREC 2008 (2008)
4. Mays, E., Damerau, F.J., Mercer, R.L.: Context based spelling correction. Inf. Process. Manage. **27**(5), 517–522 (1991)
5. Wilcox-O'Hearn, A., Hirst, G., Budanitsky, A.: Real-word spelling correction with trigrams: a reconsideration of the Mays, Damerau, and Mercer Model. In: Gelbukh, A. (ed.) CICLing 2008. LNCS, vol. 4919, pp. 605–616. Springer, Heidelberg (2008)
6. Islam, A., Inkpen, D.: Real-word spelling correction using Google web 1tn-gram data set. In ACM International Conference on Information and Knowledge Management CIKM 2009, pp. 1689–1692(2009)
7. Sonmez, C., Ozgur, A.: A graph-based approach for contextual text normalization. In: Proceedings of the 2014 Conference on Empirical Methods in Natural Language Processing EMNLP 2014, pp. 313–324 (2014)
8. Hirst, G.: An evaluation of the contextual spelling checker of Microsoft Office Word 2007(2008)
9. Németh, L.: Hunspell. Dostupno na (2010). http://hunspell.sourceforge.net/ [01.10.2013]
10. Zampieri, M., Amorim, R.: Between sound and spelling: combining phonetics and clustering algorithms to improve target word recovery. In: Proceedings of the 9th International Conference on Natural Language Processing PolTAL 2014, pp. 438–449 (2014)
11. Rusell, R.C.: US Patent 1261167 issued 1918–04-02 (1918)
12. Duran, M., Avanço, L., Aluísio, S., Pardo, T., Nunes, M.G.V.: Some issues on the normalization of a corpus of products reviews in Portuguese. In: Proceedings of the 9th Web as Corpus Workshop WaC-9, Gothenburg, Sweden, pp. 22–28, April 2014
13. De Clercq, O., Schulz, S., Desmet, B., Lefever, E., Hoste, V.: Normalization of dutch user-generated content. In: Proceedings of the International Conference Recent Advances in Natural Language Processing RANLP 2013, pp. 179–188 (2013)
14. Han, B., Cook, P., Baldwin, T.: Lexical normalization for social media text. ACM Trans. Intell. Syst. Technol. **4**(1), 5:1–5:27 (2013)
15. Andrade, G., Teixeira, F., Xavier, C., Oliveira, R., Rocha, L., Evsukoff, A.: HASCH: high performance automatic spell checker for Portuguese texts from the web. In: Proceedings of the International Conference on Computational Science, vol. 9, pp. 403–411 (2012)
16. Martins, B., Silva, M.J.: Spelling correction for search engine queries. In: Vicedo, J.L., Martínez-Barco, P., Muñoz, R., Saiz Noeda, M. (eds.) EsTAL 2004. LNCS (LNAI), vol. 3230, pp. 372–383. Springer, Heidelberg (2004)
17. Ahmed, F., Luca, E.W.D., Nürnberger, A.: Revised N-Gram based Automatic spelling correction tool to improve retrieval effectiveness. Polibits **40**, 39–48 (2009)

18. Philips, L.: The double metaphone search algorithm. C/C++ Users J. **18**(6), 38–43 (2000)
19. Avanço, L., Duran, M., Nunes, M.G.V.: Towards a phonetic Brazilian Portuguese spell checker. In: Proceedings of ToRPorEsp Workshop PROPOR 2014, São Carlos, Brazil, pp. 24–31 (2014)
20. Hartmann, N., Avanço, L., Balage, P., Duran, M., Nunes, M.G.V., Pardo, T., Aluísio, S.: A large corpus of product reviews in Portuguese: tackling out-of-vocabulary words. In: Proceedings of the Ninth International Conference on Language Resources and Evaluation LREC 2014, pp. 3866–3871 (2014)
21. Mendonça, G., Aluísio, S.: Using a hybrid approach to build a pronunciation dictionary for Brazilian Portuguese. In: Proceedings of the 15th Annual Conference of the International Speech Communication Association INTERSPEECH 2014, Singapore (2014)
22. Toutanova, K., Moore, R.C.: Pronunciation modeling for improved spelling correction. In: Proceedings of the 40th Annual Meeting on Association for Computational Linguistics, ACL 2002, pp. 144–151 (2002)
23. Pedregosa, F., Varoquaux, G., Gramfort, A., Michel, V., Thirion, B., Grisel, O., Blondel, M., Prettenhofer, P., Weiss, R., Dubourg, V., Vanderplas, J., Passos, A., Cournapeau, D., Brucher, M., Perrot, M., Duchesnay, E.: Scikit-learn: machine learning in Python. J. Mach. Learn. Res. **12**, 2825–2830 (2011)
24. Browne, K.: Snowball sampling: using social networks to research non-heterosexual women. Int. J. Soc. Res. Methodol. **8**(1), 47–60 (2005)
25. Carletta, J.: Assessing agreement on classification tasks: the kappa statistic. Comput. Linguist. **22**(2), 249–254 (1996)
26. Landis, J.R., Koch, G.G.: The measurement of observer agreement for categorical data. Biometrics **33**(1), 159–174 (1977)
27. Damerau, F.J.: A technique for computer detection and correction of spelling errors. Commun. ACM **7**(3), 171–176 (1964)
28. Brill, E., Moore, R.C.: An improved error model for noisy channel spelling correction. In: Proceedings of the 38th Annual Meeting on Association for Computational Linguistics, ACL 2000, pp. 286–293(2000)
29. van Berkel, B., Smedt, K.D.: Triphone analysis: a combined method for the correction of orthographical and typographical errors. In: Proceedings of the Second Conference on Applied Natural Language Processing, Austin, Texas, USA, pp. 77–83, February 1988

An Automatic Phonetic Aligner for Brazilian Portuguese with a Praat Interface

Gleidson Souza and Nelson Neto$^{(\boxtimes)}$

Institute of Exact and Natural Sciences, Federal University of Pará,
Augusto Correa. 1, Belém, PA 660750110, Brazil
gleidson.sousa@itec.ufpa.br, nelsonneto@ufpa.br

Abstract. The analysis of the phonetic entities of speech nearly always requires the alignment of an audio file with its phonetic transcription. However, it is an extremely labor-intensive task. An automatic alignment tool has modules that depend on the language and, while there are many public resources for some languages (e.g., English and French), the resources for Brazilian Portuguese (BP) are still limited. This work describes the development of an automatic phonetic alignment tool for BP, consisting of grapheme-to-phone converter, syllabification system and HTK-based acoustic models. This aligner is implemented and freely distributed as a plug-in of Praat. Performance tests are presented, comparing the current proposal with an existing tool.

Keywords: Phonetic alignment · Brazilian Portuguese · Pronunciation dictionary · Syllabification · HTK · Praat

1 Introduction

Automatic speech recognition (ASR) and speech synthesis (TTS) are data-driven technologies that require a relatively large amount of labeled data. As consequence, many large speech corpora have been collected for speech technology development in the recent years. And they need to be phonetically segmented with a high level of precision, i.e. the phones must be time-aligned with the sound, on risk of impairing the quality of the synthesized voice, for example. Indeed, the analysis of the prosodic structure of speech requires to know the precise position of the phonetic temporal boundaries [1]. However, manual phonetic segmentation is time-consuming, more than 13 h for a one-minute recording [2], and expensive, since it requires trained language experts.

The most widely explored phonetic alignment techniques are based either on hidden Markov models (HMM) used in forced-alignment mode or on dynamic time alignment with synthesized speech (TTS+DTW) [3]. In [4], a comparison between these two approaches has showed that in general the TTS+DTW segmentation is more accurate than HMM, however, the HMM-based phonetic aligners are more reliable. Hence, an hybrid system is proposed in [5]. The results with a Portuguese voice data suggest that the use of HMM-based along with

© Springer International Publishing Switzerland 2016
J. Silva et al. (Eds.): PROPOR 2016, LNAI 9727, pp. 374–384, 2016.
DOI: 10.1007/978-3-319-41552-9_38

TTS+DTW alignment tools can be worthy, as the former is more robust and the later is more accurate.

In this context, automatic alignment tools such as EasyAlign [6], SPPAS [7], P2FA [8] and Train&Align [9] have been developed and released. Besides a phonetic dictionary, all these tools rely on the acoustic modeling of the language with HMM. They provide the user with pre-existent speaker-independent models of each language, or models of each phoneme (monophone models) or group of phonemes (triphone models) are directly trained on the corpus to align. Then, these models are used to align an audio file with its phonetic transcription.

P2FA is an open-source automatic phonetic alignment tool based on HTK, a widespread HMM-based speech recognition toolkit [10]. It contains monophone acoustic models for American English and a Python script that can be used to do the alignment process via a command-line interface. There is also an on-line processing system on the P2FA website [11] with which the user can submit a limited size audio file and its word transcription and get back a TextGrid file compatible with the Praat, the popular speech analysis software [12]. The TextGrid contains two tiers named phone and word. The first tier holds the phonemes or speech segments and their boundaries. The alignment of words is deduced from their alignment into phonemes.

EasyAlign is also HTK-based, but its topmost layer lies within Praat, which facilitates the alignment process for a computer science non-specialist. It requires a few minor manual steps and the result is a multi-level annotation within a TextGrid composed of phonetic, syllabic, lexical and utterance tiers. It is already designed for French, English, Spanish and Taiwan Min and has been recently adapted to Brazilian Portuguese (BP) [13]. It is free distributed as a self-installable Praat plug-in only on Windows operating system, and comes with trained monophone acoustic models.

Like EasyAlign, Train&Align offers a user-friendly graphical interface implementing HTK methods and TextGrid files. However, it is proposed as a web service that can be free accessed from any platform. Its specificity is that it trains the acoustic model directly on the corpus to align, which makes it applicable to any language and speaking style. In contrast, the phonetic transcription that corresponding to the audio file must be provided to the tool, which requires a level of expertise in linguistic. The user can set some training parameters and modify the model configuration, taking into account the phonetic context. But, when using tied-state triphones, a file containing the articulatory characteristics of each phoneme must be given, which can be another difficulty.

Since the HTK tools can not be embedded in an other tool, the previous aligners assume that HTK is installed on the user's computer. However, there is no limitation on the distribution of an acoustic model created with the HTK toolkit, which can consequently be distributed under GPL license. Based on this information, SPPAS uses the open-source Julius speech recognition engine [14] to perform alignment based on HTK tied-state triphone models. The resulting alignments are a set of TextGrid files for utterance, word, syllable and phoneme segmentation. It is implemented with Python and can be executed using a graphical interface, or a console-mode, under Linux, Mac-OSX and Windows. SPPAS is currently available under GPL license for French, English, Italian and Chinese although the author claims that adding a new language is relatively easy.

A motivation of this work is to complement these previous initiatives and release an automatic phonetic aligner for BP. It is a plug-in developed for Praat, which produces a multi-tier TextGrid, including phonetic and syllabic annotations, from a sound recording and the corresponding orthographic transcription. The plug-in is made of Praat scripts, but it also includes the HVite tool (within HTK) for the alignment at the phone level and other external components. In summary, the following required tools have been developed for BP: (i) a grapheme-to-phone (G2P) converter with stress determination, (ii) scripts used for developing HTK-based tied-state triphone models and (iii) a syllabification system based on a set of linguistic rules with stress vowel indication.

Alternatively, the proposed aligner can be executed using a command-line interface. Python scripts were also implemented to automate the procedure of doing segmentation. It was tested under Linux and Windows. All these developed tools are publicly available [15].

According to [16], the most common and direct form of evaluation is comparing the automatic segmentation to a manual segmentation. Thus, aside from the description of the alignment process in details, the results of tests performed on a manually-aligned corpus are also shown, in order to measure the current proposal effectiveness. An extra comparison was made with EasyAlign [13].

One of the most usual metric to evaluate the segmentations' accuracy is called boundary-based measure, where the performance is measured as the percentage of (initial, middle, or final) boundaries which are similar in both alignments, with a certain tolerance threshold (often 20 ms), considering that the phonetic transcription is the same in both segmented corpora. Another way of evaluating the agreement between two alignments is the phone-based duration-independent measure, or Overlap Rate (OvR), that is to determine the percentage of well assigned frames, within the segment [5].

2   A Phonetic Aligner for BP

This work proposes a set of tools to automatically produce segmentations for continuous speech in BP. The result is a multi-level annotation within a TextGrid composed of phonetic, syllabic, lexical and utterance tiers. The output is illustrated as a screenshot in Fig. 1. It can be executed using scripts in a console-mode or, in order to turn it more ergonomic, as a plug-in of Praat.

The whole alignment process is a succession of four automatic steps: G2P conversion, phone segmentation, syllabification and TextGrid conversion. Figure 2 summarizes the workflow.

The first step is called phonetization, or the process of representing sounds with phonetic signs. For instance, in order to develop ASR systems, one needs a pronunciation (or phonetic) dictionary, which maps each word in the lexicon to one or more phonetic transcriptions (pronunciations). In practice, building a

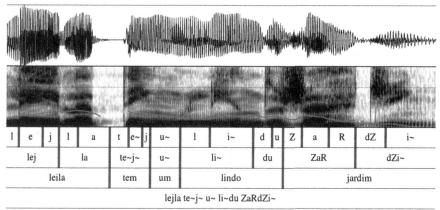

Fig. 1. The resulting TextGrid with five tiers from bottom to top: orthographic transcription, phonetic transcription, words, syllables and phonemes.

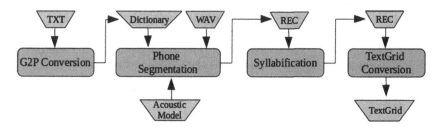

Fig. 2. The alignment process, with all acceptable inputs and outputs.

dictionary for ASR is very similar to developing a G2P module for TTS systems. In fact, a dictionary can be constructed by invoking a pre-existent G2P converter. So, this block receives as input a tokenized standard orthographic transcription and creates the corresponding dictionary.

Then, the resulting phonetic dictionary and a given audio file (.wav extension) will be used as inputs for the HTK Viterbi-based (HVite) tool in the second step. It has the function to align each word to its respective phonetic sequence, using the acoustic model. It generates a .rec file that contains the start and end times of each phoneme. The syllabification system performs the phoneme-to-syllable segmentation. Finally, the last block uses the information present in the .rec file to create a TextGrid file containing five tiers, as shown in Fig. 1. This block is a script package implemented with Python.

After this description of the alignment process, the next sections describe the tools that have been developed to compose the aligner.

2.1   G2P Conversion

This work uses a G2P converter with stress determination that is based on a set of rules described in [17]. The rules do not focus in any BP dialect and provide only one pronunciation by word. The proposed conversion is based on phonological pre-established criteria, its architecture does not rely on intermediate stages, i.e., other algorithms such as syllabic division or plural identification. There is a set of rules for each grapheme and a specific order of application is assumed. First, the more specific rules are considered until a general case rule is reached, which ends the process. Besides, the presence of a stressed vowel changes the G2P converter interpretation.

The format of each dictionary entry (i.e., the word and its phonetic transcription) suggested by HTK is illustrated by the example below:

```
leite l e j tS i sp
```

where the phone *sp* (short pause) must be added to the end of every pronunciation. Therefore the developed G2P converter deals only with single words and does not implement co-articulation analysis between words.

The rules are specified in a set of regular expressions using the C# programming language. Regular expressions are also allowed in the definition of non-terminals symbols (e.g. #abacaxi#). The rules of the G2P converter are organized in three phases. Each phase has the following function:

- a simple procedure that inserts the non-terminal symbol # before and after each word.
- the stress phase consists of 29 rules that mark the stressed vowel of the word.
- the bulk of the system, which consists of 140 rules, that convert the graphemes (including the stressed vowel mark) to 38 phones represented using the SAMPA phonetic alphabet [19].

During the research, some rules proposed by [17] were improved and others were added. Aside from the description of the added or modified rules in details, the results of tests performed on a speech corpus can be also seen in [15,18], in order to measure the G2P converter effectiveness and its influence in a ASR system performance.

2.2   Acoustic Model Training

For training an acoustic model, it is required a corpus with digitized voice, transcribed at the level of words and/or at the level of phones. In the sequel, some aspects of developing a model for BP are discussed.

The front end consists of the widely used 12 Mel-frequency cepstral coefficients (MFCCs) [20] using C0 as the energy component, and computed every 10 ms (i.e., 10 ms is the frame shift) for a frame of 25 ms. These static coefficients are augmented with their first and second derivatives to compose a 39 dimensional parameter vector per frame. Finally, the cepstral mean subtraction technique was used to normalize the MFCCs coefficients [10].

The acoustic models were iteratively refined [21]. A flat-start approach was adopted, starting with continuous single-component mixture monophone models, the HMMs were gradually improved to finally have mixtures of multiple Gaussians to model the output distributions. The set of HMMs was composed by tied-state triphones. During all the training process, the embedded Baum-Welch algorithm [22] was used to re-estimate the models.

The initial acoustic models for the 39 phones (38 monophones and a silence model) used 3-states left-right HMMs. The silence model was trained and then copied to create the tied short pause (*sp*) model with only one acoustic state. The sp has a direct transition from the entry to the exit state. After that, *cross-word* triphone models were built from the monophone models. Transition matrices of triphones that share the same base phone were tied.

Given a set of categories (also called *questions*), a decision tree specific for BP was designed for tying the triphones with similar phonetic characteristics. To illustrate, some vowels and consonants classification rules used to build the decision tree are listed below:

```
...
QS "R_V-Close" { *+i,*+e,*+o,*+u }
QS "R_V-Front" { *+i,*+E,*+e }
QS "R_Palate"  { *+S,*+Z,*+L,*+J }
QS "L_V-Back"  { u-*,o-*,O-* }
QS "L_V-Open"  { a-*,E-*,O-* }
...
```

Notice that for a triphone system, it is necessary to include questions referring to both the right and left contexts of a phone. The questions should progress from wide, general classifications (such as consonant, vowel, nasal, diphthong, etc.) to specific instances of each phone. Ideally, the full set of questions loaded using the HTK *QS* command would include every possible context that can influence the acoustic realization of a phone, and can include any linguistic or phonetic classification that may be relevant.

After tying, the number of component mixture distributions was gradually increased up to 14-Gaussians per mixture to complete the training process.

2.3 Syllabification

The proposed syllabification algorithm for BP is implemented by means of C# program language and is based on the 20 rules designed in [23], plus the two rules added by [24], including the stressed vowel mark and new improvements. All the rules are based on orthography and do not focus in any BP dialect. In fact, phonological criteria are also considered, but only the classical ones, where the grapheme sequence is admittedly represented by a single phoneme. Figure 3 illustrates the architecture of the self-contained system.

Each linguistic rule of the algorithm is basically composed of a condition to be evaluated and actions to be executed, considering that every syllable must have

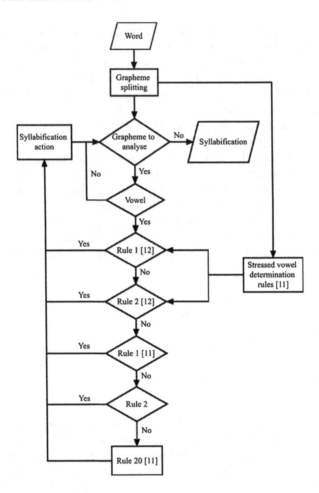

Fig. 3. Block diagram of the developed syllabification tool. There is a set of splitting rules and a hierarchical order of application is assumed, at the moment the algorithm identifies a vowel. First, the more specific rules are considered until a general case rule (Rule 20) is reached, which restarts the process. Note that some rules proposed by [23] were improved during the research, like the Rule 2, as described below.

a vowel as a nucleus. Each condition evaluates all the graphemes that surround the syllable nucleus (the vowel currently under analysis). If it is fulfilled, then the algorithm calls the method that executes the action associated to such rule to perform the required syllabification, for example, the vowel is separated from the next grapheme, or is attached to the next grapheme and separated from the subsequent graphemes, and so on. Some rules are composed of more than one action, but they also have specific conditions to help the algorithm decide which action must be taken, once the main condition is fulfilled.

The algorithms on [23,24] were subject to a baseline evaluation process with 150,000 words (and their respective syllabification) extracted from the Dicio database [25]. During this phase, the analysis of the problems gave the input for some improvements on the rules proposed by [23]. In particular, the Rule 19 was improved to treat specific hiatus (vowel + vowel) not considered in the previous analyses. Since this kind of vocalic sequence has a strong presence in the Portuguese lexicon, this modified rule is an important contribution.

A summary of the suggested modifications can be seen in [15,26], as well as some results achieved with experiments that evaluate the current syllabification tool effectiveness and its influence in a TTS system performance.

3   Experimental Results

In this experiment, the automatic alignment was estimated on the basis of the manual segmentation. Hence, the hand-aligned test corpus was made of 181 utterances recorded by a male speaker for a total duration of 7 min and 8 s. It contains 1167 words and 4887 phonetic segments (the pauses were discarded).

Then, the HVite tool was called to align each utterance to its phonetic sequence, which was provided by the G2P converter. The tied-state triphone models were trained on the basis of about 160 h of unaligned multi-speaker audio, according to the procedure described in Sect. 2.2. The reference corpus was not used to train the acoustic model.

Table 1 proposes detailed alignment performances depending on the delta range between manual and automatic boundaries, by using the time-location of the initial of each phoneme. Despite the fair comparison, it was observed that the tools achieved equivalent alignment rates, where 57.27 % and 59.30 % of the differences between the boundaries lie within 20 ms for the current proposal and EasyAlign, respectively.

In order to compare the results achieved by the proposed aligner with other tool, the utterances were also segmented using EasyAlign [13], which is currently the only one that exists for BP, as far as we know. However, it should be noted that the test corpus was used to train the acoustic model provided by EasyAlign, which is a major advantage.

The alignment accuracy was also evaluated considering the Overlap Rate (OvR) [5]. Given a segment, a reference segmentation and the segmentation to be evaluated, OvR is the ratio between the number of frames that belong to that segment in both the segmentations and the number of frames that belong to the segment in one segmentation, at least. For example, if a phone duration in the reference segmentation differs considerably from its duration in the other segmentation, the OvR quantity takes a very small value. According to [5], a segment with a OvR of 0.75 is considered well segmented.

Figure 4 shows the accuracy of the two evaluated alignment tools. The x-axis is the percentage of incorrectly assigned frames ((1-OvR)-100 %) and the y-axis is the percentage of phones that has a percentage of incorrectly assigned frames lower than the value given in the x-axis. The performances were again similar, with a slightly better OvR for EasyAlign.

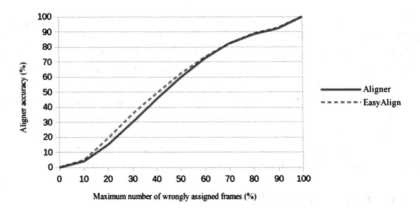

Fig. 4. Alignment accuracy (or OvR evaluation) of the two tested tools. (Color figure online)

Regarding the syllabic alignment, at present there is, to our knowledge, no generally available corpus manually aligned to evaluate such a syllabification. We will of course be pleased to perform this evaluation as soon as we can obtain one. However, a comparison between the two syllabification algorithms was performed in terms of correctly syllabified words (or word accuracy), which was obtained manually, using the Dicio database [25] as reference.

Interestingly, the current syllabification tool is observed to clearly outperform EasyAlign. The results obtained on the test corpus were respectively 96.06 % and 66.92 %. The explanation for this may be that both algorithms are rule-based and assume that all the syllables have a vowel as a nucleus, but only in EasyAlign the sonority principle is used to split the consonant clusters, in which the pauses are also used as syllabic boundaries. Therefore, EasyAlign commonly failed trying to determine the syllabic boundaries between words, putting together phonemes from different words in the same syllable.

Table 1. Percentage of boundary time differences for Manual vs. Aligner and Manual vs. EasyAlign comparisons on a BP continuous speech corpus.

	≤10 ms	≤20 ms	≤30 ms	≤40 ms	≤50 ms
Aligner	31.34 %	57.27 %	72.94 %	82.81 %	87.98 %
EasyAlign	32.35 %	59.30 %	72.33 %	82.25 %	86.53 %

4   Conclusions

This paper presented a tool to perform automatic phonetization, alignment and syllabification for BP. This multi-platform phonetic aligner works under the

Praat software and is freely available on-line. Besides, two kinds of evaluation were done: boundary-based and segment-based. We have shown that the results obtained on an external test corpus are comparable to those of an existing tool (EasyAlign for BP) that used the test corpus to train its acoustic model. It is due to the generalization achieved by the developed HMM-based speaker-independent acoustic model. Concerning the syllabification procedure, the proposed system provided the best rates with an increase of nearly 30 % compared to EasyAlign-tied in terms of words accuracy.

As there are limitations on the distribution of the HTK toolkit, one needs to previously install the HVite tool. In order to avoid it, the open-source Julius engine could be used to perform the phonetic segmentation. Future works also include expanding the speech database, aiming at reaching the performance obtained by alignment for English and French, for example. The phonetic dictionary refinement is another important issue to be addressed, considering the existing dialectal variation in Brazil. In parallel, adding new features to the software, for example, the possibility to automatically create the orthographic transcription from the audio file using the speech recognizer and an option to train new models on the corpus to align or even on other corpora.

References

1. Brognaux, S., Roekhaut, S., Drugman, T., Beaufort, R.: Automatic phone alignment. In: Isahara, H., Kanzaki, K. (eds.) JapTAL 2012. LNCS, vol. 7614, pp. 300–311. Springer, Heidelberg (2012)
2. Schiel, F., Draxler, C.: The production of speech corpora. Technical report, Bavarian Archive for Speech Signals (2003)
3. Figueira, L., Oliveira, L.C.: Comparison of phonetic segmentation tools for european Portuguese. In: Teixeira, A., de Lima, V.L.S., de Oliveira, L.C., Quaresma, P. (eds.) PROPOR 2008. LNCS (LNAI), vol. 5190, pp. 252–255. Springer, Heidelberg (2008)
4. Kominek, J., Bennett, C., Black, A.W.: Evaluating and correcting phoneme segmentation for unit selection synthesis. In: Proceedings of Eurospeech, pp. 313–316 (2003)
5. Paulo, S., de Oliveira, L.C.: Automatic phonetic alignment and its confidence measures. In: Vicedo, J.L., Martínez-Barco, P., Muñoz, R., Saiz Noeda, M. (eds.) EsTAL 2004. LNCS (LNAI), vol. 3230, pp. 36–44. Springer, Heidelberg (2004)
6. Goldman, J.P.: Easyalign: an automatic phonetic alignment tool under Praat. In: Proceedings of Interspeech, pp. 3233–3236 (2011)
7. Bigi, B., Hirst, D.: Speech phonetization alignment and syllabification (SPPAS): a tool for the automatic analysis of speech prosody. In: Proceedings of Speech Prosody, pp. 1–4 (2012)
8. Yuan, J., Liberman, M.: Speaker identification on the SCOTUS corpus. In: Proceedings of Acoustics, pp. 5687–5690 (2008)
9. Brognaux, S., Roekhaut, S., Drugman, T., Beaufort, R.: Train & Align: a new online tool for automatic phonetic alignment. In: IEEE Workshop on Spoken Language Technology, pp. 416–421 (2012)
10. Young, S., Ollason, D., Valtchev, V., Woodland, P.: The HTK Book. Cambridge University Engineering Department, version 3.4 (2006)

11. Penn Phonetics Lab Forced Aligner (P2FA): University of Pennsylvania. phon.chass.ncsu.edu/cgi-bin/step7.cgi
12. Boersma, P., Weenink, D.: Praat: doing phonetics by computer (version 6.0.06) [computer program] (2015). http://www.praat.org
13. Goldman, J.P., Miranda, M.A., Stein, C.C., Auchlin, A.: EasyAlign for Brazilian Portuguese: a (semi) automatic segmental tool under Praat. In: Proceedings of the VIIth GSCP International Conference: Speech and Corpora, pp. 116–120 (2012)
14. Lee, A.: The Julius Book, Edition 1.0.3 - rev.4.1.5 (2010)
15. FalaBrasil Project: Federal University of Pará. laps.ufpa.br/falabrasil/
16. Bigi, B.: The SPPAS participation to Evalita 2011. In: Workshop on Evaluation of NLP and Speech Tools for Italian, pp. 1–6 (2011)
17. Silva, D., de Lima, A., Maia, R., Braga, D., de Moraes, J.F., de Moraes, J.A., Resende Jr., F.: A rule-based grapheme-phone converter and stress determination for Brazilian Portuguese natural language processing. In: VI International Telecommunications Symposium, pp. 550–554 (2006)
18. Neto, N., Patrick, C., Klautau, A., Trancoso, I.: Free tools and resources for Brazilian Portuguese speech recognition. J. Braz. Comput. Soc. **17**, 53–68 (2011)
19. SAMPA: Computer Readable Phonetic Alphabet. phon.ucl.ac.uk/home/sampa/
20. Davis, S., Merlmestein, P.: Comparison of parametric representations for monosyllabic word recognition in continuously spoken sentences. IEEE Trans. ASSP. **28**, 357–366 (1980)
21. Woodland, P., Odell, J.J., Valtchev, V., Young, S.: Large vocabulary continuous speech recognition using HTK. IEEE Int. Conf. Acoust. Speech Signal Process. **2**, 125–128 (1994)
22. Welch, L.R.: Hidden Markov models and the Baum-Welch algorithm. IEEE Inf. Theory Soc. Newsl. **53**, 10–12 (2003)
23. Silva, D., Braga, D., Resende Jr., F.: Separação das sílabas e determinação da tonicidade no Português Brasileiro. In: XXVI Simpósio Brasileiro de Telecomunicações, pp. 1–5 (2008)
24. Monte, A., Ribeiro, D., Neto, N., Cruz, R., Klautau, A.: A rule-based syllabification algorithm with stress determination for Brazilian Portuguese natural language processing. In: 17th International Congress of Phonetic Sciences, pp. 1418–1421 (2011)
25. Dicionário Online de Português, 7Graus, Portugal. www.dicio.com.br
26. Neto, N., Rocha, W., Souza, G.: An open-source rule-based syllabification tool for Brazilian Portuguese. J. Braz. Comput. Soc. **21**, 1–10 (2015)

Design and Analysis of a Database to Evaluate Children's Reading Aloud Performance

Jorge Proença[1,2(✉)], Dirce Celorico[1], Carla Lopes[1,3], Miguel Sales Dias[4,5],
Michael Tjalve[6], Andreas Stolcke[7], Sara Candeias[4], and Fernando Perdigão[1,2]

[1] Instituto de Telecomunicações, Coimbra, Portugal
{jproenca,direcelorico,calopes,fp}@co.it.pt
[2] Department of Electrical and Computer Engineering, University of Coimbra, Coimbra, Portugal
[3] Polytechninc Institute of Leiria, Leiria, Portugal
[4] Microsoft Language Development Centre, Lisbon, Portugal
{miguel.dias,t-sacand}@microsoft.com
[5] ISCTE – University Institute of Lisbon, Lisbon, Portugal
[6] Microsoft and University of Washington, Seattle, WA, USA
michael.tjalve@microsoft.com
[7] Microsoft Research, Mountain View, CA, USA
andreas.stolcke@microsoft.com

Abstract. To evaluate the reading performance of children, human assessment is usually involved, where a teacher or tutor has to take time to individually estimate the performance in terms of fluency (speed, accuracy and expression). Automatic estimation of reading ability can be an important alternative or complement to the usual methods, and can improve other applications such as e-learning. Techniques must be developed to analyse audio recordings of read utterances by children and detect the deviations from the intended correct reading i.e. disfluencies. For that goal, a database of 284 European Portuguese children from 6 to 10 years old (1st–4th grades) reading aloud amounting to 20 h was collected in private and public Portuguese schools. This paper describes the design of the reading tasks as well as the data collection procedure. The presence of different types of disfluencies is analysed as well as reading performance compared to known curricular goals.

Keywords: Reading aloud performance · Child speech · Speech corpus · Reading disfluencies

1 Introduction

The use of automatic speech recognition technologies to analyse reading performance gains prominence as an alternative to any kind of manual or 1-on-1 evaluation. Usually, teachers have to spend a considerable amount of time on the task of manually assessing a child's reading ability. Automatic evaluation of literacy or reading ability (not necessarily of children) is always related to detecting correctly read words, or optionally detecting what kind of mistakes are made. Additionally, there are several systems oriented to improve the literacy of an individual [1, 2], ideally denoting and warning

© Springer International Publishing Switzerland 2016
J. Silva et al. (Eds.): PROPOR 2016, LNAI 9727, pp. 385–395, 2016.
DOI: 10.1007/978-3-319-41552-9_39

about reading errors that occur. Computer-Assisted Language Learning (CALL) is the area of research that focuses on this subject, allowing a self-practice or an oriented training of the language. These systems are most often created for foreign language learning [1, 3], and are therefore targeted at adults or young adults for whom speech technologies are significantly mature. Nevertheless, for children, there are also applications that deal with the improvement of reading aloud performance, such as reading tutors. Some projects aim at creating an automatic reading tutor that follows and analyses a child's reading, such as LISTEN [4], Tball [5], SPACE [6] and FLORA [7]. Most of these applications are helpers, e.g., by highlighting words in a sentence as they are correctly pronounced. The present work is carried out in the scope of the LetsRead project whose overall goal is to have an application that can automatically evaluate the reading aloud performance of European Portuguese (EP) children from 6 to 10 years old (1st–4th grades), and not necessarily provide feedback to them, but to their teachers and tutors.

There are currently no computer assisted applications for EP that automatically evaluate the reading aloud performance of children. Even for other languages, this automatic evaluation is a developing field, and the focus is on reading of isolated words rather than longer sentences [8, 9]. To carry out the goals of the LetsRead project, it was necessary to create a large new corpus of EP children's speech with utterances of reading tasks that are rich with common disfluencies that children commit while reading. There are some children's speech databases for EP, such as SPEECON with rich sentences [10]; ChildCAST [11] with picture naming; the Contents for Next Generation (CNG) Corpus targeting interactive games [12] and [13] with child-adult interactions. However, these databases do not exhibit the required samples of disfluent reading speech. Since children's speech has different characteristics from adult speech (such as fundamental frequency, formant frequency variability, vowel duration variability, etc. [14, 15]), special care is needed to adapt or create robust acoustic models that target children [16, 17].

This paper describes the careful design of the reading tasks as well as the collection procedure of the LetsRead database. Part of the data was verified and annotated manually including tagging of several types of disfluencies. An analysis of reading performance with a comparison to adult speakers is also presented.

2   Reading Tasks

The Portuguese government has defined certain Curricular Goals (CG) with qualitative and quantitative objectives per grade for reading aloud [18]. Some of these objectives include target reading speed of words per minute on different tasks. With the analysis of curricular goals in mind, the reading of sentences and pseudowords was the target material to be collected (described in the following subsections). It was decided not to include reading of isolated words, as the required time for a session could become too long and the child's performance is likely to decrease with extended sessions. The pseudoword reading task provides an objective analysis of reading skills. With sentences, plenty of reading disfluencies can be collected from which the overall reading performance of a child can be evaluated. Each child was presented with a reading task where they were asked to read aloud twenty sentences and ten pseudowords. Forty reading tasks were established (10 per grade) to balance repetition and diversity of the data. At a later stage, these were shortened to 5 tasks per grade, to reinforce repetition. The vocabulary of the set of sentences and pseudowords comprises a total of 2721 words. The distribution of the material for the different grades is described below.

2.1   Sentences: Selection and Difficulty Level

A large set of sentences was extracted from children's tales and school books of the level of the target group (6–10 years old, 1st–4th grades). They are mostly short sentences, although the maximum length is 30 words. Twenty sentences were included in each reading task (for a recording session of one child). The first concern for distributing sentences along the grades was to maintain a good representation of all phones, so that acoustic models with significant quality can be built with the data. The other main concerns to build appropriate reading tasks were to maintain the same average difficulty within a grade (with a rising average difficulty from 1st to 4th grades) and to have sentences of varying difficulty in a task (overlapping distributions of difficulty for the grades). Furthermore, it is necessary to capture all types of reading disfluencies to have examples for training, so the difficulty cannot be too low, although a balance must exist so as not to make the tasks too hard.

   A parameter of difficulty was developed to classify sentences according to phonological constraints. Although it would be ideal to also relate a word's difficulty to its age-of-acquisition or familiarity, not all words of the proposed reading tasks were present in available lexical databases such as ESCOLEX [19], and it was not possible to consider such features. The proposed parameter of difficulty is based on the method described in [20] where sentences are evaluated in terms of phonological complexity and variety. All words were split into syllables and a difficulty level was assigned to each syllable, determined from these rules: the length of the syllable; the multiple pronunciation of some graphemes (e.g. <mãe> [mˈɐ̃j] and <bem> [bˈɐ̃j]); the ambiguous pronunciation of consonant clusters (e.g. <prever> [prəvˈeɾ] or <florescer> [fluɾəʃˈeɾ]) and vocalic encounters (<candeeiro> [kɐ̃diˈɐjɾu] or <veem> [vˈeɐ̃j]). Since each syllable has a given minimum difficulty, the length of the sentence also contributes to difficulty.

2.2   Pseudowords Creation

Pseudowords (such as <traba> [tɾˈabɐ], <impemba> [ĩpˈɐ̃bɐ] or <culenes> [kulˈɛnəʃ]) represent non-existing or nonsense words which can be used to evaluate morphological and phonemic awareness. A novel method for the creation of pseudowords was developed. Existing tools such as Wuggy [21] take as input existing words and output pseudowords that differ in one or two syllables to the original words. This creates pronounceable words that are similar to existing words (such as <sapado> from <sapato>). The proposed method creates pseudowords without the starting point of valid words while maintaining full pronounceability. It should create unfamiliar words and the difficulty of reading them should be slightly higher than familiar or existing words. The aim was to create pseudo-words of two, three and four syllables. First, the most frequent syllables in each position for words with those number of syllables were extracted from a large lexicon of European Portuguese, CETEMPúblico [22]. Then, words of two or more syllables are created randomly from a set of the most frequent syllables. Words that have syllabic combinations that do not respect pronounceability rules are deleted as are words that exist in the lexicon. The difficulty score for a pseudoword is calculated by the same method described above for sentences. The distribution of the pseudowords along the reading tasks is also similar to sentences, promoting a varying difficulty and a rising average difficulty along the grades.

3   Data Collection

The corpus of children reading aloud was collected at 2 private and 9 public schools in urban centres and periphery areas of the central Coimbra region with children that attend primary school, aged 6 to 10 years old. A specific application was developed in which the sentences are displayed in a large font size on a computer screen simultaneously with the start of recording. This means that there is no practice time to influence performance. A screenshot of the application can be seen in Fig. 1 as well as an example of the recording environment. The recordings were performed in school classrooms chosen for their low reverberation and noise. The children were asked to read aloud a set of 20 sentences and 10 individual pseudowords. A lapel Lavalier microphone (Shure WL93) was used as the main recording device, accompanied by a standard table top PC microphone as backup (Plantronics Audio 10). The background noise could not always be completely controlled but was mostly low, also because the main recording micro-phone did a good job at filtering out background noise.

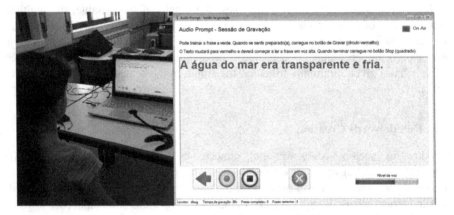

Fig. 1. Example of the recording environment (left) and software (right).

4 Corpus and Disfluencies

The collected database consists of around 20 h of recorded speech from 284 children, 147 female and 137 male, distributed from the 1st to the 4th grade with 68, 88, 76 and 52 children, respectively. A set of 104 children's speech utterances of pseudowords and sentences has been fully and manually annotated and these children (46 male and 58 female) are equally distributed among the 4 grades (26 per grade). Speech from an additional 100 children was annotated only for the pseudoword reading task. The annotated data amounts to approximately 6 h and 15 min of speech. The annotated speech exhibits a large variety of disfluencies that represent the most common types of errors in reading aloud by children. Based on [23], the rules for the annotation and labelling procedure were defined and several types of disfluency were identified:

- PRE – False starts that are followed by the attempted correction (pre-corrections, multiple can occur). Example: for prompt "*grande espanto*" [gr̃ədə iʃpẽtu], utterance is "*grande espa espanto*" [gr̃ədə **'iʃpɐ** iʃp'ẽtu].

- SUB – Substitution or severe mispronunciation of a word. Example: for prompt "*voava em largos círculos*" [vu'avɐ ẽj l'arguʃ s'irkuluʃ], utterance is "*voava em lares sicos*" [vu'avɐ ẽj **l'arəʃ s'ikuʃ**].

- PHO – Small mispronunciation of a word, usually with a change in one phone or a phone extension (EXT, marked with the symbol [:]). Example: for prompt "*A Lena chegou a casa, da escola*" [ɐ l'enɐ ʃəg'o ɐ k'azɐ dɐ iʃk'ɔlɐ], utterance is "*A Lena chegou a casa, da escola*" [ɐ l'enɐ **ʃə:g'o** ɐ k'azɐ dɐ **ɛʃk'ɔlɐ**].

- REP – Repetition of a word (multiple repetitions may occur). Example: for prompt "*Ele já me deu*" ['elə ʒ'a mə d'ew], utterance is "*Ele, ele já me deu*" ['elə **elə** ʒ'a mə d'ew].

- INS – An inserted word that is not part of the original sentence. Example: for prompt "*mas também dizem*" [mɐʃ tẽbẽj d'izẽj], utterance is "*mas também me dizem*" [mɐʃ tẽbẽj **mə** d'izẽj].

- DEL – The word was not pronounced (deletion). Example: for prompt "*onde morava uma velha*" ['õdə mur'avɐ **'umɐ** v'ɛʎɐ], utterance is "*onde morava velha*" ['õdə mur 'avɐ v'ɛʎɐ].

- CUT – The word is cut, usually in the initial or final syllable, but not corrected later. Example: for prompt "*dá água ao papagaio*" [d'a 'agwɐ aw pɐpɐg'aju], utterance is "*dá água ao papaga*" [d'a 'agwɐ aw **pɐpɐg'a**].

- PAU (…) – Intra-word pause, when a word is pronounced syllable by syllable and silence occurs in between. The symbol [...] can also appear in other disfluency events denoting a pause. Example: for prompt "*formosa e bonitinha*" [furm'ɔzɐ i bunit'iɲɐ], utterance is "*formosa e boni...tinha*" [furm'ɔzɐ i **buni...t'iɲɐ**].

Silence and noise events such as breathing, labial and background noise were also annotated. Extensions and intra-word pauses may occur simultaneously with other disfluencies and are marked with [:] and [...] inside phonetic transcriptions. The number of occurrences for each type of disfluency and their percentage of total uttered words in the database are presented in Table 1 for each of the 4 grades.

Table 1. Distribution of disfluency types in sentences for each of the four grades and in pseudowords (number of events and % of total uttered words).

Tags	Sentences					Pseudowords
	1st grade	2nd grade	3rd grade	4th grade	Total	Total
PRE	295 (7.4 %)	278 (5.7 %)	281 (4.4 %)	302 (4.1 %)	1156 (5.1 %)	318 (15.6 %)
SUB	182 (4.6 %)	149 (3.1 %)	215 (3.4 %)	208 (2.8 %)	754 (3.3 %)	263 (12.9 %)
PHO	214 (5.4 %)	169 (3.5 %)	203 (3.2 %)	143 (1.9 %)	729 (3.2 %)	476 (23.3 %)
REP	122 (3.1 %)	89 (1.8 %)	129 (2.0 %)	161 (2.2 %)	501 (2.2 %)	4 (0.2 %)
INS	30 (0.8 %)	42 (0.9 %)	42 (0.7 %)	65 (0.88 %)	179 (0.8 %)	20 (1.0 %)
DEL	5 (0.1 %)	14 (0.3 %)	16 (0.3 %)	50 (0.68 %)	85 (0.4 %)	3 (0.2 %)
CUT	11 (0.3 %)	15 (0.3 %)	29 (0.5 %)	27 (0.37 %)	82 (0.4 %)	2 (0.1 %)
EXT :	256 (6.5 %)	145 (3.0 %)	212 (3.3 %)	73 (1.0 %)	686 (3.0 %)	431 (22.7 %)
PAU...	179 (4.5 %)	126 (2.6 %)	102 (1.6 %)	65 (0.9 %)	472 (2.1 %)	251 (13.1 %)

Some interesting phenomena can be observed, such as 1st grade children being the ones that exhibit more intra-word pauses and extensions (due to slower reading), and 4th grade children having more insertions and deletions (due to faster reading). Furthermore, the defined false start type (PRE) is the most common disfluency for sentences, whereas in pseudowords mispronunciations are more common since there are fewer attempts to correct unknown words. Unexpectedly, children did not use filled pauses when trying to read aloud as teen and adults do in spontaneous speech [24], and instead pause with silence while thinking about how to read.

5   Reading Performance

5.1   Reading Speed

With annotated data, a simple analysis of the reading performance of each individual child can be done. A common metric is to evaluate reading speed considering only correctly read words, which is defined as Words Correct Per Minute (WCPM) [25]. The average values of WCPM per grade of 80 children of our corpus at the end of school year are shown in Table 2, side-by-side with the target curricular goals. A large inter-grade overlap of the distributions is observed, showing a variability in reading performance of different children, although the average does increase per grade. Figure 2 displays this behaviour with a boxplot of the distributions of WCPM, showing one clear outlier for the third grade. On data of adults and elderly speakers reading [26, 27], average words per minute are 130.3 ± 17.8 and 118.6 ± 21.7 respectively (on average, there are lower reading speeds for people above 60 years). Comparing these values to the observed child performance, there may still be expected improvement from 4th grade children, although some perform as well as adults. For sentence reading, the difference from average WCPM to curricular goals increases in absolute terms along the grades, and these lower WCPM values may be explained by the difficulty of the reading tasks. It can be concluded that the suggested increase of difficulty along the grades could be too steep to directly evaluate CG as intended, and, for overall reading ability evaluation, this difficulty needs to be taken into account. For pseudowords, although there are no CG for the third and fourth grades, average WCPM values are significantly lower than

CG, suggesting that the created pseudowords (based on joining common syllables and not on existing words) are of high difficulty.

Table 2. Per grade Mean and Standard Deviation of measured Words Correct per Minute (WCPM), Curricular Goals (CG) of WCPM and relative difference of WCPM to CG, for sentences and pseudowords reading tasks.

Grade	Words in Sentences			Pseudowords		
	WCPM	CG	WCPM-CG	WCPM	CG	WCPM-CG
1st	59.7 ± 18.1	55	+8.5 %	18.8 ± 8.0	25	−24.8 %
2nd	85.2 ± 22.9	90	−5.3 %	26.7 ± 8.4	35	−23.7 %
3rd	97.1 ± 23.5	110	−11.7 %	26.1 ± 6.5	−	
4th	110.4 ± 22.7	125	−16.7 %	34.9 ± 9.6	−	

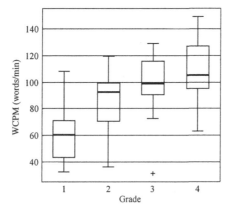

Fig. 2. Median and quartiles boxplots of Words Correct per Minute (WCPM) for sentence reading tasks for each of the 4 grades.

The defined curricular goals can be a starting point to appraise a child's reading ability. However, these do not take into account factors such as task difficulty or type of disfluencies and other ways to qualify reading performance should be considered. One possibility is to gather the opinion of experts and teachers, asking them to quantitatively rate the reading aloud performance of children (by listening to recordings of reading tasks). Their subjective opinion will be based on the several aspects of reading (speed, fluency, number of mispronunciations, etc.), and if these parameters can be quantified, there can be a correlation between the human score and a weighted average of the parameters. This is the premise of building an overall reading ability score that is well correlated with the opinion of expert evaluators [8, 9].

5.2 Pseudoword Performance

To further analyze children's performance on the task of reading individual pseudowords, the additional annotation of 100 children is considered, where they read 10 individual pseudowords each. This task differs substantially from sentence reading as morphological and phonemic awareness are the factors that influence a good performance on reading unknown words. Several interesting metrics can be extracted here, which will contribute to the overall reading performance. First, the reaction time of starting to read the word (the time between the start of recording and first try of uttering the word) reflects how fast the child is confident on reading the entire word or the first syllable, especially for first graders. However, it is not considered if the word is read correctly or not, and there are children with fast reaction times who do make several mistakes. Still, the average reaction time decreases along the grades as observed in Table 3 and Fig. 3, with only a small increase from third to fourth grades.

Table 3. Mean and standard deviation per grade of pseudoword reading reaction times (in seconds), number of uttered words with any kind of disfluency event (including extensions and intra-word pauses) and number of incorrect words.

	1st grade	2nd grade	3rd grade	4th grade
Reaction Time (s)	1.65 ± 0.83	1.35 ± 0.43	1.14 ± 0.23	1.19 ± 0.35
Number of disfluent words (out of 10)	6.54 ± 2.89	3.23 ± 2.32	2.96 ± 1.87	2.70 ± 2.24
Number of incorrect words (out of 10)	4.29 ± 2.33	2.31 ± 2.06	2.19 ± 1.57	2.17 ± 1.92

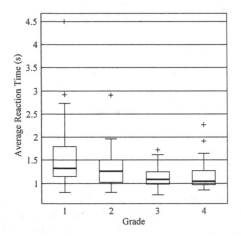

Fig. 3. Median and quartiles boxplots of average Reaction Times for the pseudoword reading task for each of the 4 grades.

Also in Table 3, the number of words that had any disfluency event is quantified. For the first grade, the average of 6.5 disfluent words out of 10 is much higher than other grades. Still, this measure is not identical to incorrect words (also presented in Table 3), since only phone extensions or intra-word pauses may occur.

6   Conclusions

A large corpus of children reading aloud was collected and analysed in terms of disfluencies and reading performance in both sentence and pseudoword reading tasks.

Various types of disfluent events were observed with the most prominent being false starts and mispronunciations. Curiously, hesitations with filled pauses were not identified. The analysed reading speed metrics fall averagely close to curricular goals per grade, although the difficulty of the tasks given to children is apparently higher for third and fourth grades than what may be expected for curricular goals evaluation. To be able to give an overall score of reading performance based on reading speed and types of disfluencies committed, a difficulty metric has to be considered as a parameter.

The main goal of the LetsRead project is to automatically analyse the overall performance of children on reading aloud tasks. For that goal, great care and effort must be taken to be able to automatically detect disfluency events and analyse disfluent reading, which are the next steps to be explored. Other necessary efforts include the optimization of the difficulty metric given to sentences and pseudowords, and obtaining the opinion of teachers on the overall reading performance of children of the LetsRead dataset, so the influence of all the parameters that contribute to reading ability can be adjusted.

Acknowledgements. This work was supported in part by Fundação para a Ciência e Tecnologia under the projects UID/EEA/50008/2013 (pluriannual funding in the scope of the LETSREAD project), and Marie Curie Action IRIS (ref. 610986, FP7-PEOPLE-2013-IAPP). Jorge Proença is supported by the SFRH/BD/97204/2013 FCT Grant. We would like to thank João de Deus, Bissaya Barreto and EBI de Pereira school associations and CASPAE parent's association for collaborating in the database collection.

References

1. Abdou, S.M., Hamid, S.E., Rashwan, M., Samir, A., Abdel-Hamid, O., Shahin, M., Nazih, W.: Computer aided pronunciation learning system using speech recognition techniques. In: INTERSPEECH (2006)
2. Probst, K., Ke, Y., Eskenazi, M.: Enhancing foreign language tutors – in search of the golden speaker. Speech Commun. **37**(3–4), 161–173 (2002)
3. Cincarek, T., Gruhn, R., Hacker, C., Nöth, E., Nakamura, S.: Automatic pronunciation scoring of words and sentences independent from the non-native's first language. Comput. Speech Lang. **23**(1), 65–88 (2009)
4. Mostow, J., Roth, S.F., Hauptmann, A.G., Kane, M.: A prototype reading coach that listens. In: Proceedings of the Twelfth National Conference on Artificial Intelligence, Menlo Park, CA, USA, vol. 1, pp. 785–792 (1994)
5. Black, M., Tepperman, J., Lee, S., Price, P., Narayanan, S.: Automatic detection and classification of disfluent reading miscues in young children's speech for the purpose of assessment. In: presented at the Proceedings of Interspeech, pp. 206–209 (2007)
6. Duchateau, J., Kong, Y.O., Cleuren, L., Latacz, L., Roelens, J., Samir, A., Demuynck, K., Ghesquière, P., Verhelst, W., hamme, H.V.: Developing a reading tutor: design and evaluation of dedicated speech recognition and synthesis modules. Speech Commun. **51**(10), 985–994 (2009)
7. Bolaños, D., Cole, R.A., Ward, W., Borts, E., Svirsky, E.: FLORA: fluent oral reading assessment of children's speech. ACM Trans. Speech Lang. Process. **7**(4), 16:1–16:19 (2011)

8. Black, M.P., Tepperman, J., Narayanan, S.S.: Automatic prediction of children's reading ability for high-level literacy assessment. Trans. Audio, Speech and Lang. Proc. **19**(4), 1015–1028 (2011)
9. Duchateau, J., Cleuren, L., hamme, H.V., Ghesquière, P.: Automatic assessment of children's reading level. In: Proceedings of the Interspeech, Antwerp, Belgium, pp. 1210–1213 (2007)
10. ELRA: ELRA - ELRA-S0180: Portuguese Speecon database. http://catalog.elra.info/product_info.php?products_id=798. Accessed 06 May 2015
11. Lopes, C., Veiga, A., Perdigão, F.: A european portuguese children speech database for computer aided speech therapy. In: Caseli, H., Villavicencio, A., Teixeira, A., Perdigão, F. (eds.) PROPOR 2012. LNCS, vol. 7243, pp. 368–374. Springer, Heidelberg (2012)
12. Hämäläinen, A., Rodrigues, S., Júdice, A., Silva, S.M., Calado, A., Pinto, F.M., Dias, M.S.: The CNG corpus of european portuguese children's speech. In: Habernal, I., Matoušek, V. (eds.) Text, Speech, and Dialogue, pp. 544–551. Springer, Heidelberg (2013)
13. Santos, A.L., Généreux, M., Cardoso, A., Agostinho, C., Abalada, S.: A corpus of European Portuguese child and child-directed speech. In: Proceedings of the Ninth International Conference on Language Resources and Evaluation (LREC 2014), Reykjavik, Iceland (2014)
14. Hämäläinen, A., Cho, H., Candeias, S., Pellegrini, T., Abad, A., Tjalve, M., Trancoso, I., Dias, M.S.: Automatically recognising European Portuguese children's speech. In: Baptista, J., Mamede, N., Candeias, S., Paraboni, I., Pardo, T.A., Volpe Nunes, MdG (eds.) PROPOR 2014. LNCS, vol. 8775, pp. 1–11. Springer, Heidelberg (2014)
15. Lee, S., Potamianos, A., Narayanan, S.: Acoustics of children's speech: developmental changes of temporal and spectral parameters. J. Acoust. Soc. Am. **105**(3), 1455–1468 (1999)
16. Hämäläinen, A., Candeias, S., Cho, H., Meinedo, H., Abad, A., Pellegrini, T., Tjalve, M., Trancoso, I., Dias, M.S.: Correlating ASR errors with developmental changes in speech production: a study of 3–10-year-old European Portuguese children's speech. In: Proceedings WOCCI 2014 – Workshop on Child Computer Interaction, Singapore, pp. 7–11 (2014)
17. Potamianos, A., Narayanan, S.: Robust recognition of children's speech. IEEE Trans. Speech Audio Process. **11**(6), 603–616 (2003)
18. Buescu, H.C., Morais, J., Rocha, M.R., Magalhães, V.F.: Programa e Metas Curriculares de Portugês do Ensino Básico. Ministério da Educação e Ciência, May 2015
19. Soares, A.P., Medeiros, J.C., Simões, A., Machado, J., Costa, A., Iriarte, Á., de Almeida, J.J., Pinheiro, A.P., Comesaña, M.: ESCOLEX: a grade-level lexical database from European Portuguese elementary to middle school textbooks. Behav Res Methods **46**(1), 240–253 (2014)
20. Mendonça, G., Candeias, S., Perdigao, F., Shulby, C., Toniazzo, R., Klautau, A., Aluisio, S.: A method for the extraction of phonetically-rich triphone sentences. In: Proceedings of the International Telecommunications Symposium (ITS), São Paulo, Brazil, pp. 1–5 (2014)
21. Keuleers, E., Brysbaert, M.: Wuggy: a multilingual pseudoword generator. Behav Res Methods **42**(3), 627–633 (2010)
22. Rocha, P., Santos, D.: CETEMPúblico: Um corpus de grandes dimensões de linguagem jornalística portuguesa. In: Presented at the PROPOR, pp. 131–140 (2000)
23. Candeias, S., Celorico, D., Proença, J., Veiga, A., Perdigão, F.: HESITA(tions) in Portuguese: a database. In: ISCA, Interspeech Satellite Workshop on Disfluency in Spontaneous Speech - DiSS, KTH Royal Institute of Technology, Stockholm, Sweden, pp. 13–16 (2013)
24. Veiga, A., Celorico, D., Proença, J., Candeias, S., Perdigão, F.: Prosodic and phonetic features for speaking styles classification and detection. In: Torre Toledano, D., Ortega Giménez, A., Teixeira, A., González Rodríguez, J., Hernández Gómez, L., San Segundo Hernández, R., Ramos Castro, D. (eds.) IberSPEECH 2012. CCIS, vol. 328, pp. 89–98. Springer, Heidelberg (2012)

25. Hasbrouck, J., Tindal, G.A.: Oral reading fluency norms: a valuable assessment tool for reading teachers. Reading Teacher **59**(7), 636–644 (2006)
26. Pellegrini, T., Hämäläinen, A., de Mareüil, P.B., Tjalve, M., Trancoso, I., Candeias, S., Dias, M.S., Braga, D.: A corpus-based study of elderly and young speakers of European Portuguese: acoustic correlates and their impact on speech recognition performance. In: INTERSPEECH, pp. 852–856 (2013)
27. Hämäläinen, A., Avelar, J., Rodrigues, S., Dias, M.S., Kolesiński, A., Fegyó, T., Németh, G., Csobánka, P., Lan, K., Hewson, D.: The EASR Corpora of European Portuguese, French, Hungarian and Polish Elderly speech. In: Proceedings of the Ninth International Conference on Language Resources and Evaluation (LREC 2014), Reykjavik, Iceland (2014)

Author Index

Printed in the United States
By Bookmasters